Advances in
MICROBIAL
PHYSIOLOGY

Advances in
MICROBIAL PHYSIOLOGY

edited by

A. H. ROSE

School of Biological Sciences
Bath University
England

D. W. TEMPEST

Laboratorium voor Microbiologie
Universiteit van Amsterdam
Amsterdam-C
The Netherlands

Volume 16
1977

ACADEMIC PRESS
London New York San Francisco

A Subsidiary of Harcourt Brace Jovanovich, Publishers

ACADEMIC PRESS INC. (LONDON) LTD.
24/28 Oval Road
London NW1

United States Edition published by
ACADEMIC PRESS INC.
111 Fifth Avenue
New York, New York 10003

Copyright © 1977 by
ACADEMIC PRESS INC. (LONDON) LTD.

All Rights Reserved

No part of this book may be reproduced in any form by photostat, microfilm, or any other means, without written permission from the publishers

Library of Congress Catalog Card Number: 67-19850
ISBN: 0-12-027716-6

Printed in Great Britain at the Spottiswoode Ballantyne Press by
William Clowes and Sons Limited
London, Colchester and Beccles

Contributors to Volume 16

A. J. BATER, *Department of Microbiology, University College, Newport Road, Cardiff CF2, 1TA, Wales*

W. T. COAKLEY, *Department of Microbiology, University College, Newport Road, Cardiff CF2, 1TA, Wales*

M. E. J. HOLWILL, *Physics Department, Queen Elizabeth College, London W8 7AH, England*

A. L. KOCH, *Division of Biological and Medical Research, Argonne, Illinois 60439, America*

A. W. LINNANE, *Department of Biochemistry, Monash University, Clayton, Victoria 3168, Australia*

D. LLOYD, *Department of Microbiology, University College, Newport Road, Cardiff CF2, 1TA, Wales*

B. E. B. MOSELEY, *Department of Microbiology, University of Edinburgh, School of Agriculture, Edinburgh EH9 3JG, Scotland*

P. NAGLEY, *Department of Biochemistry, Monash University, Clayton, Victoria 3168, Australia*

K. S. SRIPRAKASH, *Department of Biochemistry, Monash University, Clayton, Victoria 3168, Australia*

E. WILLIAMS, *Department of Microbiology, University of Edinburgh, School of Agriculture, Edinburgh EH9 3JG. Scotland*

Contents

Some Biophysical Aspects of Ciliary and Flagellar Motility

MICHAEL E. J. HOLWILL

I. Introduction	1
II. Structure	3
III. Patterns of Movement	7
IV. Some Hydrodynamic Considerations	13
A. Basic Concepts	13
B. Single Organelles	16
C. Densely Packed Organelles	18
D. Effects of Size	22
V. Functions of Axonemal Structures	24
A. General Considerations of Function	24
B. The Sliding Filament Model	25
C. Active Bending Model	33
VI. Mechanochemical Aspects	34
VII. Control of Flagellar and Ciliary Motion	40
VIII. Concluding Remarks	43
References	44

Does the Initiation of Chromosome Replication Regulate Cell Division?

ARTHUR L. KOCH

I. Introduction	50
II. Models for the Regulation of Cell Division and Chromosome Replication	53
III. The Necessity for Correlation of Events in Cell Cycles of Related Individuals	54
IV. The Role of DNA Initiation in Regulating Cell Division	57
V. Computer Simulation of the Cell Cycle Based on the Deterministic Principle	58
VI. Pulse Autoradiography of Slowly Dividing Cells: The Experimental Basis for Computer Simulation	63
A. Analysis of Autoradiographic Grain Count Data	65

VII. Experiments Concerning the Variation in the Initiation of Chromosome Replication	69
A. Other Observations	71
VIII. Possible Experimental Objections	72
A. Remaining Admonitions	74
IX. Variability of the Time Between Nuclear Division and Cell Division	76
X. Temporal Accuracy of DNA Synthesis Cycle	80
XI. Computation of the Fraction of Cells of a Size Class Engaged in DNA Replication	81
XII. Precision of Initiation of Chromosome Regulations in *Myxococcus xanthus*	90
XIII. Conclusions	93
XIV. Acknowledgements	95
References	95

Repair of Damaged DNA in Bacteria

B. E. B. MOSELEY and E. WILLIAMS

I. Survival Value of DNA Repair Mechanisms	100
II. Radiation and Chemical Damage to DNA	101
A. Ultraviolet Radiation Damage	101
B. Ionizing Radiation Damage	102
C. Mitomycin-C and Psoralen-plus-Near-Ultraviolet Radiation Damage	105
D. The extent of Repair of DNA Damage	106
III. The Isolation of Mutants Defective in DNA Repair	109
A. Mutations Leading to Increased Sensitivity of *Escherichia coli* to Radiation and/or Chemical Damage	111
IV. Photoreactivation Repair	121
A. The Substrate for Photoreactivating Enzyme	122
B. The Photoreactivating Enzyme (PRE)	122
C. The Enzyme–Substrate Complex	124
D. Photoreactivation as an Analytical Tool	125
V. Excision Repair	125
A. Methods for Studying Excision Repair	126
B. Incision (*uvrA, uvrB*)	127
C. Pre-excision (*uvrC*)	128
D. Excision (*polA*)	128
E. Re-insertion of Bases	130
VI. Post-Replication Recombination Repair	131
A. Methods for Studying Post-Replication Recombination Repair	133
B. Gap Formation Opposite Pyrimidine Dimers	133
C. Gap Repair	135
D. Recombination-Deficient Repair (Genetic Control of Post-Replication Repair)	135
VII. The Repair of Ionizing Radiation Damage	137
A. Type III Repair (*recA, recBC, exr, dnaB, ror, lig*)	137
B. Type II Repair (*polA, lig*)	138
C. Type I Repair (*lig?*)	139
D. The Contribution of Types I, II and III Repair in the Irradiated Wild-Type Cell	140
E. The Repair of DNA Double-Strand Breaks	141

VIII. The Repair of Cross-Link Damage	142
IX. The Mutagenic Consequences of Repair	144
A. Photoreactivation	145
B. Excision Repair	145
C. Recombination	146
D. The Inducible Nature of Ultraviolet-Radiation-Induced Mutagenesis in *Escherichia coli*	147
E. Repair-Dependent and Independent Mutagens	148
X. Summary	149
References	150

Structure, Synthesis and Genetics of Yeast Mitochondrial DNA

PHILLIP NAGLEY, K. S. SRIPRAKASH and ANTHONY W. LINNANE

I. Introduction	158
II. Structure and Physical Properties of mtDNA in Respiratory-Competent Yeast	158
A. *Saccharomyces cerevisiae* and *Saccharomyces carlsbergensis*	158
B. Other Yeast Species (*Petite*-Negative Yeasts)	170
III. Synthesis of mtDNA	173
A. mtDNA Synthesis *in vivo*	174
B. mtDNA Synthesis *in vitro*	175
C. mtDNA Replication and the Mitochondrial Membrane	178
D. Cellular Origin of Components Involved in mtDNA Synthesis	180
E. Level of mtDNA and Number of mtDNA Molecules in Cells of Different Strains	184
F. Synthesis of mtDNA During the Cell Cycle	187
G. Effects of Changes in Cell Physiology on the Synthesis of mtDNA	187
IV. Petite Mutants of *Saccharomyces cerevisiae*—Molecular Aspects	193
A. Structure and Organization of mtDNA in *Petite* Cells	193
B. General Aspects of *Petite* Induction	203
C. Mechanism of *Petite* Induction by Ethidium Bromide	205
D. *Petite* Negativity of Yeasts other than *Saccharomyces cerevisiae* and Closely Related Species	214
V. Mitochondrial Genetics in Yeast	216
A. Criteria of Mitochondrial Inheritance	217
B. Genetic Determinants on mtDNA	217
C. Transmission and Recombination of Mitochondrial Genes in Genetic Crosses	230
D. Genetic and Physical Map of Yeast mtDNA	245
E. Suppressiveness in *Petite* Mutants	256
References	263

Disruption of Micro-organisms

W. T. COAKLEY, A. J. BATER and D. LLOYD

I. Introduction	279
II. Strength of Cell Walls in Relation to Structure	281

	A. Cell Outer Layers	281
	B. Relationship Between Strength of Cell Wall and Breakage Method Employed	286
III.	Principles of Cell Breakage	288
	A. Osmotic Lysis	288
	B. Basic Hydrodynamics	289
	C. Cells in Flow Systems	291
	D. Mechanical Methods for Cell Disruption	295
IV.	Lability of Cell Extract Components	305
	A. Thermal Lability	305
	B. Chemical Inactivation	306
	C. Mechanical Comminution	307
	D. General Approach to Optimization of the Activity of Extracts	324
V.	Controlled Breakage of Eukaryotic Micro-organisms	328
	A. Controlled Mechanical Breakage	328
	B. Disruption after Pretreatment of Organisms	331
VI.	Concluding Remarks	334
	References	335
	Author Index	343
	Subject Index	359

Some Biophysical Aspects of Ciliary and Flagellar Motility

MICHAEL E. J. HOLWILL

Physics Department, Queen Elizabeth College, London, W8 7AH, England

I. Introduction	1
II. Structure	3
III. Patterns of Movement	7
IV. Some Hydrodynamic Considerations	13
A. Basic Concepts	13
B. Single Organelles	16
C. Densely Packed Organelles	18
D. Effects of Size	22
V. Functions of Axonemal Structures	24
A. General Considerations of Function	24
B. The Sliding Filament Model	25
C. Active Bending Model	33
VI. Mechanochemical Aspects	34
VII. Control of Flagellar and Ciliary Motion	40
VIII. Concluding Remarks	43
References	44

I. Introduction

Cilia and flagella of eukaryotic cells are motile organelles which serve to produce relative motion between a cell and its environmental liquid. In sessile organisms, such as monads and sponges, flagellar action is responsible for the production of food-bearing currents which flow over the cell surfaces, while many protozoa use cilia or flagella to propel themselves through the fluid in search of food. The patterns of beat of cilia and flagella are varied and variable, as will be described, and yet the ultrastructure of these organelles remains remarkably constant in form and dimension wherever they are found, be it the single flagellum of a trypanosome or an individual cilium from the multiciliated epithelium of the mammalian lung. There is, in fact, no definitive way to differentiate between cilia and flagella, and the distinction between them has arisen for historical reasons. Various proposals have been made for a collective name for these organelles, but none has been widely adopted so that, in the present chapter, the commonly accepted usage of the two terms will be employed.

During their motion, cilia and flagella form and propagate bends along their length. The energy for this action is derived from chemical processes within the cell, and it is of fundamental interest to enquire how the chemical energy is converted, with no intermediate heating stage, into the mechanical work required to bend the system against internal and external resistances to movement. Study of this mechanochemical system has attracted the attention of scientists from a variety of disciplines, ranging from pure biology through biochemistry and biophysics to applied mathematics. Information obtained by application of the methods and techniques of the several sciences has led to considerable advances in our knowledge and understanding of the basic operation and function of cilia and flagella.

An organelle moving through a fluid dissipates the energy derived from the chemical process responsible for the ciliary or flagellar deformation. The energy can be computed using the equations of hydrodynamics, and sets a lower limit on the chemical energy which must be made available to the organelle during motion. Elucidation of the nature of the chemical reaction is the province of the biochemist, whose techniques have been used to expose the motile machinery to chemical manipulation and hence to obtain useful information about it. Changing the environmental conditions, such as temperature and pressure, of intact cells has also yielded results which have been related to a chemical process occurring within the system. The behaviour of the internal machinery which bends a cilium or flagellum influences the shape adopted by the organelle, and a detailed study of beat patterns, together with ultrastructural studies involving the electron microscope, can provide valuable mechanistic data.

Many aspects of the motion of cilia and flagella were reviewed comparatively recently in an excellent book of some 500 pages (Sleigh, 1975a). In the present chapter, it would be impossible, and unnecessary, to present the information contained therein, but it is the aim of the author to summarize the essential material and indicate the important advances which have been made in the past few years.

II. Structure

Motile cilia and flagella are long, thin organelles (for dimensions, see Table 1) with a circular cross section containing, within a membrane, an array of nine doublet microtubules surrounding a pair of single tubules (Fig. 1), which is maintained by a network of fine

TABLE 1. Dimensions associated with cilia, flagella and their motion

Length (μm)	Diameter (nm)	Beat frequency f (Hz)	Wavelength λ (μm)	Amplitude a (μm)	Propulsive speed (μm s^{-1})
5–200	about 200	1–100	5–100	1–20	1–1000

linkages (e.g. Warner, 1975); the entire assembly is known as the axoneme. The microtubules extend the entire length of a cilium or flagellum and, in cross section, each doublet is oriented so that its axis of symmetry makes an angle of about 10° with the tangent (at the doublet) to a circle drawn through the doublets. Using a high-voltage

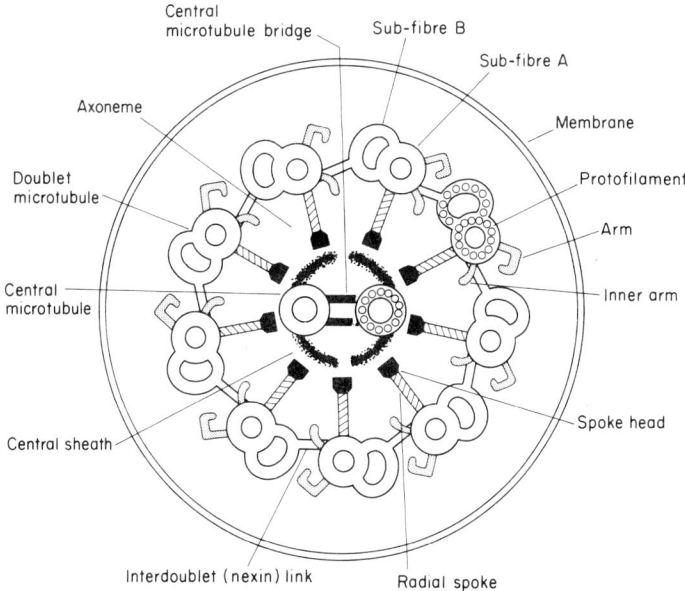

FIG. 1. Arrangement of microtubules and linking structures within a flagellum. After various authors, particularly Satir (1974), Warner and Satir (1974) and Warner (1975).

electron microscope, Gibbons (1975) reports that, in flagella with rigor waves obtained by abruptly removing ATP from a preparation, some twisting of the axoneme is observed, although he indicates that this may be a preparation artefact. A doublet consists of two parts, a complete microtubule (the A sub-fibre or microtubule) containing 13 protofilaments and the B sub-fibre which has fewer protofilaments and shares part of the wall of the A tubule. The skew of the doublets is such that the B sub-fibres contain the outermost points of the axoneme section, these points lying on a circle of diameter about 0.2 μm, while the innermost points are on the A sub-fibre and lie on a circle of approximately 0.15 μm. The spacing between adjacent doublets is between 17.5 and 20 nm and, into this space, extend the paired dynein arms which are attached to the A fibre. Viewed from tip-to-base along the axoneme, the dynein arms, which are responsible for a considerable fraction of the ATPase activity of the axoneme of all organisms so far examined for this feature, point in a clockwise direction. In most electron micrographs, the dynein arms do not extend to touch the neighbouring B sub-fibre but, if a preparation is made with ATP abruptly removed from the system, attachment of the arms to the B-fibre is observed (Gibbons, 1975).

Even when the flagellar membrane is removed by chemical procedures, the 9 + 2 axonemal complex retains its integrity due to the presence of fine linkages between adjacent sub-fibres. The linkages have been described as connecting the inner dynein arm to the neighbouring fibril (e.g. Kiefer, 1970), but Stephens (1975) has shown that axonemes with the dynein arms and B sub-fibres removed retain circumferential linkages, suggesting that the latter join neighbouring A sub-fibres. It appears likely that the interdoublet link (as it has been called by Warner, 1970) and the nexin link (Stephens, 1970) are identical structures which prevent lateral separation of the doublets, and which may be involved in the mechanical process which underlies motility. In longitudinal sections, the dynein arms and the interdoublet linkages have the same periodicity, the repeat distance being 16–20 nm depending on the organism, but insufficient information is available from such micrographs to specify the precise relationship between the two structures. The two rows of dynein arms appear to be in lateral register since, in longitudinal sections sufficiently thin to contain only one row, the same periodicity is observed.

Also attached to each A sub-fibre but directed radially inwards towards the central fibrils is another filamentous structure, the radial spoke, which is about 38 nm long, 5 nm in diameter and is terminated by an electron-dense head at, or close to, the central sheath, a structure to be discussed later in this review (Section V, p. 28; Warner and Satir, 1974). In longitudinal sections, the spokes are found in repeating groups of two or three along the A sub-fibre, each group spaced an average distance of 86 nm apart. In structures from organisms such as *Elliptio*, where spokes occur in groups of three, it is found that the spoke separations within a group are unequal, with the wider spacing towards the ciliary base. The arrangement of spokes within the axonemal matrix is such that they lie in a helical configuration, but it is important to emphasize that there is no helical structure joining the spokes together. Functional aspects of the radial spokes will be discussed in a later section (p. 28).

In earlier work on ciliary and flagellar fine structure, a helically wound sheath surrounding the central tubules was described by several authors (e.g. Gibbons and Grimstone, 1960; Pedersen, 1970) but, in recent years, observations on certain organisms have revealed an alternative structure, which, however, is still referred to as the central sheath. The structure consists of projections from each of the central fibrils, and the projections may occur in one or two rows along each central tubule. In *Elliptio* cilia, for example, there are two rows of projections to each tubule; within a given row, the projections are parallel to each other and inclined at an angle to the tubule axis (Fig. 2). According to the microtubule observed, the two rows form an erect or inverted chevron pattern, such that the ends of the projections from adjacent microtubules are close together, and may be connected (Figs. 1, 2). In transverse section, the projections appear to be curved, so that, in *Elliptio* cilia, joined chevrons would create a circle inclined to the axoneme axis. The curved nature of the projections has not, however, been established unequivocally. The longitudinal spacing of the projections is 14.3 nm in *Elliptio* cilia, and it is probably of mechanistic importance that this spacing is just one-sixth of the radial spoke-group repeat distance. This point will be discussed in more detail later in this review (p. 28).

The membrane which surrounds the axoneme is an extension of the cell membrane and sometimes carries appendages such as the mastigonemes (thin hair-like processes) of the algal flagellates. Freeze-fracture studies of the flagellar membrane (Gilula and Satir, 1972) have revealed two important patterns of membrane particles, presumably protein in character. In one arrangement, the particles lie in

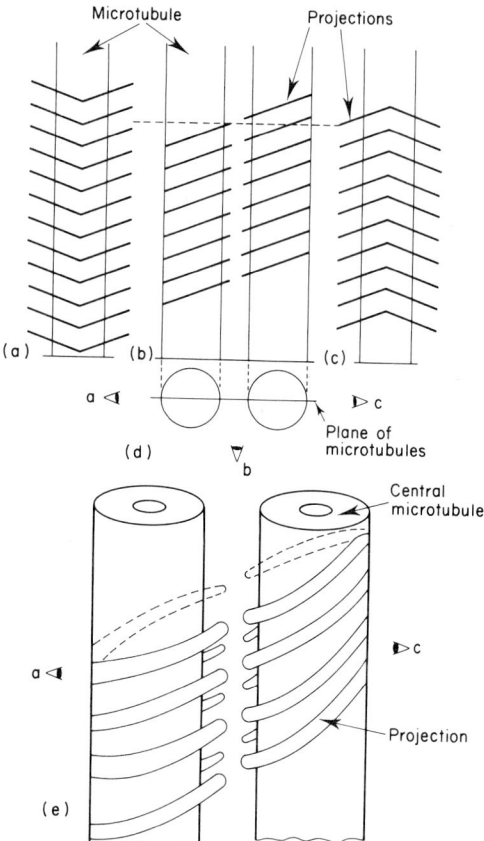

FIG. 2. The "central sheath"structure, showing the arrangement of projections from the central microtubules of *Elliptio* cilia. (a) and (c) show the "chevron" appearance of projections when viewed in the plane of the microtubules; (b) shows a pseudohelical appearance when viewed from a direction normal to the plane of the microtubules; (d) depicts a cross-section through central microtubules to indicate viewing directions for (a), (b) and (c); (e) indicates the three-dimensional impression of structure. After Warner and Satir (1974) and Satir (1974).

longitudinal rows which correspond in position to the underlying doublets, and may indicate some form of connection between the membrane and the doublets, although such a linkage has not yet been directly observed. The other pattern is the ciliary necklace, consisting of between two and six rows of membrane particles, which occurs just distal to the basal plate on which the central pair of tubules end. Linkages are observed to connect the membrane and the A tubule in the region of the necklace, and the arrangement of particles is consistent with the hypothesis that they represent the membrane-attachment sites of the linkages.

III. Patterns of Movement

The motility of cilia and flagella has been studied by a number of techniques, including high-speed cine-photography, stroboscopic observation and multiple flash photography (see e.g. Holwill, 1967; Baba and Hiramoto, 1970; Brokaw, 1965; Sleigh, 1964; Takahashi and Murakami, 1968). The results of many experiments indicate that the beat patterns of cilia and flagella can be broadly classified in two groups, namely undulatory and tonsate. Typical magnitudes of the various parameters associated with the motion of these organelles are shown in Table 1 (p. 3). A further division of undulatory movement can be made since both two- and three-dimensional waveforms are observed. It is a common practice in dealing with the hydrodynamic aspects of the problem to assume the planar waveform to be sinusoidal (Gray and Hancock, 1955; Holwill and Sleigh, 1967) and the three-dimensional undulation to be helical (Holwill and Burge, 1963; Chwang and Wu, 1971) although even superficial examination of film records indicates that these are poor approximations to the true shape. Critical analysis of the photographs shows that some waveforms can be satisfactorily matched by a pattern consisting of circular arcs linked by straight lines (Brokaw and Wright, 1963; Brokaw, 1965)—the so-called arc-line waveforms. Rather similar in form to the arc-line wave is the meander shape which is adopted by a flexible, elastic rod with freely hinged ends separated by a distance less than the length of the rod. From a mechanistic view point, it is important to establish which, if either, of these patterns is adopted by real cilia and flagella. The existence of the meander wave might suggest that elastic properties of the system are of importance in influencing flagellar motility, while an arc-line pattern could indicate a two-state bending mechanism, i.e. one in which the bending elements are deformed to a specific curvature, the other in which they are straight. As shown in Fig. 3, the arc-line and meander are very similar shapes and, although it is theoretically possible to differentiate between them (Silvester and Holwill, 1972), application of the method to photographic records presents difficulties, because of such factors as poor resolution, and is

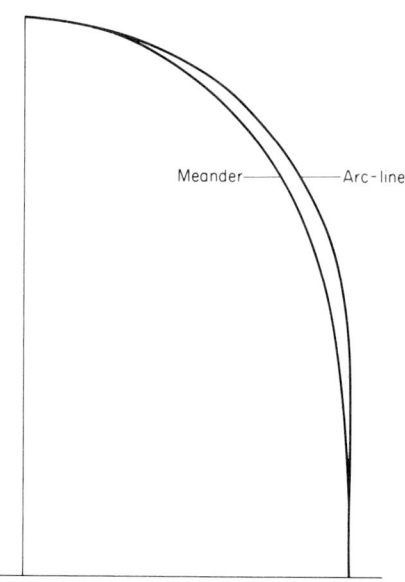

FIG. 3. Arc-line and meander waveforms of the same amplitude and wavelength, showing their close spatial relationship.

only now showing some indications of success (N. R. Silvester, M. E. J. Holwill and D. N. Johnston, unpublished results).

The majority of cells propagate bends from the base towards the tip of an associated cilium or flagellum. Only in a few cases, notably the trypanosomes, is it observed that, during normal movement, waves pass along the flagellum from the tip towards the base (Holwill, 1965, 1966b). In the case of the trypanosomes, and possibly for other organisms also (but in these cases wave analysis has not been undertaken), the direction of wave propagation can, under certain circumstances, be reversed. The reversal usually occurs when the organism encounters adverse conditions, such as an obstruction or an increase in viscosity, and can be regarded as an avoiding reaction. It is interesting to note that there appear to be no gross structure differences between the trypanosome flagellum and that of an organism which propages waves unidirectionally (Burnasheva *et al.*, 1968) although there may be subtle differences in fine structure associated with the ability to reverse the direction of wave propagation.

Waves propagated along a smooth flagellum induce forces which cause a free-swimming organism to move in the direction opposite to that of wave propagation. In certain cases, notably the ochromonad flagellates, the organism is observed to swim in the direction of the propagated wave. The flagellum of such a cell carries thin hairs or mastigonemes, often arranged in two rows on either side of the flagellum, which are believed to be responsible for the apparently reversed thrust. According to Jahn *et al.* (1964), during motility the flagellum beats in a plane which contains the mastigonemes. Reference to Fig. 4 shows that, as a wave crest passes, the mastigonemes execute a tonsate movement which provides a thrust in the direction opposite to that produced by the flagellar shaft. Provided that the mastigonemes are sufficiently long and rigid, their net induced force will be greater than that of the shaft, so that propulsion of an organism will occur in the direction of the wave. This qualitative explanation of mastigoneme

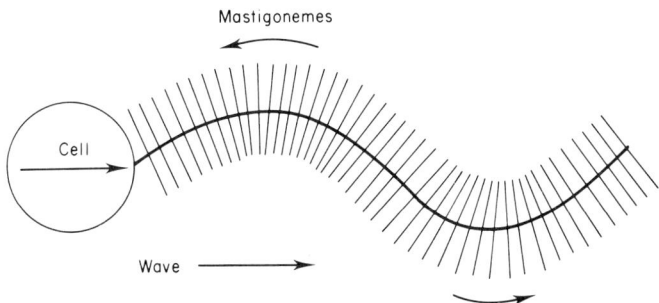

FIG. 4. Mechanism of action of mastigonemes.

action has been shown to give results consistent with experiment when subjected to hydrodynamic analysis (Holwill and Sleigh, 1967), but the mechanism has been challenged by Bouck (1971, 1972) on the basis of electron microscope observations. To be effective, the action described above requires that the mastigonemes lie in the plane of the flagellar wave, whereas Bouck suggests that a more reasonable arrangement, judged from his electron micrographs, is with the mastigonemes perpendicular to the plane of beat. Hydrodynamic analysis of the latter system indicates that reversed propulsion would not occur if the mastigonemes were passive, although Bouck (1972) has suggested a model based on freely hinged, elastic mastigonemes which he claims would produce the observed effect. Discussion of the model by Bouck is, however, rather brief and there is insufficient detail for one to understand precisely how he envisages the behaviour of the system.

Investigation of the fluid-flow patterns about *Ochromonas* flagella (Holwill and Peters, 1974) show them to be consistent with the model proposed by Jahn *et al.* (1964) although it is possible that alternative mechanisms might give the same pattern.

The tonsate movement characteristic of most cilia, but also observed for certain flagella, usually involves a more or less planar effective stroke (shown in idealized form in Fig. 5), followed by a recovery stroke which may occur in a plane or be three-dimensional (see Sleigh, 1975b for references). During the effective stroke, the organelle remains almost straight, although there may be a small degree of curvature

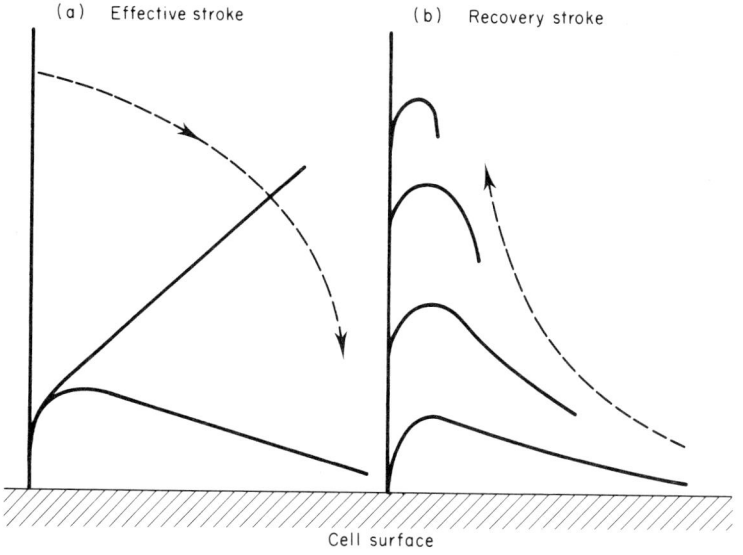

FIG. 5. Effective (a) and recovery (b) strokes of a cilium.

which is convex in the direction of the beat, and bends about its base. The recovery stroke involves propagation of the basal bend towards the tip in such a way that the organelle returns to the position it held at the beginning of the effective stroke (Fig. 5). There is usually no abrupt transition between the two phases of the beat, which is therefore observed as a continuous cyclic movement, but the abfrontal cilium of *Mytilus* sp. often moves very slowly and sometimes stops during its "effective" stroke (Sleigh and Holwill, 1969). Three-dimensional movement of cilia during their recovery strokes has been described (e.g. Parducz, 1967; Aiello and Sleigh, 1972; Machemer, 1975) for a number of organisms, such as species of *Paramecium, Opalina* and *Mytilus*. In many organisms, the bend propagated in the recovery stroke is twisted to the right or left (with reference to the effective stroke direction) so that the cilium shaft is moved parallel to the cell surface, although under conditions of increased viscosity the cilium executes a nearly helical beat.

Although the motion of an individual cilium has been described, it is unusual for a cilium to occur singly on a cell. Usually, a cell surface is covered by a large, regularly arranged field of cilia, packed so that individual organelles are about one micrometre apart, a separation considerably less than their length. Independent motion of such densely packed cilia is difficult to conceive and, in fact, interciliary coordination occurs to produce metachronal waves which pass over the cell surface much as, to use a well-worn analogy, waves of bending pass over a field of corn blown by the wind. As Machemer (1975) points out, the analogy is essentially pictorial, since the bending units in the corn field are passive while cilia bend actively. Within the metachronal pattern, cilia are synchronized in a direction parallel to a wave crest, while adjacent cilia suffer a maximum phase shift in a direction at right angles to the crests. There is considerable variation among organisms in the relative directions of the effective stroke of individual cilia and of propagation of the metachronal wave generated by the field of organelles. The two directions may be the same (symplectic metachrony), opposite (antiplectic metachrony) or at right angles (diaplectic metachrony; Fig. 6). There are clearly two possible versions of diaplectic metachrony, where the effective stroke occurs either to the right or to the left of an observer viewing in the direction of wave motion; the two are known as dexioplectic and laeoplectic metachrony, respectively. These four combinations were orginally described by Knight-Jones (1954) but relationships intermediate between these have been revealed by the critical examinations of Machemer (1975), who has proposed a "compass rose" notation to specify the intermediates. For example, one would describe a wave with an effective stroke at 45° to the wave direction as dexio-symplectic or laeo-symplectic, with appropriate notation for finer subdivisions.

Patterns of metachrony and their origins have long interested biologists, and two main concepts for wave formation have emerged, one based on a neuroidal system, the other on mechanical coupling by viscosity between adjacent organelles. It is currently believed that

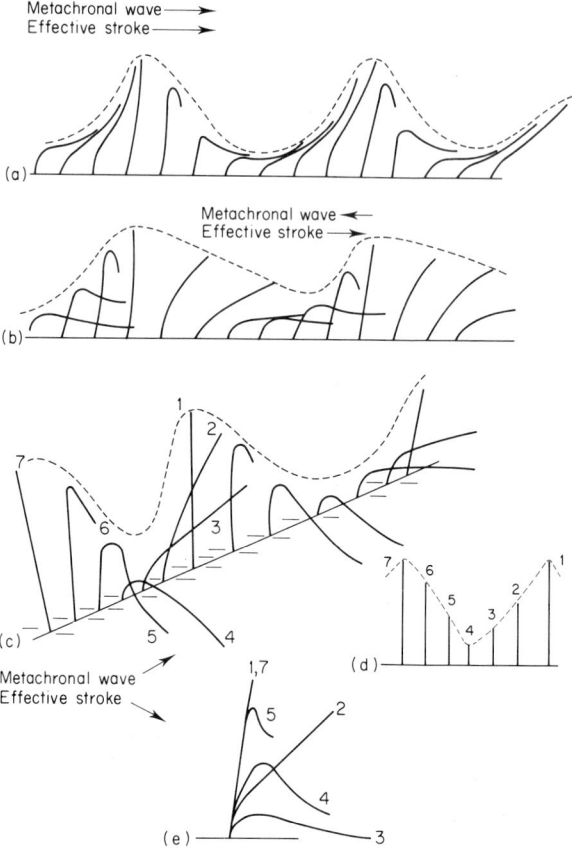

FIG. 6. Metachronal wave patterns. In (a) the wave pattern is symplectic; (b) antiplectic; and in (c), (d) and (e) dexioplectic. Fig. 6 (d) shows the appearance of the numbered cilia in (c) viewed from a direction normal to the wave propagation; (e) appearance of cilia viewed in wave direction.

metachronal wave motion is best explained by a system which involves transmission of force through the fluid from one organelle to its neighbours. This type of mechanism can be studied by exposing the field of cilia to media of different viscosities, since then the magnitude of the viscous coupling will be altered, and modifications of the wave pattern might be expected. Following his studies of *Paramecium* spp., Machemer (1972) presented an hypothesis for metachrony in which it is assumed: (a) given an assembly of oscillators (cilia) which generate fluid forces in a common direction, metachronal co-ordination will occur; (b) the type of metachronism generated is that which allows the least obstructed operation of the individual oscillators: (c) synchronization will occur in the direction of strongest viscous coupling: (d) the metachronal wavelength is a measure of the difference between the coupling forces along and perpendicular to the line of synchronization. The importance of viscous forces in metachrony has been demonstrated by a series of elegant experiments by Sleigh (1972) who used mechanical devices to stimulate or arrest the movement of *Pleurobrachia* comb plates. It was found that, if a needle was used to arrest a comb plate in such a position that the fluid flow induced by it had no effect on the neighbouring plate, metachronal waves were not observed. A similar result was obtained when a thin celluloid film was used to block fluid flow from one plate to its neighbour. In protozoa and some ctenophores, metachronal waves appear to be generated by interciliary viscous coupling forces, but there are instances among the Metazoa where blocking the metachronal wave at some point does not prevent its transmission beyond that point, thus suggesting the possible existence of a neural mechanism of co-ordination (e.g. Tamm, cited by Sleigh 1975c). It is clear that further experimentation is required to elucidate the true origin of metachronism in all organisms under a wide variety of conditions.

IV. Some Hydrodynamic Considerations

A. BASIC CONCEPTS

The forces created by the movements of cilia and flagella are derived from interactions between the flagellar surface and the fluid environment, and are used either to propel a free-swimming cell or to create fluid flow over a stationary cell surface. An understanding of these forces is of importance to an investigation of motility since, for example, a knowledge of the character of the interactions between cilia would provide support for either the mechanical or the neural mechanisms which have been proposed as the basis for metachrony. From a knowledge of the hydrodynamic interactions, it is possible to compute the energy dissipated against the external forces, and hence to specify the minimum amount of chemical energy which should be made available by the cells for motility. At its simplest, the hydrodynamic description of a cilium or flagellum involves the behaviour of a cylinder moving through a fluid. Since a complete solution to the case of a thin cylinder moving through a fluid is not possible in closed form, it is clear that motile organelles, consisting of actively bending cylinders, pose a complex problem. Fortunately, certain approximations can be made and it is possible to obtain expressions for the forces and energies involved in motility with reasonable confidence that they correctly reflect the magnitudes of these variables for the motile organelles.

When a body moves through a fluid, two types of force act upon it in addition to forces due to fluid pressure; they are *inertial* and *viscous* forces. Inertial forces arise when the body imparts momentum to the fluid, as, for instance, happens when a fish passes a wave along its body, or a ship rotates its propeller. To conserve momentum, the body acquires a velocity in the direction opposite to that of the driven fluid. Viscous forces are derived from essentially frictional effects between the body surface and the fluid. The layer of fluid in contact with the body surface remains stationary relative to the body, so that a velocity gradient is set up in the neighbourhood of a motile system. Molecular interactions within the fluid give rise to the shearing force which is deemed to be due to the viscosity of the fluid. The general analysis of a body moving through a fluid requires that both types of force be considered. Certain situations arise, however, where one of the two types of force dominates so that the other may be neglected by comparison. An order of magnitude ratio of the two forces is given by Reynolds Number (Re) defined by the relation:

$$\mathrm{Re} = \frac{l v \rho}{\eta} = \frac{\text{inertial forces}}{\text{viscous forces}} \qquad (1)$$

where v and l are, respectively, the velocity and a characteristic length of the body while ρ and η pertain to the fluid and are the density and coefficient of viscosity. The choice of v and l depends upon the particular problem to hand, but it is clear that, for large Reynolds numbers, inertial forces are greater than viscous ones, while the reverse is true for small Reynolds numbers.

In the case of a cilium or flagellum, the velocity to be used should be characteristic of the oscillatory motion of the organelle, rather than of the translational motion of the entire cell. During a cycle of undulatory or tonsate motion, the velocity of an element of an organelle varies considerably, but order of magnitude values appropriate to an evaluation of Reynolds number can be obtained by multiplying the oscillation frequency (f) by the amplitude (a) of the undulatory wave or by the length of a cilium executing a tonsate beat. The dimension of most relevance to force calculations is the diameter of an organelle rather than its length. Using the fluid properties of water to typify most cell environments, it is found that the Reynolds number for individual eukaryotic cilia and flagella lies in the range 10^{-4}–10^{-3}, although higher values are found for compound cilia such as the comb plates of ctenophores (Lighthill, 1969; Holwill, 1973, 1975). It is thus apparent that, when individual cilia and flagella are considered, the viscous forces are considerably greater than the inertial ones which may be neglected by comparison. Because the system should be considered *in toto*, the Reynolds number of the propelled cell body is also of importance. The sizes and velocities are such that the appropriate Reynolds number is perhaps an order of magnitude greater than for the flagellum, so that the drag forces acting on the cell are also predominantly viscous in character.

The equations of creeping flow (e.g. Curle and Davies, 1968) are therefore applicable to this situation but, while the equations can be set up and the boundary conditions specified, it is not possible to obtain analytical solutions for situations which correspond to ciliary or flagellar motion (for a mathematical discussion of this topic, see Lighthill, 1975). Approximate solutions have, however, been obtained and that due to Gray and Hancock (1955) has been widely used due to its inherent simplicity. In their work, these authors made use of force co-efficients C_N, C_L which are the forces per unit length per unit velocity acting on a cylinder moving normal and parallel to its length. The application of this concept to flagellar motility involves the assumption that it is possible to treat each element of an actively bending system as though it were part of a long, straight cylinder. While this is clearly open to question, it is to some extent justified by the successful prediction of swimming speeds for many organisms. The forces acting on an element δS of the flagellum are thus:

$$\delta F_N = -C_N V_N \delta S \qquad (2)$$

and

$$\delta F_L = -C_L V_L \delta S \qquad (3)$$

where δF, V are the forces and velocities in the directions indicated by the subscript. If the flagellar or ciliary beat pattern can be specified analytically, eqs. (2) and (3) can be summed over the length of an organelle to give an estimate of the propulsive force. In some cases (e.g. Gray and Hancock, 1955; Holwill and Sleigh, 1967), sinusoidal waves have been used to represent an undulating flagellum, although in principle the summation envisaged above can be obtained numerically for any general specified waveform. This latter procedure is best executed by computer using a programme of the type suggested by Brokaw (1970). The thrust generated by the organelle is equated to the drag of the cell body to give an equation for the propulsive velocity of the system in terms of the wave parameters. For a system which is propelled by three-dimensional waves, the lateral motion of the cylindrical elements creates a torque about the axis of the system (e.g. Chwang and Wu, 1971). In an equilibrium condition, this torque is balanced by counter-rotation of the cell body, and it can be shown theoretically that propulsion is not possible by the propagation of helical waves along an isolated filament.

B. SINGLE ORGANELLES

Experimental studies of organisms propelled by undulating organelles give velocities which compare favourably with the predicted values (e.g. Holwill, 1975). From the equations, it is possible to show that for an undulatory system the propulsive efficiency, expressed as the ratio of the power needed to move the cell to that required to maintain the oscillation, passes through a maximum for a wave shape corresponding to a value for $2\pi a/\lambda$ of about unity (Pironneau and Katz, 1975). For a large number of organisms, the wave shapes are close to the optimum form (e.g. Holwill, 1966a). In those experiments where the viscosity of the fluid environment is altered, the wave parameters are changed but in such a way that the wave shape remains close to the optimal value. Also, the equations of Chwang and Wu (1971) predict that, for optimal conditions, the head of an organism propelled by a helically undulating filament should have a diameter between 15 and 40 times that of the filament, thereby providing a further opportunity for experimental comparison. The diameter of most smooth eukaryotic flagella is about 0.2 μm, so that head diameters of between 3 μm and 8 μm are expected. Such a range is found for a number of protozoa, but many have head diameters considerably larger than this. Some of these organisms are known to bear, on the flagellar membrane, additional components which may increase the effective diameter of the flagellum during motion. Little information is at present available in relation to this particular problem but, as Pitelka and Schooley (1955) have noted, the hairs on the flagellum of *Euglena viridis* would give an effective filament diameter of at least 0.4 μm, thus giving a larger optimum cell diameter than quoted for naked flagella. Although the body of this species of *Euglena* is not spherical, it is hydrodynamically equivalent to a sphere of diameter 20 μm, based on viscous drag calculations, and so is close to the optimum range. If the criterion of equivalent viscous torque is applied, the diameter of *Euglena* sp. falls within the specified range. It would be interesting to examine other flagellates with this type of waveform and comparatively large cells to determine whether accessory structures have a hydrodynamically optimizing role.

The equations derived by Chwang and Wu (1971) also predict the rate at which the cell body should rotate in terms of the frequency of a helical propeller. Again, little experimental information is available, but the figures for *Euglena viridis* (Holwill, 1966b) give a predicted value of order 10 for the ratio of the flagellar beat frequency to the rotational frequency of the head. This represents order of magnitude agreement with the observed value (about 12) which is satisfactory in view of the hydrodynamic modelling necessary in the case of *Euglena* sp. Yoshida *et al.* (1975) and Shimada *et al.* (1975) carried out a study on the motion of flagellated bacteria in which they were specifically concerned with the relationship between propulsive speed and the rotational frequency of the cell body using an analysis based on previously derived equations (Holwill and Burge, 1963; Chwang and Wu, 1971). The hydrodynamic model required for flagellated bacteria involves a number of assumptions which are known to be unsatisfactory, but which are necessary if a solution is to be obtained, so it is encouraging that these authors obtained considerably better than order of magnitude agreement between theory and experiment.

As already indicated (Section IVA, p. 13), a major use of the hydrodynamic equations is to predict reliably the energy dissipated externally by a flagellum. The above discussion indicates that there is considerable agreement between theoretical predictions and experimental observations when propulsive speeds and rotational frequencies are considered, and so provides some measure of confidence that the assumptions upon which the arguments are based are reasonable. It is, however, important to recognize that the analyses described so far are based on components, in specific directions, of the forces which act on the system. It is conceivable that the actual force distribution along a flagellum, for example, could be different from that assumed theoretically and yet give the same integrated effect for the propulsive force and torque. Calculations of the power dissipated clearly depend on the force distribution considered, and it is therefore desirable to test the theory with further experimental results, preferably dependent on the complete force distribution in an unambiguous manner. The fluid flow generated by a motile organelle satisfies this criterion, since the flow field for a given force distribution is unique, and as it can be mapped experimentally using, for instance, polystyrene spheres in the medium, provides a critical test for the theory. Preliminary results of such observations have been reported (Lunec, 1975) and indicate considerable discrepancies between the predicted and experimental flow fields about flagella. Further work is required on this problem to ascertain whether the lack of agreement stems from a basic hydrodynamic misrepresentation of the system or from experimental parameters which have not been adequately considered.

C. DENSELY PACKED ORGANELLES

It should, in principle, be possible to use the equations developed for single organelles to produce an analysis appropriate to large aggregates of cilia or flagella. The difficulties encountered in this method are great, and first approaches to the problem of metachronal waves on ciliated organisms were made using an "envelope" technique (Blake, 1971a, b, c) in which the ciliary field is modelled by a continuous membrane with undulations which correspond to the metachronal wave. From Fig. 6 (p. 12), it is apparent that the envelope model is satisfactory for symplectic metachrony (Fig. 6a) but the problem of antiplectic metachronal waves (Fig. 6b), having considerable fluid flow across the "envelope", requires an alternative approach, which involves a consideration of the behaviour of individual organelles in the field.

In modelling symplectic metachrony by an undulating envelope, consideration is given to the fact that the tips of cilia in the field execute both lateral and transverse movements. This type of motion cannot therefore be modelled satisfactorily by a sinusoidal oscillation of a sheet, a problem which has been studied by a number of authors (e.g. Taylor, 1951; Reynolds, 1965; Tuck, 1968). The appropriate model requires a sheet which is assumed to vibrate transversely and longitudinally, so that the general path traced by an element of the sheet is a closed curve; if both the vibrations are simple harmonic, the curve will be an ellipse (Fig. 7). The theoretical analysis pertaining to this model is developed by Blake (1971b), but it will be sufficient here to note the main conclusions and compare them with the available experimental evidence. A major result to emerge is that, for a freely swimming organism when the excursion of an element of the sheet is predominantly transverse, the direction of movement of the organism

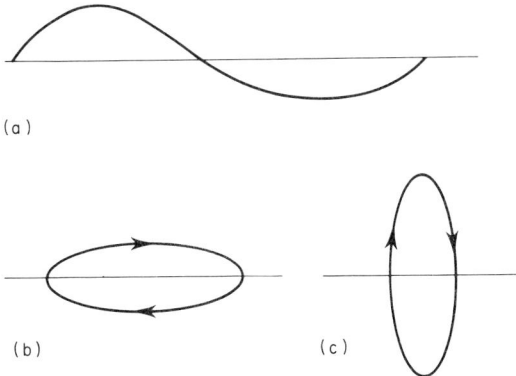

FIG. 7. Wave shape of envelope used to model metachronal waves (a). Figs (b) and (c) represent paths traced by an element of the envelope; (b) predominantly longitudinal motion, (c) predominantly transverse motion.

is opposite to the propagation direction of the metachronal wave; when an element executes a longitudinal vibration, the directions of wave propagation and of propulsion are the same. The metachronal waves on *Opalina* sp. are roughly symplectic, with the ciliary tip (corresponding to the envelope element) executing a predominantly transverse path. As predicted, *Opalina* sp. swims in the opposite direction to the wave motion. The envelope model for antiplectic metachrony is, as noted on page 18, not satisfactory for quantitative assessment, but it is interesting that it would correspond to a predominantly longitudinal particle oscillation. As required by the theory, species of *Paramecium*, which bear waves with an antiplectoid appearance, swim in the same direction as its propagated waves. In these organisms, the direction of swimming is opposite to that of the effective strokes of the individual cilia. For large amplitude waves, the maximum propulsive velocities for the envelope model are about half the wave speed. Species of *Opalina* move with a velocity between 0.3 and 0.6 times the wave speed, in reasonable agreement with the prediction (Blake and Sleigh, 1975).

In more general models for predicting the behaviour of large fields of cilia, Blake (1972) and Keller *et al.* (1975) consider a system of cylinders attached at one end to a plane surface and arranged in a regular array. The difference between the two models lies in assumptions concerning the fluid flow experienced by the cilia. Blake (1972) assumes that each cilium is exposed to a steady unidirectional flow created by the remainder of the ciliary field, while Keller *et al.* (1975) consider the more realistic condition that the fluid flow is varying. A qualitative interpretation of the models can be obtained by considering the fluid regions about a cilium to be divided into near-field and far-field parts. Near-field effects concern the behaviour of fluid within the field of cilia, while far-field effects contribute to the propulsive velocity of an organism. For an element of a cylinder on an infinite plane surface, the near-field is a sphere of radius $0.5h$ where h is the height of the element above the surface. Hence, in the effective stroke, an individual cilium influences a relatively large volume of fluid in terms of near-field effects but this volume is considerably reduced during the recovery stroke, when much of the cilium is close to the surface (Fig. 8). By considerations of the near-field effects, a qualitative assessment of the fluid flow for a field of cilia can be obtained. Near the surface to which the cilia are attached, near-field effects due to both the effective and recovery strokes contribute, so that an oscillatory motion of the fluid is to be expected in this region. Further from the surface, the fluid is affected by the effective stroke only so that uni-directional flow should be observed. More detailed hydrodynamic analysis of both models suggests that there should be a net back-flow of fluid close to the surface to which the cilia are attached. Sleigh and Aiello (1973) have performed a detailed experimental study of fluid flow around bomb plates of *Pleurobrachia* spp., and report no back-flow in the region expected. However, the Reynolds number appropriate to this organism is of order unity so that the low Reynolds number approximation made in the analysis may not be valid. Reasonable agreement between theory and experiment is provided by the work of

Jahn and Votta (1972) and of Cheung and Winet (1975), who obtained the velocity profile of the fluid surrounding species of *Paramecium* and *Spirostomum*. No back-flow close to the surface of *Paramecium* was reported, but it was observed for *Spirostomum*. Other features of the experimentally determined flow are close to the predicted behaviour.

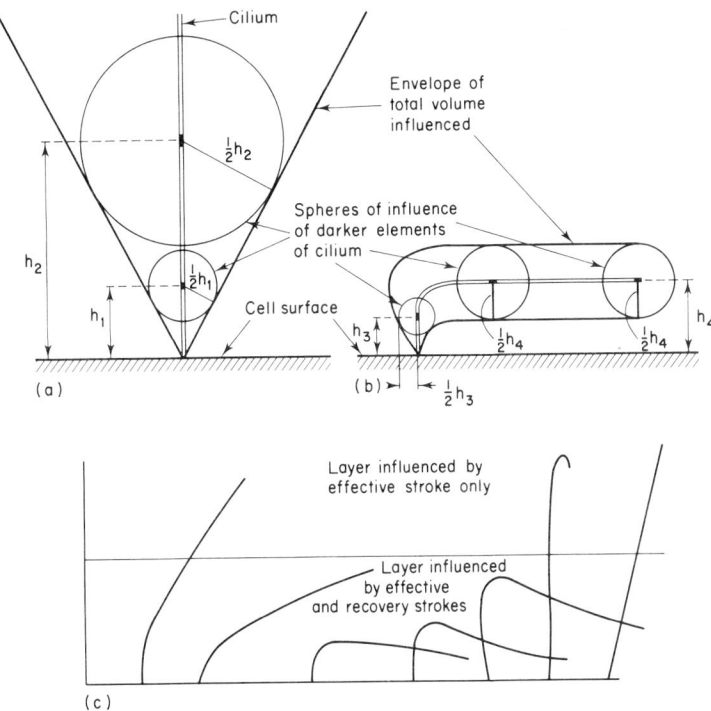

FIG. 8. Diagram showing how the sphere of influence results in large volumes of fluid being affected during the effective stroke (a), while smaller volumes are affected in the recovery stroke (b). Fig. 8(c) shows how the near-field effects contribute to fluid flow within the ciliary layer. Modified from Blake (1972).

Further observations of this type are needed to test the detailed predictions of the theories.

The analyses can also be used to obtain values for the propulsive velocity for comparison with experimental observation. In the case of symplectic metachrony, good agreement is found between the two types of model while, as noted on page 18, the physical behaviour of the antiplectic system is satisfactorily represented only by a model where individual cilia are considered. The predicted values of the swimming speeds of species of *Paramecium* (dexioplectic metachrony) and *Opalina* (symplectic) are of the same order of magnitude as the observed speeds, which are 1000 μm s^{-1} and 50 μm s^{-1}, respectively (e.g. Sleigh and Blake, 1976). The predicted swimming speed for *Pleurobrachia* sp. is between 10 and 50 mm s^{-1} and agrees well with the observed value (75 mm s^{-1}) although this may be fortuitous since the assumptions made do not strictly apply to this cell.

Brennen (1974, 1975) has developed an alternative approach to the problems of ciliary propulsion based on the observation that the unsteady components of the fluid flow caused by the cilia are confined to a very thin layer of fluid close to the organism, provided the metachronal wavelength is much less than the overall dimension of the organism. Outside the oscillatory boundary layer, a steady fluid flow is created by the uniform motion of the cell through the fluid. Clearly the flow fields must match at the junction of the oscillating boundary layer with the region of steady flow. From equations which describe the motion of fluid in the boundary layer and the fact that the net force acting on the swimming cell is zero, the translational velocity is obtained. This technique is applicable to finite models and allows the effect of shape on propulsive performance to be evaluated. While the model can, in principle, be applied to ciliary fields executing quite general motions, Brennen (1974, 1975) has so far considered only cases which can be represented by an envelope model, with the further restriction that the wave amplitude is small compared to its wavelength. Agreement with experiment is reasonable for several organisms, and the results have allowed an equation to be predicted for the large amplitude situation. The velocities calculated from this equation show considerably better agreement with experiment than the small-amplitude case.

D. EFFECTS OF SIZE

Comparative studies on the motility of ciliates and flagellates of different sizes have recently been reported (Sleigh and Blake, 1977; Holwill, 1977). The sizes of cells propelled by large numbers of cilia range from about 20 μm (*Uronema* sp.) to 50 mm (*Pleurobrachia* sp.) while, for eukaryotic organisms propelled by a few undulating flagella, the body size varies from a few micrometres to perhaps 100 μm. In absolute terms, the velocities of many species of ciliated cells are roughly the same so, when the speed is expressed in terms of body lengths per second, it decreases with increasing size and a plot of scaled speed against body length is linear (Sleigh and Blake, 1977; Fig. 9). Two organisms, namely species of *Opalina* and *Pleurobrachia*, have parameters which do not lie on the line. With both organisms, this may

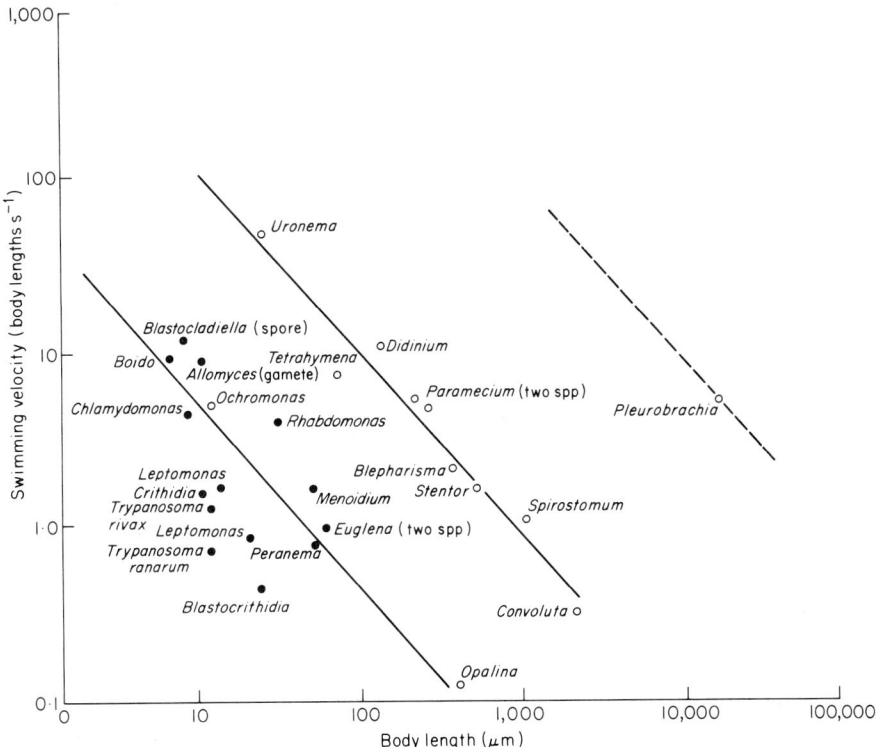

FIG. 9. Linear relationships between body length and swimming speed for a number of ciliates and flagellates (some results from Blake and Sleigh, 1977). ○ indicates relationship for ciliates; ●, flagellates.

be due to different methods of propulsion; in *Pleurobrachia* inertial, as well as viscous, forces may be involved while *Opalina* propagates symplectic metachronal waves in contrast to the other organisms many of which are known to sustain diaplectic metachrony. Data available for flagellates also produce a line, essentially parallel to that for the majority of ciliates, which includes the point for *Opalina* sp. The significance of the linear behaviour is not clear, but is no doubt associated with the similar propulsive mechanisms used in the linearly related organisms. It may also be relevant that a graph of total ciliary length (i.e. the product of the number of cilia and the length of an individual organelle) against the swimming speed in body lengths per second is linear.

Results of interest also emerge when the hydrodynamic efficiency of the system is considered. For ciliated cells, it appears that the most efficient are those which lie in the smaller size range, from 25 μm to 250 μm. Larger organisms tend to expend more energy since they bear many more cilia, but this effect is negated to some extent by a decrease in ciliary size and beat frequency. For cells propelled by undulating flagella, maximum hydrodynamic efficiency is attained throughout the size range, which may be extended to include bacteria at the lower end and small worms at the upper (Holwill, 1977). From the few results which have been presented in relation to scaling, patterns of similar behaviour emerge, and it may be that a more extensive comparative study of ciliates and flagellates could reveal important information about the hydrodynamic and biological behaviour of these cells.

V. Functions of Axonemal Structures

A. GENERAL CONSIDERATIONS OF FUNCTION

Patterns of ciliary and flagellar motility vary widely from one species to another, and even among organisms of the same species, but the 9 + 2 microtubular arrangement of the axoneme is common to them all and is practically invariant in dimension and arrangement. Any mechanism which purports to explain bend formation and propagation in terms of the behaviour of the structural components of an organelle must be able to explain the diversity of motile patterns observed. Two basic models for the mechanism of bending have been evolved, one based on the idea that local contractions occur in certain structures, thereby producing a bending force (e.g. Bradfield, 1955), the other assuming that relative sliding of filaments produces the necessary stresses of deformation (e.g. Satir, 1975a). Costello (1973a, b) recently described a theory which relies on local active bending of the peripheral doublets and is thus similar in type to the first mechanism to which I referred. It has long been assumed that the outer nine doublet tubules are intimately associated with motility, but it is only in recent years that detailed electron-microscope studies have allowed a functional interpretation of other axonemal structures to be made. Information about the axonemal functions has also been obtained in computer simulation of flagellar movement (e.g. Brokaw, 1972a, b, c).

The forces generated by the axoneme are sustained by chemical energy provided by the cell, and the observed bending is a macroscopic manifestation of conformational changes in macromolecules within the flagellum. Coupling between the chemical reactions and the axonemal structure therefore exists, and it is not unreasonable to suppose that aspects of flagellar motion reflect, directly or indirectly, the behaviour of the underlying mechanochemical reaction. As it is known that changes in the external environment, such as temperature and pressure, alter the rate of chemical reactions, the effects on motility of changing these parameters have been studied with a view to describing some aspects of the reaction associated with bending.

B. THE SLIDING FILAMENT MODEL

1. *Experimental Approaches*

As implied by its name, the sliding filament model of ciliary or flagellar action generates forces by interactions which tend to produce relative sliding of the axonemal microtubules. It is apparent that, if sliding of inextensible fibrils occurs during ciliary motility then, if a single bend exists on the organelle, the tip of a fibril on the concave side of the bend will project beyond one on the convex side by an amount which depends on their separation. In elegant experiments which provided the experimental basis for development of the model, Satir (1965, 1968) sectioned the tips of cilia of the freshwater mussel (*Elliptio* sp.) fixed in both effective and recovery phases, and found quantitative agreement with the predictions. In longitudinal sections, the radial spokes provide useful markers which allow relative sliding of the doublet fibres to be determined. Sections containing the basal plate allow corresponding spoke groups on separate fibres to be identified, and it is found that their spatial relationship through a bend is entirely consistent with the sliding of inextensible fibres; no evidence for contraction was obtained. More direct evidence for the sliding model has been obtained by Summers and Gibbons (1971) who studied the behaviour of sea-urchin spermatozoa which had been demembranated by addition of the detergent Triton X-100. Addition of ATP to this preparation causes the extracted organelles to propagate waves and propel themselves through the fluid environment. If the detergent-treated flagella are exposed briefly to trypsin, it is found that the radial spokes and interdoublet links are digested, and subsequent exposure of the system to ATP induces the microtubules to slide past each other, thereby causing the axoneme to become longer and thinner. The extruded tubules tend to coil, suggesting a release of tension inherent in the intact system but, once all of the doublets are separated, the individual fibrils show no tendency to move.

By analogy with the system of sliding filaments in muscle, the dynein arms attached to one doublet microtubule are believed to interact with the neighbouring B sub-fibre in the presence of ATP to produce conformational changes such that shearing forces are established between the two. Until recently, there was no direct evidence that the dynein arms do attach to sub-fibre B, but Gibbons (1975) has shown that, if the ATP concentration in the medium surrounding a demembranated sample is rapidly decreased from about 10 mmol m^{-3} to 0.1 mmol m^{-3}, a substantial fraction of the dynein arms form cross bridges between adjacent fibrils. This procedure causes the flagella to stop with stationary waveforms (rigor waves) which are similar in shape to the waves borne by the re-activated axoneme during motion. On adding ATP in concentrations of less than 10 mmol m^{-3} to a suspension of flagella with rigor waves, the flagella straighten, thus suggesting that the dynein cross bridges are responsible for maintaining the wave shape in the rigor state. In support of this idea is the observation (Gibbons and Gibbons, 1974) that addition of sufficient trypsin to destroy the nexin linkages and the radial spokes does not cause relaxation of the rigor waves if ATP is not added. Gibbons (1975) reports that the elastic properties of the rigor system are qualitatively similar to those of a model of the axoneme constructed from flexible curtain wires cross linked in a bent configuration. Both systems are more resistant to bending than to twisting, and it may be relevant to a detailed examination of the system that the axoneme can be twisted through an angle of 180° by a current of fluid which has a velocity about five times *slower* than the relative velocity of flagellum and fluid obtaining during normal motility.

Relative filament sliding, of itself, cannot induce bending but would give rise to compressive or tensile forces within the flagellum. It is necessary to have a system which transmits the shearing forces to the flagellar structure in such a way that bending moments are created. The appropriate conditions would exist if free sliding were prevented in a region of the flagellum to be bent since, in the absence of free sliding and provided the structure could withstand the compressive thrusts, bending moments would be generated. Free-filament sliding at the proximal end of a flagellum is generally assumed to be impossible because of the basal structures, so that shearing forces developed by interdoublet interaction along the length of a flagellum would cause the organelle to bend into a circular arc. Localized bending could occur by restricting the bending moments to a limited portion of the organelle or if the flexural rigidity of the flagellum is variable.

Initiation of a bend at the base of a flagellum could be achieved by active sliding in a limited proximal region of the flagellum, thereby confining the active bending moment to this region, with fibres in the distal part able to slide freely but passively. The same effect could be attained by active-shear development along the length of the flagellum, with a smaller flexural rigidity at the proximal region than over the remainder of the organelle. To initiate bends at the distal end of a flagellum, as observed in trypanosomes, would require similar interactions to those already described, although the presumed absence of sliding at the proximal end of the flagellum would require the sliding generated through a bend to be accommodated distal to it. Once initiated, the propagation of a bend would require the local active bending moment or stiffness change to travel along the flagellum.

In structural terms, the forces required to produce the bending moment are derived from shearing forces developed by interactions between the dynein arms and the peripheral doublet microtubules. Local bending moments would be created by active interactions over a short region of the flagellum, while changes in the flexural rigidity would be the result of interfibrillar linkages, which might be the dynein arms, the nexin links or the radial spokes. Of these three, the dynein arms and the radial spokes both appear to behave in a manner consistent with a transduction mechanism, but little is known about the properties of the nexin links. As already noted, cross linking by the dynein arms is sufficient to maintain a rigor wave (Gibbons and Gibbons, 1974) and may be used to sustain an actively propagating bend.

Alternatively, the stiffness of the flagellum may be controlled by the radial spokes which link the peripheral doublets to the central fibrils. The work of Warner and Satir (1974) on cilia of *Elliptio* sp. indicates that the spoke heads undergo a cyclic motion involving attachment to and detachment from the central sheath complex. The angular arrangement of the spokes varies according to whether the region of the cilium examined is straight or bent; in straight regions, practically all of the spokes remain normal to the long axis of the flagellum while, in bent regions, all of the spokes are tilted (Fig. 10). The tilted spokes could be the result of relatively large forces being created between the spoke heads and central sheath, thereby restricting free sliding and inducing the bent configuration. Within the straight regions, the spoke head–central sheath interactions could be significantly weaker than in the bent region, so that the spoke heads are functionally detached and free filament sliding occurs. This conclusion is supported in studies by Goldstein and Pivonka (1975) who used the change in angle of a flagellum relative to the base to infer the displacement of the microtubules within the flagellum. In addition to concluding that there is no rigid cross linking within the straight regions, they also suggest that bent regions can tolerate substantial extrinsic sliding without disruption or effect on propagation. In mechanistic terms, the restriction on active filament sliding could occur on either side of a bend, but it may be significant that on the concave side the spoke tilt direction is consistently such that the spoke head always lies proximal to its site of origin on the A sub-fibre. On the convex side of a bend, the spoke tilt usually occurs with the spoke head distal to its A-fibre origin, although the opposite is sometimes observed. The variable patterns can be explained in terms of active sliding of some filaments in a particular direction on one side of the axoneme, with passive sliding on the other. The development of bends of opposite curvature on the same organelle would require asynchronous active sliding of different tubules, a phenomenon which could also explain certain observed spoke patterns.

As developed so far, the sliding filament model of ciliary bending involves active relative sliding between peripheral doublets. The forces required to generate the sliding originate from interactions between

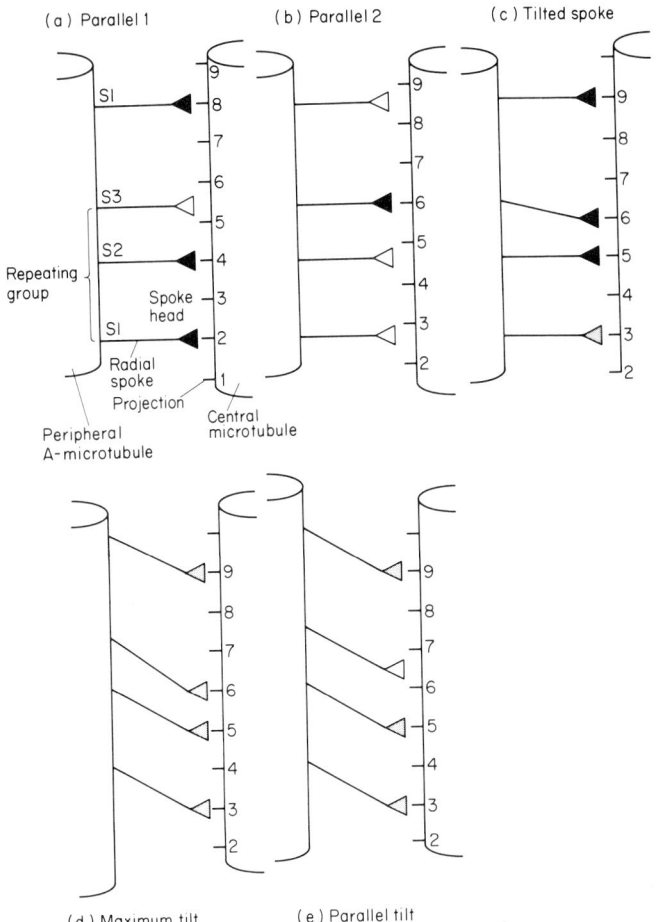

FIG. 10. Arrangements of spokes and central-sheath complex obtaining (a), in straight and bent region (b), in straight regions only (c), in bent regions only (d, e). Radial spokes with shaded heads are directly opposite a central-sheath projection, and may be functionally attached. Open heads lie between projections and are functionally unattached. The "vernier" arrangement of the spokes and projections ensures that, if required, the central sheath and outer doublets can remain permanently attached. The sequence (a) to (e) represents a possible sequence of cyclic spoke action. Modified from Warner and Satir (1974) whose nomenclature convention is used.

the dynein arms and the fibres in the presence of ATP, while the necessary bending moments are produced when sliding is restricted by attachment of the radial spokes to the central sheath. The function of the interdoublet links in sliding is not clear. If they are permanently attached to fibres at both ends then, to allow the degree of sliding observed, these links must stretch by an apparently unreasonable amount. It is possible that, like the radial spokes, the nexin links are detachable at one end, and may be associated in a functional way with the inner dynein arms (Warner and Satir, 1974). There is, however, some evidence that, in the axonemes of *Sciara* sperm (which do not conform to the 9 + 2 pattern), considerable extension of the interdoublet links is possible (Dallai *et al.*, 1973) so that permanent attachment of the nexin links may be feasible, thereby avoiding the need to postulate cyclic attachment and detachment behaviour.

The possibility that the axonemal fibrils twist during movement may have implications for the underlying mechanism. Satir (1963) reported a rotation of the fibres in gill cilia of *Elliptio* sp., which are known to execute a three-dimensional beat pattern (Aiello and Sleigh, 1972). A similar effect has been reported for sea-urchin sperm (Gibbons, 1975) an organism known to have a planar wave form. Twisting in this axoneme appears to be localized at the junction between curved and straight regions, regarding the flagellum to be of the arc-line form. Alternate twists have opposite senses, so that twisting does not accumulate over the flagellar length. The functional reason for the twisting of the outer doublets is not apparent, although as Gibbons (1975) remarks this behaviour would decrease the amount of sliding needed to produce a bend of given angle. It has generally been accepted, though the evidence is not strong, that those cilia and flagella with a planar beat are oriented during motion so that the plane of motion is normal to a sheet which contains the central fibrils. This would appear to be the optimal arrangement in mechanical terms, and is apparent in the high-voltage electron micrographs of Gibbons (1975) through a bend. In the straight regions of the axoneme, the central fibrils appear to be twisted to lie in a plane at about 45° to the bending plane. It is interesting to note that, in *Ochromonas* sp., there is some dispute as to the behaviour of the central microtubules during motility. The mastigonemes of this organism are attached to the membrane at sites contained by the plane of the central fibrils. If the mastigonemes are passively effective in their action, they should lie in the plane of beating (Holwill and Peters, 1974) which would imply that the fibres themselves are contained by the beat plane. Bouck (1971, 1972) argues the opposite case, that, since the central fibrils are probably normal to the beat plane, the mastigonemes must operate in a manner somewhat different from that generally accepted (see Section III, p. 9). Further studies are needed to ascertain the true configuration of the fibrils during motility *in vivo*.

2. Computer Simulation

Brokaw (1972a, b, c) has adopted an alternative approach to the study of a sliding filament model by considering the behaviour of the mechanism in computer-simulated models. The technique corresponds to the fabrication in real space of a sliding filament model, but the parameters of such a model would be many, and it would not be an easy matter to study responses to changed internal and external conditions. Computer simulation allows all of the variables of the system to be changed at will and thus, in principle, permits a thorough investigation of a sliding filament model under a variety of conditions. The responses of the model can then be compared with appropriate experimental results, and modifications made where necessary. The use of a computer provides information about the dynamic behaviour of a sliding filament system, in contrast to the electron microscope studies already described, which show the arrangement of the fibrils at a particular moment in time. The two approaches are thus complementary and are proving of value in a detailed analysis of ciliary mechanisms of movement.

The computer program models the flagellum as a series of connected straight-line segments, hinged at their ends to allow changes in angular orientation. Assuming the viscous, elastic and active bending moments to be balanced at each joint, equations can be written which, together with appropriate boundary conditions, are solved to obtain the rates of bending at each joint. This information allows the configuration of the model after a small time interval to be determined, so that a repeated application of the analysis reveals the behaviour of the system as a function of time. The model is somewhat simpler than the intact axoneme in that it contains only two parallel inextensible fibrils which maintain a constant spacing along the length of the model and between which active sliding is assumed to occur.

The external viscous bending moments can be computed with reasonable confidence using the approach of Gray and Hancock (1955; see Section IV, p. 15), but the effects of the internal resistances to motion have to be estimated. Brokaw (1972a, b, c) assumes that, in the most general case, internal resistances to bending can be both viscous and elastic and originate from bending and shearing motions within the filament. To overcome the various "passive" bending moments, the flagellum must generate active moments along its length (e.g. Machin, 1958, 1963), and this, of course, is the function of the sliding filaments. In his computer model, Brokaw (1972a) assumes that active bending is controlled by the curvature of the flagellum, and he has found that the models obtained on solving the moment equations spontaneously initiate and propagate waves of a form very similar to those of a real flagellum.

The parameters of wavelength, amplitude and frequency used to describe a flagellar wave are each influenced predominantly by different components of the computer-simulated model, although the effects on individual wave parameters cannot be completely ascribed to a single property of the model. To limit the wave amplitude, it is necessary to introduce non-linear elastic terms (i.e. it is assumed that the elastic components do not obey Hooke's law) the magnitudes of which determine, to a large extent, the wave amplitude generated by the model. The wavelength of the model is controlled by the value of the internal viscous resistances relative to the viscous drag experienced externally by the system. It is found that the wavelength is more sensitive to changes in the internal viscous bending resistance than to the shear resistance. Beat frequency is influenced by interaction between the viscous bending moments (internal and external), and the component of the active moment in phase with the viscous moments. The computer simulated model tends to maximize its frequency (to a value determined by the interaction of viscous moments), a result which may reflect a frequency-determining requirement near the base of a real flagellum, where the viscous moments are less than in the middle region. The wave shapes generated by the model are similar in form to those on real flagella, with the possible exception of regions near the flagellar base, and it is interesting that the computer system spontaneously initiates waves. It is not necessary to have a special wave-initiating component in the flagellum base. Real flagella are also known to have this capability. Flagella isolated from *Crithidia oncopelti* by laser microbeam or mechanical micromanipulation techniques are able to continue beating although they do not contain basal structures (Goldstein *et al.*, 1970; Holwill and McGregor, 1974).

A further development of the model has incorporated a cyclic activity of cross bridges linking the two filaments (Brokaw and Rintala,

1975). The cross-bridge cycle involves attachment and detachment phases each of which can be assigned a rate constant which, in the general case, depends on the local curvature of the system. If both rate constants are dependent on curvature, the computer simulation gives results in qualitative agreement with certain experiments, notably the effects of viscosity on wave parameters (e.g. Brokaw and Gibbons, 1975) and the effects of dynein removal or other chemical procedures on motiltiy (e.g. Brokaw and Benedict, 1968; Brokaw and Gibbons, 1973). The model does not, however, satisfactorily predict the ATP response of chemically extracted axonenemes. Better agreement is obtained if a recovery stage is included in the cross-bridge cycle, so that three rate constants are required to describe the activity. On the basis of the computer simulated model, realistic waveforms can be generated and the results of some experiments qualitatively matched. Since there is incomplete agreement with experimental observation, some of the assumptions relating to internal operation of the model require modification. For example, in one of the cross-bridge models, a cross link is expected to stretch to between 120 nm and 180 nm, which is significantly longer than the dynein arms of real flagella visualized by electron microscopy. Provided the active moment generated by the model is controlled by the local curvature of the flagellum, wave propagation will result. There is at present no evidence to suggest that such a feedback system operates during movement of real flagella, but it is possible that the radial spoke interactions already described could provide the appropriate control. The internal elastic and viscous resistances required in the control of wavelength and amplitude could result from properties of individual fibrils, from cross linking between them and from matrix interactions with the microtubules. An attempt to assign a particular interaction within the real flagellum to a specific term in the computer model would be purely speculative at this stage. There is, however, sufficient correlation between the model and the real system to expect that further detailed study of the computer formulation will provide valuable information about the internal mechanisms of flagellar motility, and give some indication of the required functions of the various axonemal components.

C. ACTIVE BENDING MODEL

Although much of the available experimental evidence is taken to support a sliding filament model of flagellar activity, Costello (1973a,

b) used some of the structural evidence to formulate a model based on active bending of the outer doublet fibrils combined with a stiffening of the central tubules. The structural evidence relies heavily on electron micrographs of macerated flagella, which show coiling of the doublet fibrils but frequently reveal the central tubules to be straight. The model assumes that both doublets and singlets can exist in an activated or a relaxed state. The activated state of a doublet involves active bending, while that of a singlet corresponds to stiffening. During motility, it is envisaged that active waves of bending and stiffening of the respective tubules pass along the axoneme, thereby producing the observed flagellar or ciliary waveforms. Ciliary motility is caused, according to this model, by successive bending of three microtubules in one lateral half of the axoneme while the central fibrils are stiffened, and a recovery stroke generated by bending of the microtubules in the other half with attendant relaxation of the central pair. In planar waves, the bending is caused as for the effective stroke, with the central fibrils remaining stiffened for the active period of the entire wavelength. Since wave propagation is a continuous phenomenon, this presumably suggests that the central fibrils are permanently stiffened. Helical movements are induced by sequential bending of the outer doublets. The model contains many features of earlier local-contraction models (see Holwill, 1966a, for references) but includes a specific role for the central pair of fibrils. The experimental evidence upon which this model is based is open to other interpretations, and it is essential that the properties of the model be sufficiently well defined to allow further tests of its mechanism to be made.

VI. Mechanochemical Aspects

I have already remarked that the energy needed for ciliary and flagellar movement is provided by chemical processes within the cell. In common with all chemical reactions, cellular chemistry is influenced by changing the environmental parameters and, in particular, reaction rates are modified by altering the temperature and pressure to which the reactants are exposed. Changing the parameters of the environment which surrounds a motile cell should provide a response which reflects the behaviour of reactions underlying motility. To establish a working hypothesis, it is necessary to relate some aspect of the flagellar wave parameters directly to a variable which is a function of the chemical reactions. A simple interpretation, and one which has had some success, is that the flagellar beat frequency can be regarded as proportional to the rate constant of some reaction within the flagellum. In elementary terms, this is not an unreasonable assumption as it is likely that a reaction cycle occurs at least once at each site along the flagellum for each wave that passes. There is some experimental evidence in support of this, since Brokaw (1967) observed that roughly one ATP molecule is used by each axonemal ATPase molecule for each beat of a sea-urchin sperm flagellum. It is thus possible that the beat frequency indicates the rate at which the cycle is executed, and so may be expressed in terms of a rate constant, or combination of rate constants, associated with the cycle. It is therefore appropriate to consider the effect on rate constants of changing the environment in which the reaction is occurring.

It is shown in the statistical treatment of reaction rates (e.g. Glasstone et al., 1941) that a first-order reaction rate, k, is related to changes in entropy $\Delta S°$ and enthalpy $\Delta H°$ occurring in the reaction by the equation:

$$k = \frac{kT}{h} \exp\left(\frac{T\Delta S° - \Delta H°}{RT}\right) \quad (4)$$

where **k** and h are, respectively, the Boltzmann and Planck constants, while T is the thermodynamic temperature. Changes in thermodynamic parameters are, in fact, related to the production of an activated complex from the primary reactants. The enthalpy term is the sum of the internal energy change, $\Delta E°$, and any work done during the process so, if a volume change $\Delta V°$ accompanies transition to the activated state, the following relation holds:

$$\Delta H° = \Delta E° + p\Delta V° \quad (5)$$

where p is the pressure to which the system is exposed. Substituting Eq. (5) in Eq. (4) and taking logarithms gives:

$$\ln\left(\frac{k}{T}\right) = \ln\left(\frac{k}{h}\right) + \frac{\Delta S°}{R} - \frac{\Delta E°}{RT} - \frac{p\Delta V°}{RT} \quad (4) \quad (6)$$

If the relation:

$$k = \alpha f \quad (7)$$

where α is a constant and f is the beat frequency, is assumed, it is clear from Eq. (6) that linear relations should obtain between $\ln(f/T)$ and

$1/T$ at constant pressure, and between $\ln(f)$ and p at constant temperature. From the intercepts and slopes of the graphs a number of thermodynamic parameters can be calculated.

A number of organisms have been studied in relation to the thermal dependence of their flagellar activity at atmospheric pressure (e.g. Holwill 1969, 1970; Holwill and Silvester, 1967) and it was found that the predicted relationships hold over a limited temperature range. Indications of a departure from linearity for marine invertebrate spermatozoa (Holwill, 1969, 1970) at the higher temperatures used were confirmed by Coakley and Holwill (1974) who performed experiments on *Crithidia oncopelti* at various temperatures and pressures. It was suggested that the results at the higher temperatures were indicative of enzyme denaturation. Over the rest of the temperature range, plots of $\ln(f/T)$ against $1/T$ show two linear portions, indicating that, for *Crithidia oncopelti* at least, the results cannot be interpreted in terms of variation in a single chemical rate constant. An interpretation which does fit the experimental results is the assumption that the enzyme is subject to reversible low-temperature denaturation and, over the temperature range considered, exists in a mixture of two states, one active, the other inactive (see e.g. Brandts, 1967). The proportion of active enzyme would thus depend on the temperature, and an analysis of the experimental results is possible if it is assumed that the beat frequency is decreased in proportion to the amount of enzyme denatured. Some support for this assumption comes from the work of Gibbons and Gibbons (1973), who found that the beat frequency of demembranated sea-urchin sperm decreased in proportion to the number of dynein arms (i.e. the amount of enzyme) removed from their sites of presumed action. The beat frequency now depends, in a relatively simple way, on two chemical constants: one, the rate constant referred to above, the other the equilibrium constant for the denaturation process. The magnitude of the thermodynamic parameters determined on the basis of this model are of the order of magnitude expected.

The relationships between beat frequency of *Crithidia oncopelti* and pressure at several different temperatures also did not conform to the simple behaviour predicted by Eq. (6), but again showed two linear regions (Coakley and Holwill, 1974). It is possible to interpret these results on the basis of reversible enzyme denaturation which occurs above a certain critical pressure for a given temperature. The linear section of the graph above the critical pressure would thus be due in part to the kinetics of an enzyme denaturation process as well as of an enzyme reaction underlying motility, while below the critical pressure the kinetics of the enzyme reaction dominate.

From their results, Holwill and Silvester (1967) found a linear relationship between changes in entropy and enthalpy associated with a variety of organisms. A basic inference of this, and one that is to be expected, is that the chemical reactions characterized by the beat frequencies of the organisms are similar, and form a homologous series with a common mechanism (e.g. Laidler, 1965). From published results, Holwill and Silvester (1967) were also able to determine activation parameters associated with muscular activity. They found that the changes in entropy and enthalpy calculated from data pertaining to striated muscle plotted onto the same linear relationship obtained for flagella, thus suggesting a basic similarity in chemical, as well as mechanistic, terms between the two systems. Both the system of muscular contraction and of flagellar action are believed to use ATP as a source of energy, and it is therefore of interest to consider the thermodynamic parameters of an ATP–ATPase complex which have been determined *in vitro* (Ouellet *et al.*, 1952). The point which corresponds to breakdown of the ATP–ATPase complex lies closest to the line describing the flagellar response, while that obtained for the complex formation is a significant distance from the line. It is accordingly tempting to speculate that the rate-limiting step within flagella is breakdown of the ATP–enzyme (ATP–dynein?) complex.

This discussion of experimental results concerns intact cells. Considerable information about flagellar mechanisms has, however, been obtained using organelles which have been treated chemically to render the membrane permeable or to remove it completely. The former situation is obtained using glycerol-based solutions while demembranation requires the use of detergents (for a review of techniques, see Goldstein, 1975). By chemically treating cilia and flagella, the axoneme is exposed to molecules in the surrounding medium and, as noted earlier, addition of ATP with the appropriate combination of ions causes the extracted system to move in a manner very similar to that observed *in vivo*. It is therefore possible to study the behaviour of the axoneme in response to changed chemical environments, assuming that the concentration of compounds in the bulk solution is that to which the axonemes are exposed.

With this type of preparation, it is possible to vary the ATP concentration to obtain information directly about the kinetics of the reactions within the axoneme and their relationship to the wave parameters. The dependence of beat frequency on ATP concentration follows Michaelis–Menten kinetics (e.g. Brokaw, 1967; Holwill, 1969) with the frequency approaching a limiting value at high concentrations of substrate. Provided the appropriate extraction procedure is followed (and the precise formulation of the solutions varies from one organism to another), the limiting frequency can be similar to that of the living organism. It was thought that the integrity of the basal structures in the extracted systems was needed for *in vivo* frequencies to be reached (e.g. Holwill, 1973) but recent work with *Crithidia oncopelti* has shown that fragments of extracted flagella which do not contain the basal region can be induced to beat at these rates (M. E. J. Holwill and J. L. McGregor, unpublished results).

Using this technique, it is also possible to assay the chemical performance of the axoneme. For instance, the rate at which ATP is hydrolysed can be determined, and it was on this basis that the relationship between beat frequency and usage of ATP mentioned earlier was determined (Brokaw, 1967). It has also been shown that the presence of a divalent cation is required for the extracted system to be motile (e.g. Gibbons and Gibbons, 1972; Douglas and Holwill, 1972), with magnesium the best activator, manganese weaker and calcium inducing but feeble motility.

A direct comparison of the behaviour of intact and extracted flagella is also possible by exposing both systems to changed physical environments. Studies of the thermal dependence of flagella activity in two species of sea urchin showed that the activation parameters of enthalpy and entropy changed by the same amount for the whole and chemically treated cells of each species (Holwill, 1969), thus giving support to the assumption inherent in many studies that the reactions giving rise to motility *in vivo* and *in vitro* are identical.

Changes in the viscosity of the environment also modify the motility of intact and chemically extracted cells, as implied in Section IV (p. 16), and considerable use has been made of this effect (see Brokaw and Gibbons, 1975, for references). Different organisms respond in varied ways to the changes. It is usual for both the wavelength and frequency to decrease as the viscosity is increased but, for some cells, the power output per beat decreases while in others it increases

(Brokaw, 1966; Rikmenspoel et al., 1973). When combined with measurements of metabolic rate at different viscosities, the indications are that the efficiencies, expressed as hydrodynamic work done against external viscous resistances for each mole of ATP hydrolysed, remain unchanged or increase significantly (Brokaw and Gibbons, 1975). In experiments which examine the effects of changing the ATP concentration and environmental viscosity, Brokaw (1974a, 1975a) found that, at low ATP concentrations when the beat frequency is about 1 Hz, changes in viscosity had little effect on the rate of beating. The beat frequency was more sensitive to changes in the viscosity at high concentrations of ATP, when the beat frequency was in excess of 10 Hz. The results are consistent with the hypothesis that, at low ATP concentrations, the beat frequency is regulated by an intrinsic rate-limiting process while the behaviour is viscosity-limited at high ATP concentrations. The mechanism can be visualized in terms of a two-step reaction sequence in the mechanochemical cycle underlying motility; one step would be sensitive to changes in ATP concentration while the other would alter in response to changing viscosity. The beat frequency (f) would then be governed by the relation:

$$1/f = 1/r_1 + 1/r_2$$

where r_1 is a function of viscosity and r_2 is directly proportional to the ATP concentration, so that the beat frequency is related, at constant viscosity, to the ATP concentration by Michaelis–Menten kinetics as experimentally observed. Experimental viscosity data can be matched by setting $1/r_1 = c\eta^\gamma$ where c and γ are constants and η is the co-efficient of viscosity of the fluid environment. A prediction, based on the control of wavelength by an internal bending viscosity, that γ should have a value of 0.25 is not satisfied. Two other models which might reasonably account for the experimental results were considered. One of these was that there might be a maximum possible rate of sliding between filaments which could be reached at the small viscous loadings prevailing at low beat frequencies; alterations in the viscosity would not then influence beat frequency. The other model considered was that the elastic properties of the flagellum might become dominant at low frequencies and hence control the frequency (see also Brokaw, 1975b). Neither of these models was able to account satisfactorily for all of the experimental results. Further developments of the model are required if the predictions are to agree with observations, but the system clearly has potential for providing information about intraflagellar processes.

In the real flagellum, motility is a direct result of coupling between chemical activity within the cell and mechanical properties of the system. Several theoretical models have been proposed which incorporate this coupling (e.g. Miles and Holwill, 1971; Lubliner and Blum, 1971; Brokaw and Rintala, 1975) and each provides predictions which correspond to certain experimental observations. The model formulated by Brokaw has been developed to the greatest degree (see Section V, p. 3) and is, perhaps, the most versatile of those which have been proposed. Each model contains a significant number of variable parameters, the choice of which depends on experiment and intuition. It is clear that a considerable amount of further experimental work is required to test the models in more detail.

VII. Control of Flagellar and Ciliary Motion

Re-orientation of many organisms in response to stimuli depends on a modification of the motion of their cilia or flagella, thus indicating that cellular control is exercised over the activity of these organelles. The change in beat pattern assumes various forms depending on the organisms considered. In species of *Paramecium* and *Opalina* and in similar protozoa, re-orientation of the organism which constitutes the avoiding reaction is brought about by altering the direction of the effective strokes of the cilia, so that the metachronal waves change their direction of propagation. The cilia can, and frequently do, change the direction of their effective strokes by 180°, but intermediate positions have been observed, especially under conditions of controlled stimulation (e.g. Naitoh and Eckert, 1975). Biflagellates, such as species of *Chlamydomonas*, which normally swim by propagating asymmetric bends along their flagella from base to tip to give the appearance of a breast-stroke action, reverse their direction of swimming by extending the flagella before the cell and propagating symmetric waves, also from base to tip (e.g. Lewin, 1952). Re-orientation in a chemotactic response of the male gametes of species of *Allomyces*, and probably of *Tubularia* sperm, is effected by the flagellum momentarily ceasing its motion and turning the head in a new direction before flagellar motility resumes (Miles and Holwill, 1969; Miller and Brokaw, 1970). In *Crithidia oncopelti* and other trypanosomes, reversal of the direction of movement is effected by changing the direction of wave propagation from the normal tip-to-base to become base-to-tip (e.g. Holwill, 1965, 1966a). Flagella, which often execute a tonsate beat involving part or all of their length, e.g. the free flagellum of species of *Peranema*, are able to alter the forms of motion to become undulatory (e.g. Lowndes, 1941). While these descriptions refer to protozoa, it is known that metazoan cilia stop their movement in response to a nervous stimulus (e.g. Mackie *et al.*, 1974; Takahashi and Murakami, 1968).

Of the systems described, ciliary reversal in *Paramecium* spp. has been the subject of the most exhaustive studies which have been recently comprehensively reviewed (Naitoh and Eckert, 1975). Other organisms which have been studied in relation to ciliary and flagellar control include bracken spermatozoids (Brokaw, 1975b), *Crithidia oncopelti* (Holwill and McGregor, 1975), and species of *Chlamydomonas* (Hyams and Borisy, 1975), *Tubularia* sperm (Miller, 1975) and *Elliptio* spp. (Satir, 1975b). The directional responses of the respective cilia and flagella are sensitive to calcium ions in all organisms. This conclusion is reached on the basis of experiments in which the ionic environment of intact and chemically demembranated organelles is altered. The ionophore A 23187, which transports divalent cations and in particular calcium across membranes, has been used to study the behaviour of *Crithidia* flagella and *Elliptio* cilia. The critical calcium concentration for motility appears to be in the region of 10^{-4} mol m^{-3} when demembranated organelles are studied. Below this concentration, normal motility is observed while, above it, the orienting response develops. It may also be significant that the concentration of calcium ion influences the symmetry of demembranated sea-urchin spermatozoa (Brokaw *et al.*, 1974). Omission of calcium ions from the extracting solution or an increase in the concentration of this ion in the re-activating solution caused asymmetric propagation while some flagella exhibited local flexing. The critical concentration for the sperm behaviour is again about 10^{-4} mol m^{-3}.

The origin of the calcium ions required to mediate the various responses also varies from one species to another. In several of the organisms discussed, calcium ions are derived from the medium external to the cell (e.g. Naitoh and Kaneko, 1973; Brokaw, 1974b; Miller, 1975; Satir, 1975b). When the appropriate signal is given, channels open in the membrane to admit calcium ions to the cell interior and, when the flagellar or ciliary response is completed, the ion is actively removed from the cell by a calcium "pump". In *Crithidia oncopelti*, it appears that there is sufficient Ca^{2+} inside the cell to allow the wave reversal to occur on stimulation, since the response occurs with the Ca^{2+}-sequestering agent EGTA in the external medium.

The site at which calcium ions are active within the flagellum is obviously of great interest. It is likely that all of the phenomena described are to be explained in terms of the same internal mechanism which affects the bend formation and propagation. From the viewpoint of the sliding filament model, calcium ions would probably either influence the active filament sliding directly or modify the transduction mechanism (i.e. probably the radial spoke–central sheath interaction). Wave reversal in *Crithidia oncopelti* and ciliary reversal in species of *Paramecium* could be induced by a reversal in the relative direction of sliding between filaments or by altering the effective stiffness of the system in the appropriate manner by controlling the attachment sites of the radial spokes. Efforts to detect a site of calcium ion action by electron microscopy have made use of a technique for producing electron-dense regions which are calcium dependent (Oschman and Wall, 1972). In the technique, large concentrations (up to 75 mol m^{-3}) of calcium ion were added to the specimen during the preparative procedures, and produced deposits in the proximal region of *Paramecium* cilia (Fisher and Keneshiro, 1975; Plattner, 1975). In view of the difference in concentrations used and those present within a living cell (perhaps 0.5 mol m^{-3}), these results must be regarded carefully, but it is tempting to suggest that, in the proximal portion of the cilia, there are calcium-affinity sites which may be involved in control of motility. It will be interesting to discover whether such sites exist in other cilia and flagella, and whether techniques can be devised to detect the location of calcium ions at much lower concentrations in these organelles.

The flagellar and ciliary actions already discussed occur in response to stimulation of an electrical, chemical or tactile nature. To respond, the motile elements within the flagellum are influenced by calcium ions admitted to or released from within the cell on receipt of a signal. The most obvious signal is an electrical one carried by the membrane and, in the case of species of *Opalina* and (particularly) *Paramecium*, extensive experiments (described by Naitoh and Eckert, 1975) have shown that the ciliary responses of these organisms are correlated with alterations in membrane potential. It is likely that electrophysiological experiments involving re-orientating responses in various organisms will also reveal correlated bio-electric membrane signals.

VIII. Concluding Remarks

This survey describes the considerable developments which have occurred in the fundamental aspects of ciliary and flagellar motility during the past decade. Many developments have been a direct result of improved observational and analytical methods, and have led to an increasingly widespread use of the axonemal system in the study of mechanochemical coupling at the cellular level. The discussion has concentrated on the more biophysical aspects of the problem, but it is clear that a knowledge of the biochemical behaviour of the microtubular macromolecules (tubulin) is necessary for a complete understanding of the system. Developments in this area have been recently reviewed by Stephens (1975) and Mohri and Ogawa (1975). Genetic control of motility has not been described, although the results of work by Kung (1971) are implicit in the control of motility in *Paramecium* spp. It has been possible to produce a mutant of this protozoon which can move in one direction only, and the results indicate that the membrane of the variant has been modified (Kung and Eckert, 1972; Kung and Naitoh, 1973). A non-motile mutant of *Chlamydomonas reinhardii* has the central fibres disrupted (Warr et al., 1966), suggesting their importance in motility, although it is believed that organelles with $9 + 0$ fibrillar complexes are motile. This is a promising area of reasearch which is now yielding valuable information about the motile process (Warr, 1974; Allen and Borisy, 1974).

Many flagellated cells do not have the $9 + 2$ fibrillar complex characteristic of the eukaryotic organelles discussed in this review. The majority of these are the sperm tails of various insects (e.g. Phillips, 1975; Afzelius, 1975) which contain microtubules in a variety of arrangements and have motility patterns which can be complex, but little critical work on the motion of these cells has as yet been undertaken. A comparative study of the motility of these cells with attendant structural correlations would provide important information of relevance to the general problems of motility in which microtubules are implicated.

REFERENCES

Afzelius, B. A., (ed.) (1975). "The Functional Anatomy of the Spermatozoon", pp. 251–339. Pergamon Press, Oxford.
Aiello, E. and Sleigh, M. A. (1972). *Journal of Cell Biology* **54**, 403.
Allen, C. and Borisy, G. G. (1974). *Journal of Cell Biology* **63**, 5a.
Baba, S. A. and Hiramoto, Y. (1970). *Journal of Experimental Biology* **52**, 675.
Blake, J. R. (1971a). *Journal of Fluid Mechanics* **46**, 199.
Blake, J. R. (1971b). *Journal of Fluid Mechanics* **49**, 209.
Blake, J. R. (1971c). *Bulletin of the Australian Mathematical Society* **5**, 255.
Blake, J. R. (1972). *Journal of Fluid Mechanics* **55**, 1.
Blake, J. R. and Sleigh, M. A. (1974). *Biological Reviews* **49**, 85.
Bouck, G. B. (1971). *Journal of Cell Biology* **50**, 362.
Bouck, G. B. (1972). *In* "Advances in Cell and Molecular Biology", (E. J. Dupraw, ed.), vol. 2, pp. 237–271. Academic Press, New York.
Bradfield, J. R. G. (1955). *Symposium of the Society for Experimental Biology* **9**, 306.
Brandts, J. F. (1967). *In* "Thermobiology", (A. H. Rose, ed.), pp. 25–72. Academic Press, London.
Brennen, C. (1974). *Journal of Fluid Mechanics* **65**, 799.
Brennen, C. (1975). *In* "Swimming and Flying in Nature", (T. Y. Wu, C. J. Brokaw and C. Brennen, eds.), vol. 1, pp. 235–251. Plenum Press, New York.
Brokaw, C. J. (1965). *Journal of Experimental Biology* **43**, 155.
Brokaw, C. J. (1966). *Journal of Experimental Biology* **45**, 113.
Brokaw, C. J. (1967). *Science, New York* **156**, 76.
Brokaw, C. J. (1970). *Journal of Experimental Biology* **53**, 445.
Brokaw, C. J. (1972a). *Biophysical Journal* **12**, 564.
Brokaw, C. J. (1972b). *Journal of Mechanochemistry and Cell Motility* **1**, 151.
Brokaw, C. J. (1972c). *Journal of Mechanochemistry and Cell Moltility* **1**, 203.
Brokaw, C. J. (1974a). *Journal of Cell Biology* **63**, 38a.
Brokaw, C. J. (1974b). *Journal of Cellular Physiology* **83**, 151.
Brokaw, C. J. (1975a). *Journal of Experimental Biology* **62**, 701.
Brokaw, C. J. (1975b). *Biological Journal of the Linnean Society* **7**, Supplement 1, 423.
Brokaw, C. J. and Benedict, B. (1968). *Journal of General Physiology* **52**, 283.
Brokaw, C. J. and Gibbons, I. R. (1973). *Journal of Cell Science* **13**, 1.
Brokaw, C. J. and Gibbons, I. R. (1975). *In* "Swimming and Flying in Nature", (T. Y. Wu, C. J. Brokaw, and C. Brennen, eds.), vol. 1, pp. 89–126. Plenum Press, New York.
Brokaw, C. J., Josslin, R. and Bobraw, L. (1974). *Biochemical and Biophysical Research Communications* **58**, 795.
Brokaw, C. J. and Rintala, D. R. (1975). *Journal of Mechanochemistry and Cell Motility* **3**, 77.
Brokaw, C. J. and Wright, L. (1963). *Science, New York* **142**, 1169.
Burnasheva, S. A., Ostrovskaya, M. V. and Yurzina, G. A. (1968). *Doklady Biological Sciences* **181**, 375.
Cheung, A. T. W. and Winet, H. (1975). *In* "Swimming and Flying in Nature", (T. Y. Wu, C. J. Brokaw and C. Brennen, eds.), vol. 1, pp. 223–234. Plenum Press, New York.
Chwang, A. T. and Wu, T. Y. (1971). *Proceedings of the Royal Society of London, Series B* **178**, 327.
Coakley, C. J. and Holwill, M. E. J. (1974). *Journal of Experimental Biology* **60**, 605.
Costello, D. P. (1973a). *Biological Bulletin* **145**, 279.
Costello, D. P. (1973b). *Biological Bulletin* **145**, 292.

Curle, N. and Davies, H. J. (1968). "Modern Fluid Dynamics", 290 pp. van Nostrand Reinhold Company, London.
Dallai, R., Bernini, F. and Guisti, F. (1973). *Journal of Sub-Microscopical Cytology* **5**, 137.
Douglas, G. J. and Holwill, M. E. J. (1972). *Journal of Mechanochemistry and Cell Motility* **1**, 213.
Fisher, G. W. and Kaneshiro, E. S. (1975). *Journal of Cell Biology* **67**, 115a.
Gibbons, B. H. and Gibbons, I. R. (1972). *Journal of Cell Biology* **54**, 75.
Gibbons, B. H. and Gibbons, I. R. (1973). *Journal of Cell Science* **13**, 337.
Gibbons, B. H. and Gibbons, I. R. (1974). *Journal of Cell Biology* **63**, 970.
Gibbons, I. R. (1975). *In* "Molecules and Cell Movement", (S. Inoué and R. E. Stephens, eds.), pp. 207–232. Academic Press, New York.
Gibbons, I. R. and Grimstone, A. V. (1960). *Journal of Biophysical and Biochemical Cytology* **7**, 697.
Gilula, N. B. and Satir, P. (1972). *Journal of Cell Biology* **53**, 494.
Glasstone, S., Laidler, K. and Eyring, H. (1941). "The Theory of Rate Processes", 611 pp. McGraw-Hill, London.
Goldstein, S. F. (1975). *In* "Cilia and Flagella", (M. A. Sleigh, ed.), pp. 111–130. Academic Press, London.
Goldstein, S. F., Holwill, M. E. J. and Silvester, N. R. (1970). *Journal of Experimental Biology* **53**, 401.
Goldstein, S. F. and Pivonka, P. R. (1975). *Journal of Cell Biology* **67**, 137a.
Gray, J. and Hancock, G. J. (1955). *Journal of Experimental Biology* **32**, 802.
Holwill, M. E. J. (1965). *Journal of Experimental Biology* **42**, 125.
Holwill, M. E. J. (1966a). *Physiological Reviews* **46**, 696.
Holwill, M. E. J. (1966b). *Journal of Experimental Biology* **44**, 579.
Holwill, M. E. J. (1967). *In* "Proceedings of the 7th International Congress on High-Speed Photography", p. 265. Helwich, Darmstadt.
Holwill, M. E. J. (1969). *Journal of Experimental Biology* **50**, 203.
Holwill, M. E. J. (1970). *Acta Protozoologica* **7**, 301.
Holwill, M. E. J. (1973). *Science Progress, Oxford* **61**, 63.
Holwill, M. E. J. (1975). *In* "Cilia and Flagella", (M. A. Sleigh, ed.), pp. 143–175. Academic Press, London.
Holwill, M. E. J. (1977). *In* "Scale Effects in Animal Locomotion", (T. J. Pedley, ed.), pp. 232–242. Academic Press, London.
Holwill, M. E. J. and Burge, R. E. (1963). *Archives of Biochemistry and Biophysics* **101**, 249.
Holwill, M. E. J. and McGregor, J. L. (1974). *Journal of Experimental Biology* **60**, 437.
Holwill, M. E. J. and McGregor, J. L. (1975). *Nature, London* **255**, 157.
Holwill, M. E. J. and Peters, P. D. (1974). *Journal of Cell Biology* **62**, 322.
Holwill, M. E. J. and Silvester, N. R. (1967). *Journal of Experimental Biology* **47**, 249.
Holwill, M. E. J. and Sleigh, M. A. (1967). *Journal of Experimental Biology* **47**, 267.
Hyams, J. S. and Borisy, G. G. (1975). *Journal of Cell Biology* **67**, 186a.
Jahn, T. L. and Votta, J. J. (1972). *Annual Reviews of Fluid Mechanics* **4**, 93.
Jahn, T. L., Landman, M. D. and Fonseca, J. R. (1964). *Journal of Protozoology* **11**, 291.
Keller, S. R., Wu, T. Y. and Brennen, C. (1975). *In* "Swimming and Flying in Nature", (T. Y. Wu, C. J. Brokaw, and C. Brennen, eds.), vol. 1, pp. 253–271. Plenum Press, New York.
Kiefer, B. I. (1970). *Journal of Cell Science* **6**, 177.
Knight-Jones, E. W. (1954). *Quarterly Journal of Microscopical Sciences* **95**, 503.
Kung, C. (1971). *Genetics* **69**, 29.
Kung, C. and Eckert, R. (1972). *Proceedings of the National Academy of Sciences of the United States of America* **69**, 93.

Kung, C. and Naitoh, Y. (1973). *Science, New York* **179**, 195.
Laidler, K. J. (1965). "Chemical Kinetics", 566 pp. McGraw-Hill, New York.
Lewin, R. A. (1952). *Biological Bulletin* **103**, 74.
Lighthill, M. J. (1969). *Annual Review of Fluid Mechanics* **1**, 413.
Lighthill, M. J. (1975). "Mathematical Biofluiddynamics", 102 pp. Society for Industrial and Applied Mathematics, Philadelphia.
Lowndes, A. G. (1941). *Proceedings of the Zoological Society of London* **113**, 99.
Lubliner, J. and Blum, J. J. (1971). *Journal of Theoretical Biology* **31**, 1.
Lunec, J. (1975). *In* "Swimming and Flying in Nature", (T. Y. Wu, C. J. Brokaw and C. Brennan, eds.), vol. 1, pp. 143–159. Plenum Press, New York.
Machemer, H. (1972). *Journal of Experimental Biology* **57**, 239.
Mackie, G. O., Paul, D. H., Singla, C. M., Sleigh, M. A. and Williams, D. E. (1974). *Proceedings of the Royal Society of London, Series B* **187**, 1.
Mohri, H. and Ogawa, K. (1975). *In* "The Functional Anatomy of the Spermatozoon", (B. A. Afzelius, ed.), pp. 161–167. Pergamon Press, Oxford.
Machin, K. E. (1958). *Journal of Experimental Biology* **35**, 796.
Machin, K. E. (1963). *Proceedings of the Royal Society of London, Series B* **158**, 88.
Machemer, H. (1975). *In* "Cilia and Flagella", (M. A. Sleigh, ed.), pp. 199–286. Academic Press, London.
Miles, C. A. and Holwill, M. E. J. (1969). *Journal of Experimental Biology* **50**, 683.
Miles, C. A. and Holwill, M. E. J. (1971). *Biophysical Journal* **11**, 851.
Miller, R. L. (1975). *Journal of Cell Biology* **67**, 285a.
Miller, R. L. and Brokaw, C. J. (1970). *Journal of Experimental Biology* **52**, 699.
Naitoh, Y. and Eckert, R. (1975). *In* "Cilia and Flagella", (M. A. Sleigh, ed.), pp. 305–352. Academic Press, London.
Naitoh, Y. and Kaneko, H. (1973). *Journal of Experimental Biology* **58**, 657.
Oschman, J. L. and Wall, B. J. (1972). *Journal of Cell Biology* **55**, 58.
Ouellet, L., Laidler, K. J. and Morales, M. F. (1952). *Archives of Biochemistry and Biophysics* **39**, 37.
Parducz, B. (1967). *International Review of Cytology* **21**, 91.
Pedersen, H. (1970). *Journal of Ultrastructure Research* **33**, 451.
Phillips, D. M. (1975). *In* "Cilia and Flagella", (M. A. Sleigh, ed.), pp. 379–402. Academic Press, London.
Pironneau, O. and Katz, D. F. (1975). *In* "Swimming and Flying in Nature", (T. Y. Wu, C. J. Brokaw and C. Brennen, eds.), vol. 1, pp. 161–172. Plenum Press, New York.
Pitelka, D. R. and Schooley, C. N. (1955). *University of California Publications in Zoology* **61**, 79.
Plattner, H. (1975). *Journal of Cell Science* **18**, 257.
Reynolds, A. J. (1965). *Journal of Fluid Mechanics* **23**, 241.
Rikmenspoel, R., Jacklet, A. C. and Orris, S. E. (1973). *Journal of Mechanochemistry and Cell Motility* **2**, 7.
Satir, P. (1963). *Journal of Cell Biology* **18**, 345.
Satir, P. (1965). *Journal of Cell Biology* **26**, 805.
Satir, P. (1968). *Journal of Cell Biology* **39**, 77.
Satir, P. (1974). *Scientific American* **231**, 45.
Satir, P. (1975a). *In* "Cilia and Flagella", (M. A. Sleigh, ed.), pp. 131–142. Academic Press, London.
Satir, P. (1975b). *Science, New York* **190**, 586.
Shimada, K., Yoshida, T. and Asakura, S. (1975). *In* "Swimming and Flying in Nature", (T. Y. Wu, C. J. Brokaw, and C. Brennen, eds.), vol. 1, pp. 31–43. Plenum Press, New York.

Silvester, N. R. and Holwill, M. E. J. (1972). *Journal of Theoretical Biology* **35**, 505.
Sleigh, M. A. (1964). *Quarterly Journal of Microscopical Science* **105**, 405.
Sleigh, M. A. (1972). *In* Essays in Hydrobiology", (R. B. Clark and R. Wooton, eds.), pp. 119–136. University of Exeter.
Sleigh, M. A., (ed.) (1975a). "Cilia and Flagella", 500 pp. Academic Press, London.
Sleigh, M. A. (1975b). *In* "Cilia and Flagella", (M. A. Sleigh, ed.), pp. 79–92. Academic Press, London.
Sleigh, M. A. (1975c). *In* "Cilia and Flagella", (M. A. Sleigh, ed.), pp. 287–304. Academic Press, London.
Sleigh, M. A. and Aiello, E. (1973). *Acta Protozoologica* **11**, 265.
Sleigh, M. A. and Blake, J. R. (1977). *In* "Scale Effects in Animal Locomotion", (T. J. Pedley, ed.), pp. 243–256. Academic Press, London.
Sleigh, M. A. and Holwill, M. E. J. (1969). *Journal of Experimental Biology* **50**, 733.
Stephens, R. E. (1970). *Biological Bulletin* **139**, 438.
Stephens, R. E. (1975). *In* "Cilia and Flagella", (M. A. Sleigh, ed.), pp. 39–76. Academic Press, London.
Summers, K. E. and Gibbons, I. R. (1971). *Proceedings of the National Academy of Sciences of the United States of America* **68**, 3092.
Takahashi, K. and Murakami, A. (1968). *Journal of the Faculty of Science, University of Tokyo* **11**, 359.
Taylor, G. I. (1951). *Proceedings of the Royal Society of London, Series A* **209**, 447.
Tuck, E. O. (1968). *Journal of Fluid Mechanics* **31**, 305.
Warner, F. D. (1970). *Journal of Cell Biology* **47**, 220a.
Warner, F. D. (1975). *In* "Cilia and Flagella", (M. A. Sleigh, ed.), pp. 11–37. Academic Press, London.
Warner, F. D. and Satir, P. (1974). *Journal of Cell Biology* **63**, 35.
Warr, J. R. (1974). *Subcellular Biochemistry* **3**, 149.
Warr, R., McVittie, A., Randall, Sir John and Hopkins, J. (1966). *Genetics Research* **7**, 335.
Yoshida, T., Shimada, K. and Asakura, S. (1975). *Journal of Mechanochemistry and Cell Motility* **3**, 87.

Note Added in Proof

Recent observations (Warner, 1976) have shown that corresponding dynein arms in the outer and inner rows are longitudinally displaced from each other by 3–4 nm, rather than being in lateral register as suggested on p. 4. Further evidence in support of the sliding microtubule hypothesis (p. 25), and information concerning the interaction between the microtubules and the dynein arms, have been obtained in elegant experiments performed by Sale and Satir (1976; and unpublished results). These authors treated isolated *Tetrahymena* cilia with detergent and trypsin to cause interdoublet sliding as described on p. 26 for sea-urchin spermatozoa, and subsequently observed their preparations in the electron microscope. Because the arrangement of spokes within a group is asymmetric (see p. 5), this feature provides a means whereby the direction of the base and tip of an organelle can be identified. With this information, Sale and Satir (1976) deduce that relative sliding occurs in one direction only, such that the microtubule to which the arms are attached pushes the adjacent microtubule towards the tip.

Experiments on waveform analysis (p. 7) have been facilitated by the development of a computer-controlled instrument which automatically tracks along a flagellar image and provides data prepared for further computer analysis (Silvester and Johnston, 1976). Preliminary results obtained using this device suggest that waves on the flagellum of *Crithidia oncopelti* are closer in form to the arc-line shape than to the meander.

Brennan (1976) has performed a more rigorous hydrodynamic analysis of the behaviour of hispid flagella than that of Holwill and Sleigh (1967) referred to on p. 9. The later work supports the earlier quantitative conclusions, and shows that the mastigonemes must be essentially rigid to be effective in producing reversed movements. If the stiffness of the mastigonemes falls below 3.5×10^{-25} Nm^2, they become ineffective and the flagellum behaves as if it were smooth. Brennan (1976) reinforces the conclusion of Holwill and Sleigh (1967) that, if the mastigonemes are passive, they should lie in the plane of beating to provide the maximum reverse thrust.

Brokaw (1975, 1976a, b) has continued to develop the computer simulation of flagellar movement discussed on p. 31. It is shown that, as an alternative to control by curvature of an organelle, if the cross-bridges between the filaments have self-oscillatory properties, propagated waves are possible. The 9 + 2 pattern of microtubules in the flagellar axoneme is shown to be an efficient means of producing planar bending in an organelle (Brokaw, 1977) and provides justification for the simplification to a two-fibre model used during the simulation analysis.

Studies on the response of flagella to pressure and temperature (p. 36) have been extended by examining the combined effects of changing these parameters and the viscosity of the medium (M. E. J. Holwill and J. Wais-Steider, unpublished observations). It is found that, for the flagellum of *Crithidia oncopelti*, as the viscosity of the medium is increased the beat frequency of the cells becomes less sensitive to the effects of pressure, until at a particular viscosity the beat frequency remains unaltered as the pressure is changed. For viscosities higher than this particular value, increasing the pressure causes an increase in the beat frequency, in direct contrast to the results obtained at lower viscosities. The results can be interpreted phenomenologically in terms of the product of two rate constants rather than the expression used by Brokaw (p. 39) in his interpretation of the combined effects of viscosity and ATP concentration on the beat frequency of sea-urchin spermatozoa.

The implication of calcium ions in ciliary reversal in *Paramecium* (p. 41) is supported by experiments involving intracellular injection of calcium (Saiki and Hiramoto, 1975). Tsuchiya and Takahashi (1976) confirm the occurrence of calcium-dependent electron-dense regions in the basal region of *Paramecium* cilia treated as described on p. 42.

Additional References

Brennen, C. (1976). *Journal of Mechanochemistry and Cell Motility* 3, 207.
Brokaw, C. J. (1975). *Proceedings of the National Academy of Sciences of the United States of America* 72, 3102.
Brokaw, C. J. (1976a). *Biophysical Journal* 16, 1013.
Brokaw, C. J. (1976b). *Biophysical Journal* 16, 1029.
Brokaw, C. J. (1977). *Journal of Mechanochemistry and Cell Motility* 4, 101.
Saiki, M. and Hiramoto, Y. (1975). *Cell structure and Function* 1, 33.
Sale, W. S. and Satir, P. (1976). *Journal of Cell Biology* 71, 589.
Silvester, N. R. and Johnston, D. N. (1976). *Journal of Physics E: Scientific Instruments* 9, 990.
Tsuchiya, T. and Takahashi, K. (1976). *Journal of Protozoology* 23, 523.
Warner, F. D. (1976). *Journal of Cell Science* 20, 101.

Does the Initiation of Chromosome Replication Regulate Cell Division?

ARTHUR L. KOCH*

Division of Biological and Medical Research, Argonne National Laboratory, Argonne, Illinois 60439

I.	Introduction	50
II.	Models for the Regulation of Cell Division and Chromosome Replication	53
III.	The Necessity for Correlation of Events in Cell Cycles of Related Individuals	54
IV.	The Role of DNA Initiation in Regulating Cell Division	57
V.	Computer Simulation of the Cell Cycle Based on the Deterministic Principle	58
VI.	Pulse Autoradiography of Slowly Dividing Cells: The Experimental Basis for Computer Simulation	63
	A. Analysis of Autoradiographic Grain Count Data	65
VII.	Experiments Concerning the Variation in the Initiation of Chromosome Replication	69
	A. Other Observations	71
VIII.	Possible Experimental Objections	72
	A. Remaining Admonitions	74
IX.	Variability of the Time Between Nuclear Division and Cell Division .	76
X.	Temporal Accuracy of DNA Synthesis Cycle	80
XI.	Computation of the Fraction of Cells of a Size Class Engaged in DNA Replication	81
XII.	Precision of Initiation of Chromosome Regulations in *Myxococcus xanthus* .	90
XIII.	Conclusions	93
XIV.	Acknowledgements	95
	References	95

* Permanent and present address: Department of Microbiology, Indiana University, Bloomington, IN 47401.

49

I. Introduction

It is universally agreed that the cell cycle of eucaryotes is composed of an orderly sequence of steps; that is, the cell must proceed through the many stages that comprise each of G_1, S, G_2, and the four discernible phases of mitosis. It is believed that each stage only starts at the completion of the previous stage. Perhaps the only variation in this theme is the reversible transition from this sequence into and out of a special cell compartment, designated G_0, in which cells do not progress through the stages but remain for an indefinite time in limbo. The cell cycle in procaryotes appears to differ from this pattern. This was apparent during earlier days when five independent studies showed that the time spent in DNA synthesis was nearly 100% of the cell division cycle (Schaechter et al., 1959; McFall and Stent, 1959; Abbo and Pardee, 1960; Young and Fitz-James, 1959; Pachler et al., 1965). In fact, it was reported that DNA synthesis was continuous throughout the cell cycle. We know now that, even for cells growing rapidly, DNA synthesis is not continuous, but is composed of an overlapping chromosome replication process, and that the actual gaps only can be detected in slowly growing cultures (Lark, C., 1966; Kubitschek et al., 1967; Helmstetter, 1967). Modern estimates of the mean fraction of the cycle spent in the synthesis of a complement of DNA are shown in Table 1, using the nomenclature shown in Table 2. In the enteric procaryote, DNA synthesis spans most of the cell cycle at fast and moderate growth rates, and a number of reports suggest that this is so even in very slowly growing cultures. But this difference between pro- and eucaryotes in the proportion of time spent in DNA synthesis only stresses one of the basic differences between the cell cycles of these two major classes of organisms. A second difference is that caryokinesis and cytokinesis are not closely connected in time in procaryotes; while in eukaryotes, nuclear division precedes cell division by a few minutes—in some cases only by an infinitesimal fraction of the cell cycle. In contrast to eucaryotes, Schaechter et al. (1962) showed that, in bacteria, the nuclear division event in a number of cases occurred one-half a cell cycle before cell division. The average age during the cycle at which a round of DNA synthesis stops depends on the growth rate. Cooper and Helmstetter (1968) presented evidence that at moderate to fast growth rates, the average length of time required to replicate the chromosome ($\overline{C}$), and the average delay before division ($\overline{D}$), are each approximately constant and independent of the average cell division

	Thymine requirement	Strain	$\bar{C}/\bar{T}$ $\bar{T}$ faster than 45 min	$\bar{C}/\bar{T}$ $\bar{T}$ slower than 45 min	Range of doubling time (min)
From Millipore filtration analysis—Cooper and Helmstetter (1968)	Thy$^+$	B/r	$41/\bar{T}$	0.65	
From composite of tracer and analytical technique—data summarized in Chai and Lark (1970)	Thy$^-$	15/T$^-$	$40/\bar{T}$	0.63–$7.49/\bar{T}$	22–270 (6)[d]
From analytical data alone Kubitschek and Freedman (1971)	Thy$^+$	B/r	$47/\bar{T}$	$47/\bar{T}$	20–1200 (49)
From autoradiography[a]					
Chai and Lark (1970)	Thy$^-$	15T$^-$		$0.67 = 80/120$	120
Forro (unpublished)	Thy$^-$			$0.50 = 70/140$	140
Pierucci and Zuchowski (1973)	Thy$^+$			$0.65 = 78/120$	43,120
From gene frequency in balanced growth					
Bird et al. (1972)[c]	Thy$^-$	K12	$27.5/\bar{T}$		25, 39
Masters and Broda (1971)[c]		K12/B/r	$28/\bar{T}$		20–25 (3)
Chandler et al. (1975)	Thy$^-$, Thy$^+$	K12	$40/\bar{T}$	$40/\bar{T}$	22–220 (8)
From mutation studies of synchronized cells—Hohlfeld and Vielmetter (1973)	Thy$^+$	K12	30/42		42
From increase in DNA after initiation block—Zaritsky and Pritchard (1971)[b]	Thy$^-$	K12, 15T$^-$	48/48, 52/53		48, 52
From rate stimulation after DNA inhibitor is removed Zaritsky and Pritchard (1971)[b]	Thy$^-$		45/50	40/60, 47/52–60	40, 64 52–60

[a] Values calculated this paper.
[b] Values in the presence of deoxyguanosine.
[c] Values calculated by Painter (1974).
[d] Values in parenthesis are the number of essentially different growth rates studied.

TABLE 2. Abbreviations used to define the parameters of the cell cycle of individuals and populations

	Size, mean size critical size at	Interval and mean interval between	Age and mean age at
Birth	$b, \bar{b}, b_c$		0
		$B, \bar{B}$	
Initiation of DNA synthesis	$c, \bar{c}, c_c$		
		$C, \bar{C}$	$B + C, \bar{B} + \bar{C}$
Termination of DNA synthesis	$f, \bar{f}, f_c$		
			$T = B + C + D$
Nuclear division	$n, \bar{n}, n_c$	$D, \bar{D}$	$\bar{T} = \bar{B} + \bar{C} + \bar{D}$
Cellular division	$d, \bar{d}, d_c$		

time ($\bar{T}$). Thus, nuclear division occurs earlier and earlier in the cell cycle as the growth rate increases. Cooper and Helmstetter (1968) also presented evidence that, at slow growth rates, chromosome replication is completed two-thirds of the way through the cell cycle, but Table 1 shows that this view is far from universally held.

The Cooper–Helmstetter model explicitly assumes that the procaryote proceeds step by step through the cell cycle, as does the eucaryote, but is capable of working on several cycles at once: each cycle an orderly sequence like all the others, simply displaced in phase. However, the facts are also consistent with a radically different mode of operation in which cell division does not depend on the previous completion of chromosomal replication a definite time before. Rather, multiple signals responding to different levels of macromolecular accumulation may independently trigger separate phases such as chromosome replication and cell division. I believe that the question of how many independent signals (one, two, or more) is much more important than the current controversy concerning the position of the gap in DNA synthesis relative to the cell division cycle. Consequently, in this paper, I do not choose to debate whether $\bar{B}$ (the average gap before DNA synthesis) or $\bar{D}$ (the average gap after DNA synthesis) is longer under any specified growth conditions. I do wish to question whether size of cells engaged in initiation of DNA replication has smaller variance than the size of cells in the act of cytoplasmic division, and whether the time intervals between various stages of the cell cycle are distributed with large or small variance for cells within a balanced growing population. Although these variances of size and times can only be estimated indirectly, they, and the regulation of events of the cell cycle, are evidently connected, and I feel that the available evidence that I have assembled, together with the computer modelling, imply at least two sites in the cell cycle at which cues are taken from some measures of the cell's macromolecular synthesis to trigger parts of the process.

II. Models for the Regulation of Cell Division and Chromosome Replication

The sizes of cells in a population in the act of either cellular or nucleoid division varies only 9–15% in many circumstances (Koch and Schaechter, 1962). However, while the cell division process is relatively precise in terms of cell size, it is less precise in terms of cell age. The standard deviation of the age at division (life-length) is quite large; typically, it is about 22–27% of the mean doubling time (see Koch and Schaechter, 1962). This is to say that the range of ages during which the central 95% of the cells divide is at least eight-tenths of the mean doubling time, $\bar{T}$. The variation in nucleoid division is similarly broad.

The model of Koch and Schaechter (1962), based on the data of Schaechter et al. (1962), assumed that nuclear division resulted when the cell achieved one critical size (n_c) and that cell division is tied to them achieving a second, but larger, critical size (d_c). (For definitions of these quantities see Table 2.) In the model of Cooper and Helmstetter (1968), cell division was alternatively tied to a fixed lag, $\bar{D}$ minutes long, after termination of chromosome replication.

It is much more logical to assume that initiation of chromosome replication, rather than termination or nuclear division, is a point of cellular control. That initiation of chromosome replication takes place when the cells achieve a threshold mass (c_c) is implicit in the Cooper–Helmstetter model and explicit in the models of Donachie (1968) and of Pritchard et al. (1969), and the later papers from Helmstetter's laboratory. In fact, the essential part of Donachie's original model was that c_c does not even change as the culture medium is changed sufficiently to result in large changes in the growth rate.

Although there has been a good deal of research in this decade on the biochemical mechanism of the many processes involved both in DNA replication and in cell division, there has been little progress either experimentally or conceptually concerning the regulation of these processes in response to cellular growth. The only recent original suggestions about these matters are firstly that of Helmstetter (1974), who argues that cell membrane synthesis may have a direct role in this regulation, and secondly that of Rosenberg *et al.* (1969) who proposes that the control may be mediated in a negative fashion.

III. The Necessity for Correlation of Events in Cell Cycles of Related Individuals

In the long run, the fundamental constraint regulating nuclear and cell division, and the initiation of chromosome replication, in procaryotes in many circumstances has to be the ability of the cells to carry out, as rapidly as possible, biosynthetic processes under the existing environmental conditions. This is primarily dependent on the product of the rates of two types of processes: first, on the optimized rate for the ability of the protein of the cell to function enzymatically and structurally in energy and biosynthetic metabolism; and second, on the optimized rate for the function of protein synthetic machinery to create more of that machinery and other protein. There is no reason to believe that either of these processes would vary significantly from cell to cell. (Experimental measurements concerning this will be discussed below.) Rather the specific rate of these processes would be expected to be essentially the same in any two cells, and not subject to statistical fluctuation, since the numbers of reactants per cell is large: that is, there are tens of thousands of ribosomes, and of the individual species of tRNA, protein synthesis factors, and enzyme molecules per cell. Even though there are fewer molecules of each mRNA species, many are under negative control so that a just sufficient amount are produced for the needed output of protein, while the need is dependent on kinds of molecules present in many copies. Consequently two cells of the same size in the same growing population should increase at very nearly the same rate, independent of chance fluctuation, and at division should partition each kind of these critical entities without important chance fluctuation in proportion to the volume of each daughter cell.

The synthesis of macromolecules (particularly RNA and protein) require much more of the cells resources, by far, than do the combined processes of DNA replication and cell division, whose control is under consideration here. It would therefore be very poor economy to have these major cellular processes limited by any minor processes in terms of utilization of cell resources, as important as the processes may be. Further, if the processes of chromosome initiation and cell division are necessary, but subservient to these major processes, then in balanced growth the events of cell division would be expected to be more dependent on size than on cell age, although the dependency should itself depend on the nutritional circumstances, and other conditions. Our measurements (Schaechter et al., 1962) show that this is indeed the case, and that the coefficient of variation of the age of cells in the act of division (of the order of 20%) is approximately twice the coefficient of variation of length of rod-shaped enteric organisms in the act of division (10%), measured on the same sample of cells. Other measurements, before and since, are consistent with the hypothesis that size at division varies less than age at division.

Had the cell a division mechanism inherently consisting of a series of stages in the cell cycle, each to be timed from the completion of the last stage independently of the cell size at that time, then one would have expected the cell size at division to have a larger (or equal) coefficient of variation than that for the age distribution, since random sources of uncorrelated variation only add. Consequently any mechanism which does not lead to negative correlations between parts of a cell cycle or between cycles of related individuals is excluded. A chance fluctuation in timing of one phase must sooner or later lead to a fluctuation of opposite sign at some later phase. This logic eliminates several previously considered pure branching processes such as those proposed by Rahn (1931) and Kendall (1948), and demands statistical models containing deterministic elements.

A logical consequence of any "deterministic" model for variation-free protoplasm growth, with cell division processes subservient to it via variation-producing processes, is that they predict a negative correlation of the life-lengths of mother cells with their daughters (or later descendents), and a positive correlation of life-length of sister cells in those circumstances where cell division partitions cytoplasm equally between daughters. If cell division was controlled fundamentally by the cells' size, then one would expect an additional delay to not only lengthen its life-length but to shorten the life-length of both its daughters. The simplest theory predicted that the mother–daughter correlation be −0.5 and the sister–sister correlation be +0.5. The experimental data (ours and others) indicate that both types of correlation were more positive than predicted. However, it could be shown theoretically that the effect of non-uniformity in growth conditions, either spatially or temporally, would tend to make these correlation coefficients more positive, and therefore the experimentally observed values represented upper bounds, and do support the theory that cell size, or some related property, rather than timing from some event of the cell cycle, serve a primary role in triggering critical events of the cell cycle.

Later analysis (Koch, 1966) of the cinematographic studies of Hoffman and Frank (1965) of the distributions of combined life-length, of lines of descent spanning nine generations, fully supported a deterministic model. Data drawn from the experiment of Kubitschek (1962) gave full support to the thesis (Koch, 1966; Kubitschek, 1967) that the variance of the collective time between one cell division and another several generations later does not increase indefinitely with increasing number of generations. If the life-length in one generation is uncorrelated with that in succeeding generations, one expects that the variance will continue to increase in direct proportion to the number of generations.

New data relevant to this point are now available from the thesis of Merijean Kelley (1974) which showed that cells synchronized by a modification of the technique of Cutler and Evans (1966) remain synchronized for up to 24 hours (21 generations). In her experiments the synchronized cells were diluted, and then later again diluted into growth medium by large enough factors to discount cell–cell interaction for maintaining the synchrony. Cutler and Evans (1966) had observed the maintenance of synchrony for four generations. Failure to lose synchrony implies very strongly that there were no local variations in specific rates of biosynthesis, and that the cell division process apportions the cytoplasm extremely evenly between every pair of daughters.

These observations together form a strong basis for the assumption that the autocatalytic properties of macromolecular biosynthesis are the deterministic part of the control of the cell division process. On the other hand, other much less precise processes must take their cue from the amount of one or more macromolecules in the cell and, with some random fluctuation, trigger certain of the discontinuous stages in the genetic replication and/or the morphological parts of the cell division processes. The ultimate aim of this paper is to guess whether there are a number of such triggers for different phases of the cell cycle, or whether there is only one triggering event from which the other events of the cell cycle are precisely timed.

IV. The Role of DNA Initiation in Regulating Cell Division

As mentioned above, Donachie (1968) and, independently, Pritchard (1968) suggested that there was only a single triggering event and that was the initiation of the chromosome replication. Donachie showed that if the critical size for the initiation (c_c) was invariant with growth rate, that $\bar{m}$, the mean size of cells in the culture, would be a particular function of doubling time, dependent on $\bar{C} + \bar{D}$. This relation can be expressed as:

$$\bar{m} = kc_c 2^{(\bar{C}+\bar{D})/\bar{T}},$$

where k is a constant for any stable size distribution (and is equal to ln 2 if cell mass grows exponentially, and T has no variation within the population; see Koch and Schaechter, 1962). The quantity $\bar{C} + \bar{D}$ can be calculated from the dependence of average mass per cell at any two growth rates for which c_c is constant.

With this relationship, estimates of $\bar{C} + \bar{D}$ shown in Table 3 have been calculated from several bodies of data in the literature. The first

TABLE 3. Estimate of $\bar{C} + \bar{D}$ through the use of the Donachie (1968) relationship[a]

Origin of data and organism	Doubling time range (minutes)	$\bar{C} + \bar{D}$ (minutes)
Schaechter et al. (1959) *Salmonella typhimurium*	300–25	62
Helmstetter and Cooper (1968) *Escherichia coli* B/r	60–20	64
Dennis and Bremer (1974) *Escherichia coli* B/r	90–24	82
Kubitschek (1974) *Escherichia coli* B/r	600–22	44

[a] The relationship is based on the assumption that the critical size of the cell at the time of initiation of chromosome replication is growth rate independent.

and last sets of data are consistent with the constancy of c_c and a constant value of $\bar{C} + \bar{D}$, independent of growth rate. The data from Kubitschek's laboratory require either that the theory that c_c be constant is wrong, or that C or D, or both, are very much shorter than usually assumed, even by Kubitschek. The data from Helmstetter's laboratory are consistent with both a constant c_c and a constant $\bar{C} + \bar{D}$, with the value of the latter sum in agreement with his model for moderate to fast growth. The first three are calculated from the dependency of dry mass, as assessed turbidimetrically, per countable cell. The data from Kubitschek's laboratory are based on the modal cell size as measured by his modified Coulter counter.

Taken together these data do support Donachie's model for moderate to rapidly growing cultures. In this region there is a unanimity of result that $\bar{C} + \bar{D}$ is of the expected order of 60–70 minutes, and is growth rate independent if c_c is assumed constant. No such confidence can be expressed for slow growth rates: c_c must change if $\bar{C} + \bar{D}$ change as some claim, and it must vary in special ways for the theories espoused by others to be valid.

It seems to me that the evidence that c_c is invariant with respect to growth is not sufficient to build an elaborate theory, as has been done by Pritchard and his school. On the other hand, the theory is sufficiently appealing that I have prepared a computer simulation program to test in detail the autoradiographic experiments discussed below against the prediction of a variety of models, including initiation-controlled cell cycles. The program, and its results, are presented in the following two sections.

V. Computer Simulation of the Cell Cycle Based on the Deterministic Principle

The distribution of numbers of individuals, drawn from synchronously growing balanced culture through the cell cycle, can be approximated to a high degree of accuracy with a model for normal "balanced" growth based on the assumption of the four simple, and in many cases, quite realistic postulates of Koch and Schaechter (1962). These are: (1) Protoplasm grows deterministically, with essentially no variation from cell to cell—on a per unit protoplasm basis. I shall make the somewhat stronger assumption that all the cells grow with the same growth rate constant throughout each individual growth cycle. This is for computational convenience, but the important and critical part of the assumption is that cells born small grow proportionately slower than do the cells that were born large. This accords with our experimental findings (Schaechter et al., 1962) and has never, to my knowledge, been questioned. (2) The cell divides when it achieves a mass appropriate to the growth medium and strain. In this paper, the average size at birth is set arbitrarily and without loss of generality at 1.0. (3) Although there is some mechanism determining the size at division, fluctuation from this norm does occur, and whether this results from significant variation in the biology of the cell division process, or from our inability to identify those cells that have just actually divided, is not clear (and is immaterial, at least initially, for the present discussion). The fluctuations are small, and it will be assumed that they are Gaussian with a small coefficient of variation, usually about 10–15%. The distribution of cell sizes of the growing population is quite insensitive to the assumed shape of the size distribution at division. (4) The final assumption concerns the variation of size between newborn sister cells. In this section it will be assumed that the partition is precise. Evidence that this is valid for certain enteric organisms has been presented above (p. 55).

A computer simulation was set up in which the computer systematically chooses sizes at birth (b) and division (d) ranging from that size which is three standard deviations below the mean to three standard deviations above, in steps of 0.1 of the assumed standard deviation. All combinations of sizes at birth and division are treated. For each, the contributions to the intermediate size classes of a growing population, where m ranges from b to d, are made. The size of any cell class is designated by the variable m. As mentioned, the size scale has been arbitrarily set so that $\bar{b}$ is 1 and $\bar{d}$ is 2. The presently used computer program has storage arrays which handle the distribution of cell sizes, where m ranges from 0 to 3 in steps of 0.01.

The contributions of any particular sub-population of cells with a given value of b and d decrease as m increases from b to d. During balanced asynchronous growth, there is a smaller contribution to larger cell categories because there were fewer cells belonging to this sub-population at the earlier time when the cells that now are of the larger size under consideration were formed by fission. The contribution of cells formed at a time such that a cell of a size then of b can just have grown to size m is proportional to b/m. This can be seen as follows: let us focus on the sub-population that was born at a chosen size, b, and will divide at the chosen size, d. We designate by x_{-t} the number of individuals of this class present at any earlier time t. Then the number present, x, is given by:

$$x = x_{-t} e^{-\lambda t},$$

where λ is the growth rate constant $\simeq (\ln 2)/\overline{T}$. Now we can ask how long ago cells of current size m had a size of b. On the assumption of exponential growth we can write:

$$m = b e^{\lambda t}.$$

Consequently:

$$\frac{b}{m} = \frac{x}{x_{-t}},$$

and the number of cells of size m is proportional to b/m.

There is also an additional factor of $1/m$ because the range of each category of sizes corresponds to a different range of sizes at the birth of those cells (see Koch and Schaechter, 1962).

Thus, for the range of the discrete steps of b and d, contributions to every m category of the form of a triple product of the probability of that value of b occurring multiplied by the probability of that value of d occurring multiplied by b/m^2 is computed. These are added into the appropriate storage register corresponding to the value of m. The final sums of such products accumulated in each m size category are totalled for all sizes, and then each register is divided by this total and multiplied by 100, so that the cumulative sum for the entire population will be normalized to 100%.

This method of computation is accurate, except for numerical limitations, as compared with the analytical integral equation presented in Koch and Schaechter (1962). That expression (their equation 5) is only valid, as was clearly stated, for those cases where variation in the size at division, and of new born cells, is very small. Limitation of the previous treatment has been elaborated on by Powell (1964) and by Painter and Marr (1968). Since the numerical approach adopted here overcomes these objections, this is then the first computation of the theoretical distribution of the sizes of cells in balanced growth where the computations follow accurately the hypotheses of the model.

It is appropriate, therefore, to compare the canonical or zero variation case with the new computation, and with the previously calculated values from the earlier integral equation formulation. The new results are shown in Table 4, and the distribution for certain choices for the coefficient of variation at the size at division are shown as the continuous distributions in Fig. 1. Calculations for variation in the size at division no greater than 30% can be made. Above this value the range of choices of the birth size yields a significant number of cells with a negative size. While the computation still accurately reflects the mathematical model, it no longer corresponds to a physically realistic situation.

A comparison of the first and second lines of Table 4 shows that the computer simulation accurately agrees with the predictions of the simple "inverse square distribution" also designated canonical or zero variation case (Koch and Blumberg, 1976). Comparison of the new computations with those based on the integral equation may be best made by noting that previously (Koch, 1966) the coefficient of variation for the steady state distribution of cell sizes for the case, where the size at division has a 15% coefficient of variation, was 25.69% *versus* the present 23.53%. The corresponding skewness statistics were 0.6047 *versus* 0.5869. The difference of these estimates means that some error was introduced by the method of calculation used previously.

The comparison of the theory with the 981 cells measured by Chai and Lark (1970) is also given in Fig. 1. The fit was made utilizing only the mean cell size from their data. The agreement with the theoretical model, with a coefficient of variation of the distribution of sizes at birth and division (0.1), is excellent except for the tails at the extremes of cell size. These classes are present in too large amounts to be consistent with our model. Their existence implies minor classes of cells, both extraordinarily large and small, that do not replicate their protoplasm as fast as the rest. This is a confirmation of the results of Harvey *et al.* (1967) who used the Collins and Richmond (1962) method for treating experimental size distributions obtained with a Coulter counter. They found that the rate of protoplasm synthesis was substantially proportional to the cell size, except for very large and very small cells. Both of the latter classes appeared to grow abnormally slowly. These exceptional cells represent a few percent of the total and, in many cases, probably an even smaller percentage of the DNA synthesis going on at any time.

TABLE 4. Parameters of the size distribution for balanced growth

Coefficient of variation of size at division	Mean	Coefficient of variation	Skewness statistic, γ_1	Kurtosis statistic, β_2
0^a	1.386	20.17%	0.4900^b	$+2.09^b$
0	1.3917	20.10	0.4844	+2.09
0.05	1.3854	20.55	0.5045	−0.43
0.1	1.3825	21.65	0.5463	−30.58
0.15	1.3774	23.53	0.5869	−59.58
0.2	1.3722	25.78	0.6085	−69.54
0.25	1.3623	28.77	0.5870	−66.03
0.3	1.3512	31.82	0.5377	−58.01
Chai and Lark (1970)	—	25.2	0.474	+3.01

^{a} This line is for the canonical size distribution, $\Theta(m) = c/m^2$, where only m values inside the range 1 to 2 are non-zero.

^{b} The skewness statistics, γ_1, has an expected zero value for a normally distributed sample. The kurtosis statistic, β_2, has an expected value of +3 for a normally distributed sample.

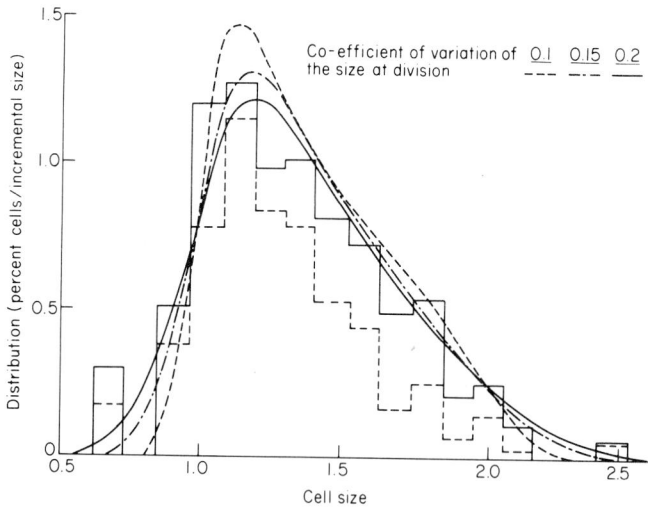

FIG. 1. Theoretical and experimental cell size distributions. The smooth curves show the theoretical distributions of cell sizes expected for balanced populations of asynchronously growing cells by the Koch and Schaechter (1962) model for the case that cell division always gives rise to equal sized daughters. The solid histogram is data reported by Chai and Lark (1970) from electron microscope measurements of 960 cells. The dashed histogram shows the distribution of cells engaged in DNA synthesis (see text).

To the extent that these two extreme classes of cell size represent pathological processes, it is reasonable to assume for the bulk of the cells that there are no variations in growth rate constant, and that the variation in size at division is quite small. For the Chai and Lark (1970) data, it is 10% or less of the mean size at division. To be overly generous on this point, computations for an assumed variation of 15% are presented below in Fig. 10 (p. 86).

A further check of applicability of this model to growing cultures comes from a comparison with the experiments of Woldringh (1976). In these experiments several thousand cells were measured in the electron microscope. For two growth rates he found a coefficient of variation of the sizes at division of 0.13. This corresponds to an expected coefficient of variation in the size of the population of 23% interpolated from Table 4. He found 24% and 25% for the two growth rates.

VI. Pulse Autoradiography of Slowly Dividing Cells: The Experimental Basis for Computer Simulation

The earliest published experiment presenting evidence concerning DNA synthesis within a population of cells in balanced growth is that of Chai and Lark (1970). These workers studied *Escherichia coli* growing slowly (120 minutes doubling time in a medium containing aspartate). At this growth rate, the gaps between chromosome replication events could be discerned. They pulse-labelled with tritiated thymine for one-thirtieth of a generation, and then examined by electron microscopy the autoradiograms produced in very thin emulsions over the fixed cells. The exposure was such that replicating cells had many grains overlaying them, and a clear discrimination could be made, they felt, of labelled cells and cells with only one or two background grains. The salient conclusion (see their Figs. 2 and 3, and our Fig. 1), but one that they did not draw, is that DNA synthesis is *not accurately phased* with growth of individual cells. Never more than 90% of the cells in any cell size class were synthesizing DNA, and never more than 60% *were not* synthesizing DNA in any cell class. Their description of their experimental technique suggests that these experiments were meticulously carried out, and that there were appreciable numbers neither of false negatives, because of Poisson statistics, nor of dead

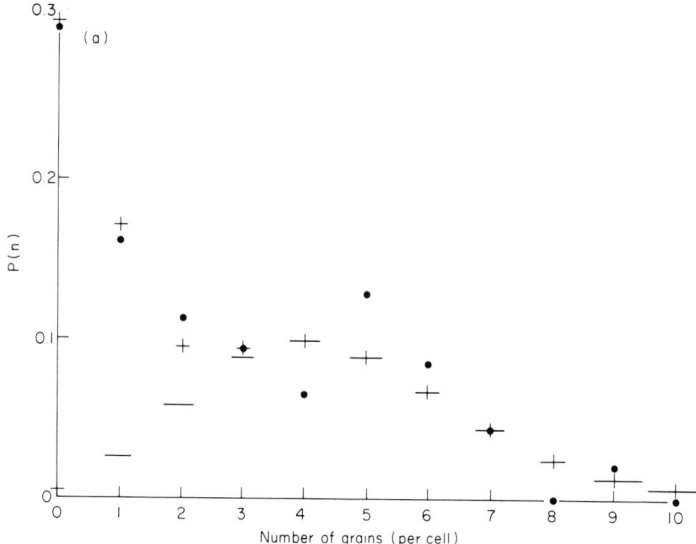

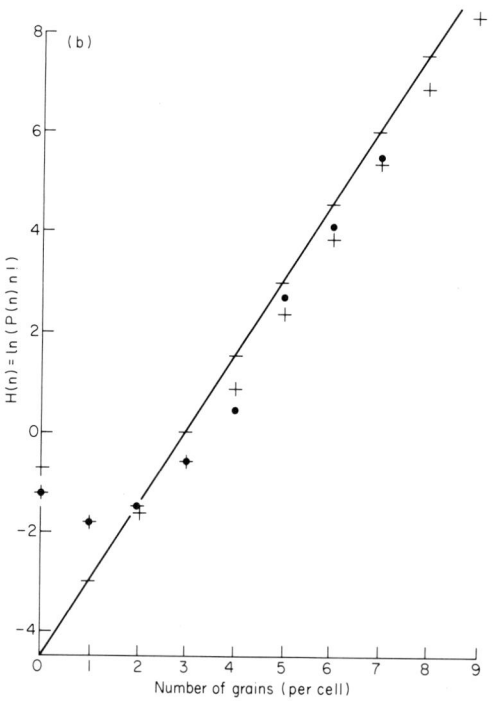

cells in the population, nor of false positives due to spurious incorporation of radio-activity.

Similar conclusions can be drawn from the heretofore unpublished experiments of Fred Forro (personal communication) who pulse-labelled *E. coli* (growing with 140 minutes doubling time in an acetate-containing medium) with purified tritiated thymine for 5% of a doubling time. The general technique used in these experiments is that of Forro and Wertheimer (1960).

A. ANALYSIS OF AUTORADIOGRAPHIC GRAIN COUNT DATA

I have analysed Forro's data, using the method of Hanawalt *et al.* (1961) of plotting $\ln[P(n)n!]$ *versus* n, where n is the number of grains over a cell and $P(n)$ is the *a posteriori* probability of observing n grains. If all cells have the same amount of radio-active thymine in their DNA, then these workers noted that the grain counts should follow the Poisson law, and this type of plot should be linear and have both a negative x-axis intercept and a slope equal to the natural logarithm of the average number of grains produced in the experimental period. Clearly this is not the case for Forro's combined data (see Fig. 2). Even for most cell size classes, the results are more consistent with the assumption that there is a group of cells with only little radio-activity contributing mainly to the null class, together with another group with more, and nearly constant, radio-activity producing the strain line portions of the graphs. To analyse these plots more readily, I have calculated what $\ln[P(n)n!]$ should be in the case of arbitrary mixtures of two homogeneous populations of cells with different amounts of radioisotope. If $\bar{x}_g$ and $\bar{x}_s$ are the mean number of grains per cell for the

---

FIG. 2. A. Distribution of autoradiographic grains. The closed circles show the observed distribution in unpublished experiments of Forro for the data when all size classes are pooled. The horizontal dashes show 52% of the distribution to be expected for a homogeneous population of cells where the mean number of silver grains per cell is 4.5. The crosses show the distribution to be expected if the remaining 48% of the population is also homogeneous with a mean number of silver grains per cell of 0.5.

FIG. 2. B. The H-transformation. The transformed data of Fig. 2A are shown here. (There is an exception in that the horizontal dash for zero class in the natural logarithm of 0.48 + 0.00578, where the first term is the contribution of the fully unlabelled gap cells and the second is the contribution of the labelled cells that just by chance did not form grains).

cells in the total gap and the cells in the synthesis period, respectively, the Poisson law can be written for each:

$$P_g(n) = \frac{\bar{x}_g^n e^{-\bar{x}_g}}{n!}$$

and

$$P_s(n) = \frac{\bar{x}_s^n e^{-\bar{x}_s}}{n!}$$

For the combined population, if G and S are the percentages of the two types of cells, we can write:

$$P(n) = (GP_g(n) + SP_s(n))/100.$$

The function employed by Hanawalt et al. (1961) is then given by:

$$H(n) = \ln[P(n)\, n!]$$

$$= \ln[((100 - S)\bar{x}_g^n e^{-\bar{x}_g} + S\bar{x}_s^n e^{-\bar{x}_s})/100].$$

A computer program was made which tabulated this function for a range of the three variables S, $\bar{x}_g$, and $\bar{x}_s$.

To make these ideas clear I show in Fig. 2a (horizontal lines) a histogram of the Poisson distribution of grains over cells expected, if every cell has the same radio-activity and the emulsion exposed such that, on average, there are 4.5 grains/cell. I have actually multiplied the Poisson values by 52% because I want to consider cases where this DNA-synthesizing population constitutes 52% of the total population. The actual population is imagined to be formed 52% from cells with a mean of 4.5 and 48% from cells containing low level of radio-activity contributing an average of 0.5 grains per cell. The contribution of these gap cells in the population augments the zero and single-grain classes to a higher degree than the classes of more densely labelled cells. These two populations together result in the combined distributions shown in Fig. 2a (crosses). Figure 2b shows the $\ln(P(n)n!)$ transformation of this hypothetical mixture (crosses). Also shown is the case where there is no adventitious radio-activity, and now the gap cells only contribute to the zeros class (horizontal lines). The solid line is the case for $S = 100\%$ and $\bar{x}_s = 4.5$. It can be seen that, except for the zeros class, the plot is virtually identical with that for $S = 52\%$, $\bar{x}_s = 4.5$, $\bar{x}_s = 0.5$. Moreover the slope is identical with that of the $n > 3$ values for the case $S = 52\%$, $\bar{x}_s = 4.5$, $\bar{x}_s = 0.5$. This means that such slopes are reliable indications of $\bar{x}_s$, and almost independent of $\bar{x}_g$, or the percent of cells engaged in DNA synthesis.

In Fig. 2a, 2b, 3 are also shown the grain counts of the 202 cells examined by Forro. In Fig. 2a and 2b they can be compared with the best fitting parameters, and in Fig. 3 the fit to values of $\bar{x}_g$ and $\bar{x}_s$ on either side of the optimum. The fit was achieved as follows.

Figure 3 shows fits to the combined data from the 202 cells examined by Forro for various choices of $\bar{x}_g$ and $\bar{x}_s$. A straight line

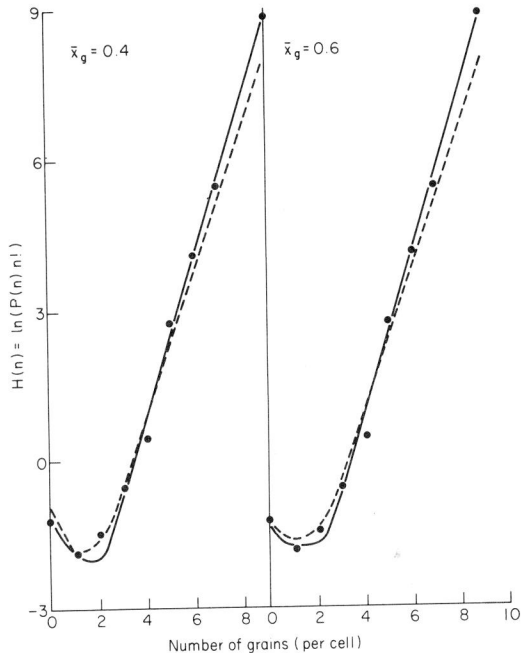

Fig. 3. H-transformation fits to Forro's autoradiographic data. The same set of data shown in Fig. 2 is reproduced in both panels. The solid lines corresponds to the best fits with $\bar{x}_s$ taken as 4 and the dashed line for $\bar{x}_s = 5$ grains per cell. The fit adopted for Figs. 2 and 4 is intermediary both between the values of $\bar{x}_g$ and $\bar{x}_s$ fitted here. In all cases $S = 52\%$ fitted the data most closely.

through the data for $n > 3$ gives a mean grain per cell estimate of 4.8 from the slope, and 5.0 from the intercept. To encompass the range, values of $\bar{x}_s$ of 5 (solid lines) and $\bar{x}_s$ of 4 (dashed lines) were tested. The tabulated values of $H(n)$ were consulted, and values of $\bar{x}_g$ were chosen that could give reasonable fits to the data with cells containing only a few grains. For $\bar{x}_g = 0.4$ (left panel), the value of S that fitted best was 52% for both choices of $\bar{x}_s$; similarly for $\bar{x}_g = 0.6$ (right panel), the value of S that fitted best was also 52%. The line that best represents the data should fall below the data points in regions where a grain number class is missing, since the corresponding value $H(n)$ is negative infinity when $P(n)$ is 0.

On this basis, I assumed an intermediate value of $\bar{x}_g$ of 0.5 and an intermediate value of $\bar{x}_s$ of 4.5 for Fig. 2 and to fit the data when partitioned by cell class. All the curves shown in Fig. 4 were drawn with those values of $\bar{x}_g$ and $\bar{x}_s$, and the table was consulted to choose the best value for S. The fit to the full data is also shown by the crosses in the

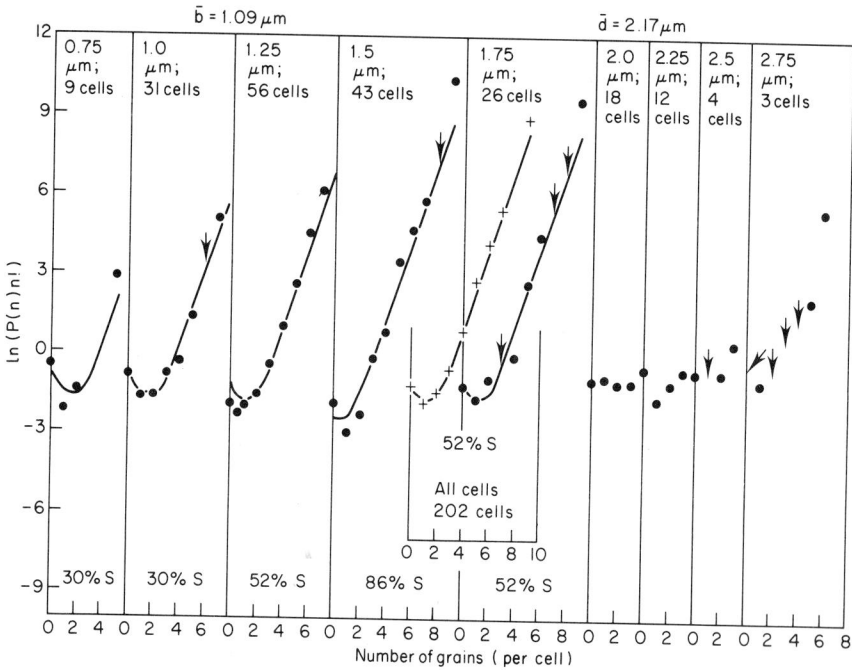

FIG. 4. H-transformation of Forro's autoradiographic data by cell size class (see text).

insert in Fig. 4. Two conclusions can be drawn from these fits. First, cells in all but the largest size class contained a significant number of individuals which were synthesizing DNA, but in all size classes there were certain cells which were not engaged in chromosomal DNA synthesis. The precision of the fit depends not only on the number of cells counted, but also on the magnitude of S. Second, those that are engaged in DNA synthesis do so at very nearly the same rate independently of the cell size (at least in the size range classes from 1.0 μm to 1.75 μm).

No fit has been made for the classes of very large cells from Forro's data. There may be an absolute gap since the plot for the combined cell classes 2.0 and 2.25 µm long could have been fitted on the assumptions that the population is homogeneous, and that all cells have one grain per cell (corresponding to a horizontal line $H(n) = -1$). In that case one could reasonably suppose that all of this one grain is background and is larger than 0.5 grains, assumed above, because the cells under consideration here are larger than average. On this assumption it follows that there is no replication going on in these very large cells.

In summary, the strong conclusion to be drawn from the analysis of the data of Chai and Lark (1970) and of Forro is that there is no size class in which 100% of the cells are engaging in DNA synthesis.

VII. Experiments Concerning the Variation in the Initiation of Chromosome Replication

The experiments discussed above did not involve synchronization; rather cells labelled during asynchronous growth were grouped together on the basis of similar size in the analysis of the autoradiographic data. If the culture is either synchronized by choosing conditions that bunch the cells in one phase at a time, or if cells of a narrow cell size range, or cell age range, are selected from the parental population, macroscopic experiments using ordinary liquid scintillation counting techniques can be done. Biochemical properties of samples of such cultures can be measured, including, of course, DNA synthesis by pulse-precursor incorporation. Contingent on the quality of synchronization obtained, and the details of the DNA cycle regulation, gaps are expected during which no radio-active DNA is formed.

In all published experiments, the observed incorporation during short pulses of thymidine or thymine, on a per cell basis, varies relatively little over the cycle of a synchronized culture, and the incorporation certainly does not vary in an all-or-none manner. This is especially noteworthy for the new experiments of Kelley (1974), who synchronized cells by the method of Cutler and Evans (1966), where the cell division steps were sharply maintained for many doublings after the synchronizing procedure had been completed and unrestricted growth had been initiated by a large dilution into growth medium.

Also worthy of note are the experiments of Gudas and Pardee (1974), in which the cells were synchronized by the technique of Mitchison and Vincent (1965) and the growth of a sample of the smallest cells was followed.

Experiments where cells are synchronized and then growth properties followed are subject to the general criticism that the synchronization procedure may perturb the subsequent growth. For this reason many workers have labelled cultures in balanced asynchronous growth and *then* subjected the cells to separation procedures. During the separation further growth is prevented, or only serves (hopefully) a role in separating cells initially in different phases.

Experiments of this latter kind, performed in a number of laboratories, also suggest that DNA synthesis is not well phased with respect to the cell cycle, where time instead of cell size is the experimental variable. In all published experiments from the laboratories of Helmstetter, Cooper, and Pritchard, where cells in balanced growth are pulse-labelled and then analysed by elution from a membrane filter, the apparent wave of the DNA label is much more diffuse than the wave of cell division. This is also true for the wave of duplication of genetic regions coding for particular gene functions (Hemstetter, 1968). In these experiments the measured quantity, in principle, should not fall to zero during a gap, so that one cannot test critically the degree to which some cells are engaged in DNA synthesis completely out of phase with the rest of the cells.

Finally, there are experiments where exponentially growing cells are pulse-labelled, then separated by size using a gradient centrifugation technique, and the distribution of the radio-active DNA label among cells of various sizes is analysed. The results from several laboratories are consistent with the belief that, in slowly growing cultures, initiations take place in cells of all sizes. In the work of Kubitschek *et al.* (1967), the ratio of the rate of DNA synthesis per cell during the period of replication, compared with the presumed gap period, is no more than 4 : 1. Our work (Blumberg and Koch, unpublished) gave similar ratios on a per cell basis. Although the observation, that the specific radio-activity in the presumptive gap was 25% of that of the presumptive synthesis period, could be attributed to contamination, or to other DNA replication processes, it also can be plausibly attributed to the initiation events occurring at an irregular size or age.

A. OTHER OBSERVATIONS

Yet another argument can be brought forward in favour of quite irregular spacing of chromosome initiation. This concerns the pattern of inheritance of radio-active DNA, as ascertained by autoradiographic methods. Two novel techniques were devised by Forro and Wertheimer (1960). In the first, radio-active cells were allowed to grow into microcolonies, and these then were layered with photographic emulsion. By this method, segregating units of radio-active DNA could be scored. In the second, but more difficult technique, the cells were separated by micromanipulation as they divided, and after several generations autoradiography was performed and the family tree was reconstructed. The results of the study of Forro and Wertheimer (1960) and those of Forro (1965) and Lark, C. (1966) are consistent with the hypothesis that recombination frequently occurs; that is, radio-activity associated at one time in a single strand of DNA is partitioned between two daughter cells.

More relevant for the present paper, there is also evidence in the data of Forro (1965), using the micromanipulation procedure, that an "inappropriate" number of conserved units (i.e., a number other than 2, 4, 8, etc.) is observed fairly frequently. Forro concluded, "Our results, particularly the occurrence of the $1n$ and $3n$ cells, do suggest, however that cell division is not a necessary or sufficient condition for the initiation of a round of chromosomal replication. Neither must cell division await the start of a new round of DNA synthesis in both daughter chromosomes."

This leads to several possibilities. First, multiple forks may exist in some chromosomes in cells growing in a medium in which a single pair of forks, or at most two completed chromosomes, is predicted by the Cooper–Helmstetter model. Second, a cell with two chromosomes may occasionally initiate chromosome synthesis in one, but not the other. Third, chromosome initiation in some cells might follow immediately after the termination of a previous round with much less than the average total gap $(\bar{B} + \bar{D})$. All these possibilities may be related to the earlier work from Lark's laboratory, summarized in Lark, K. G. (1966). Lark concluded that in cultures growing slowly in a medium containing succinate, newly formed cells have two chromosomes and that first one, and then the other, replicates before the next cell division. A consequence of Lark's hypothesis is that only one chromosome would be labelled in a short pulse, and only this one would be subject to self-destruction by radio-active decay. On this basis, Lark's hypotheses was critically disproved by Koch and Pachler (1967) who observed no predicted immunity against "suicide" of pulse-labelled succinate-grown cells. While our experiment disproved Lark's hypothesis for the bulk of the cells, it may well be possible that the data of his group were really showing that there was inaccurately scheduled DNA synthesis in a significant fraction of the cells in the population.

VIII. Possible Experimental Objections

This concludes the several lines of evidence in favour of irregular scheduling of chromosome initiation. Before considering the consequences of this, let us consider several complications and objections to the tracer experiments, both autoradiographic and otherwise.

Certainly the main objection is that some of the tritium in the TCA-insoluble fraction may not reflect chromosomal DNA replication. First, there may be contributions from radio-activity adventitiously adsorbed on the membrane filters, or on the cells. Second, there may be contributions from thymidine disphosphorhamnoside (Lark, 1963), a precursor of certain lipopolysaccharides. In principle, suitable measures and controls should obviate or correct for these two processes. Third, the commercial radio-active precursor preparation may be contaminated. Radiodecomposition, and failure of quality control by the manufacturer, should have impressed one with the necessity for purity checks. This is particularly so if the tracer is to be diluted with non-radio-active carrier, thus decreasing the specific activity of the compound, but not that of the impurity.

These processes, in general, will contribute to the radio-activity of cells both in and out of the DNA replication process, but do not reflect the replication of DNA. In addition, there are some processes other than normal chromosomal replication that lead to labelling of the cellular DNA. There may be a contribution from plasmid replication. There is, at present, no valid control of which I am aware that would measure the extent of this contribution in any specific case. Certainly, the cultures used in the autoradiography experiments of Chai and Lark (1970) and in Forro's unpublished work were of *E. coli* strain 15 T$^-$, which is known to contain at least one plasmid—a defective phage. There may be others. Exhaustive electron microscopy, and use only of strains that prove to be clean, would be the way to eliminate this possibility, but this has never been done. However, it can be argued that most plasmids are present in only a few copies (one or two per cell) and that each represents one or two percent of the chromosomal DNA. Consequently, there should be few false positives in autoradiography experiments where the pulse is about 10% of the length of time of cell synthesis.

Also, there is a contribution from repair, although three lines of evidence can be cited to suggest that, in the absence of detectable damage to DNA, this contribution is very small. Ronald Ley (personal communication) has conducted experiments with *E. coli* strains that were made to incorporate bromouracil subsequent to long-term thymine labelling. He measured the repair incorporation of bromouracil into the thymine-labelled strands by its ability to sensitize single-strand DNA to near-ultraviolet light scission. He estimated that about one in 10,000 thymine bases is replaced by bromouracil in a cell cycle. This corresponds roughly to the estimates given by Grivell and Hanawalt (1973) of 0.05% repair turnover per generation of a temperature-sensitive strain of an *E. coli* K-12 derivative at a non-permissive temperature. K. G. Lark (personal communication) finds that repair and replication can be distinguished with the use of anti-polA antibody in a cell-permeable DNA synthesizing system. Assuming that replication alone is partially inactivated by the procedures necessary to set up the cell permeable system, and that repair proceeds at the rate it would in the intact cell, he then calculates that repair is 0.3% of replication in intact cells.

An additional criticism that can be levelled at experiments with thymine auxotrophs is that DNA synthesis can be slowed by low levels of thymine. These matters are discussed critically by Pritchard (1974), and depend very much on strains, the particular mutation, and on the past history of the culture. Although the autoradiography experiments used "super-low thymine-requiring" strains, and fairly high levels of thymine, one can suppose that they reflect a partial thymine, or deoxyriboside, deficiency. At faster growth rates Zaritsky and Pritchard (1973) have shown that guanosine deoxyriboside decreases C. On the other hand, both the experiments of Chai and Lark (1970) and of Forro were done under conditions of balanced growth, and at standard growth rates for thymine-sufficient strains for given carbon sources, and at what are now presumed to be high enough levels of thymine to more than adequately maintain growth.

On the basis of these several lines of evidence I feel fairly confident, therefore, in concluding that the initiation of rounds of DNA synthesis is neither well controlled with respect to the cell size, nor to cell age. While it is clear that there is a positive correlation of these events, the correlation is nowhere near $+1$, nor is the mean relative phasing constant, even in similar growth situations.

A. REMAINING ADMONITIONS

There are two other quite different classes of *caveats* to complete the list. One has to do with the question of the detailed growth of cellular material through the cell cycle. This field has been much discussed. No one has detected a period of inhibition of cellular growth such as is seen in higher organisms at preprophase and mitosis (see Mitchison, 1971); on the other hand, one can easily imagine that small starts and stops might exist, and never have been detected with the tools at hand. Because critical information on this important question of continuity of growth is lacking, the controversy in the current literature has hinged on the much grosser debate of whether cells demonstrate linear or exponential growth. The experiments of Ecker and Kokaisl (1969) are critical in showing that the rates of both protein and DNA synthesis are in direct proportion to cell size. On the other hand, Kubitschek (1968a, b; 1971) defended the proposition that the cell volume increases at a constant rate (linear growth), although an array of papers can be cited in favour of the rate of volume increase proportional to volume (exponential growth), for the typical cell, for at least the bulk of the cell cycle (Schaechter *et al.*, 1962; Harvey *et al.*, 1967; Errington *et al.*, 1965). This question is further discussed in the paper of Koch and Blumberg (1976).

The relevant issue here is that, even if volume growth is linear and macromolecular synthesis exponential, with or without interruption, classification of cells on the basis of length is a valid measure of the cell position in the cell cycle. This follows from the fact that rod-shaped cells only get bigger and never grow smaller. It also follows from the microscopic and autoradiographic evidence that cells of increasing length (as measured in the phase or electron microscope) monotonically increase their rate of protein and RNA syntheses, and that cell division takes place over a remarkably short range of cell lengths.

The final area of doubt has to do with the continuity of chromosome replication. The synthesis of DNA occurs remarkably fast; it can be calculated that the replicating chromosome must rotate at 5000 rpm. This is an average, assuming bidirectional replication and a 40 minute $\bar{C}$ time. However, maybe it is yet faster than that, but there are intermittent stops and starts during the replication process. Table 1 (p. 51) lists $\bar{C}$ from a variety of experiments at a variety of growth rates; it can be seen that there are apparently valid experiments indicating that $\bar{C}$ is in the neighbourhood of 30 minutes. On the other hand, three out of five studies cited in the table present evidence to suggest that $\bar{C}$ lengthens under conditions of slow growth.

This lengthening could happen in several ways. The replication process might be slowed at the biochemical level. Maaløe (1961) and Maaløe and Kjeldgaard (1966) suggested that C might be a constant because replication might be limited by the speed with which a single molecule of the critical polymerase could replicate the DNA. This could well be independent of the variation in levels of the precursor triphosphates. Maaløe's suggestion antedated the discovery by Okazaki of the role of small fragments of DNA covalently linked to RNA in the replication process. One can now easily imagine that the C period could vary, depending on circumstances, on the basis of a multitude of processes: the rate of opening up the double strand helix, the length of the Okazaki fragments, the number remaining to be covalently bound into the growing chains, the size of the regions that are filled by the combined action of the DNA polymerase I and ligase. One would only expect a constant C period, independent of growth environment, if it could be established that the rate of unwinding was limited by the action of a single molecule of protein that moves with, and determines, the fork. This is an aspect of the current studies of the biochemistry of DNA replication, but as of now there is no concensus.

Second, although it is now clear that chromosome replication occurs bidirectionally, it is possible that this is not the only mode during normal growth. Certainly, during sexual transfer of DNA, replication is unidirectional. Perhaps it is possible for both modes of replication to co-exist, and that the proportion of the two modes varies as the growth conditions alter, becoming predominantly unidirectional at long cell cycle times. If replication is almost exclusively bidirectional at fast growth rates, and unidirectional at slow growth rates, this would weaken the conclusions of Chander et al. (1975) that $\bar{C}$ is 40 minutes, independent of $\bar{T}$. It could be that there is a preferred direction for the unidirectional mode. Some evidence against this possibility has been recently supplied by Rodriguez and Davern (1976), who found by autoradiography that bidirectional replication occurs during growth on succinate as well as on glucose.

Third, we could image that movement of the fork is discontinuous. It is possible that the failure to observe 100% of the cells in the synthesis phase, in any size class, in the autoradiographic experiments of Chai and Lark (1970) and of Forro (unpublished), are simply due to "internal" gaps of 5 or 10 minute duration interspersed within the C phase. Then, the calculations of C presented below from the autoradiographic experiments could be in error, and the total C period, which must include these internal gaps, must be greater than two-thirds of the cell cycle length, and DNA synthesis must be thought of as nearly continuous, on a grosser scale, with little or no time lag between the termination of one round of chromosome replication and the beginning of the next.

IX. Variability of the Time Between Nuclear Division and Cell Division

There are no data bearing on the intraculture constancy of C, or the total period $C + D$. If $C + D$ were rigorously constant, then c (the size at initiation) should be closely correlated with the size of the same cell at cell division, d; in addition, c and d should have identical coefficients of variation. If $C + D$ varies from cell to cell in the culture, then, on the assumption that the chromosome initiation controls the subsequent cell division, and the assumption basic to all statistics that random sources of error only can increase the error of a total process, c should have a smaller coefficient of variation than d. If c is rigorously constant, then similarly the size at nuclear division (n) should have a smaller coefficient of variation than d.

One can adduce that the period in the cell cycle from nuclear division to cell division fluctuates considerably. Woldringh (1976) has done serial sections of $E.\ coli$, from balanced cultures of bacteria, and examined them in the electron microscope. From three-dimensional reconstructions, he identified the number of nuclear bodies in a cell, as well as whether the cell was in the act of cell division. His findings show that nuclear division takes place in cells in the central third of the cell size range. These data have been used to construct Figs. 5a and 5b. They show on probit graph paper the cumulative nuclear division events (●) and cumulative cell division events (+). The open circles show nuclear division data corrected for cell division depletion of each size class. For the case of cells growing with a doubling time of 32 minutes, in glucose casamino acid medium (Fig. 5a), the median lengths at nuclear division and at cell division correspond to a 15-minute average period between these two events. The slopes correspond to coefficients of variation of 23% and 14% of the mean doubling time, for nuclear and cell division, respectively.

In Fig. 5b the results with cells growing in a glycerol-containing medium, with a doubling time of 60 minutes, are presented. The difference in the mean size of the cells engaged in the two processes corresponds to a time of 7.2 minutes. The slopes correspond to coefficients of variation of 11% and 14% for nuclear and cell division, respectively.

In Figs. 5a and 5b the lines through the nuclear division data have been drawn through the corrected data between ± one probit from the 50% point (0 probit). These results are only approximate because they represent the analysis of only those few cells belonging to the particular length class. Estimates of $\bar{D}$ in the literature are usually of the order of magnitude of 20–30 minutes for cells growing in this range of doubling times. The results are consistent with our previous findings (Schaechter et al., 1962) using the technique of Mason and Powelson (1956). For *E. coli*, growing with doubling times of 28 and 32 minutes, we give data from which one can calculate that the interval from nuclear division to cell division has a mean period of 11.0 and 12.6 minutes, respectively. Similar results were reported with other enteric organisms. As I have plotted Woldringh's data, the coefficient of variation for nuclear division was somewhat larger than that for cell division, in one case, and smaller in the other. In our previous work they were nearly the same for the *E. coli* and *Salmonella typhimurium* cultures, ranging from 6.3 to 9.9% for nuclear division and from 8.5 to 14% for cell division.

Two conclusions can be drawn from these comparisons. First, the same stage in the cycle is identified by separation of bodies observed in the phase-contrast microscope, by the Mason–Powelson technique, as

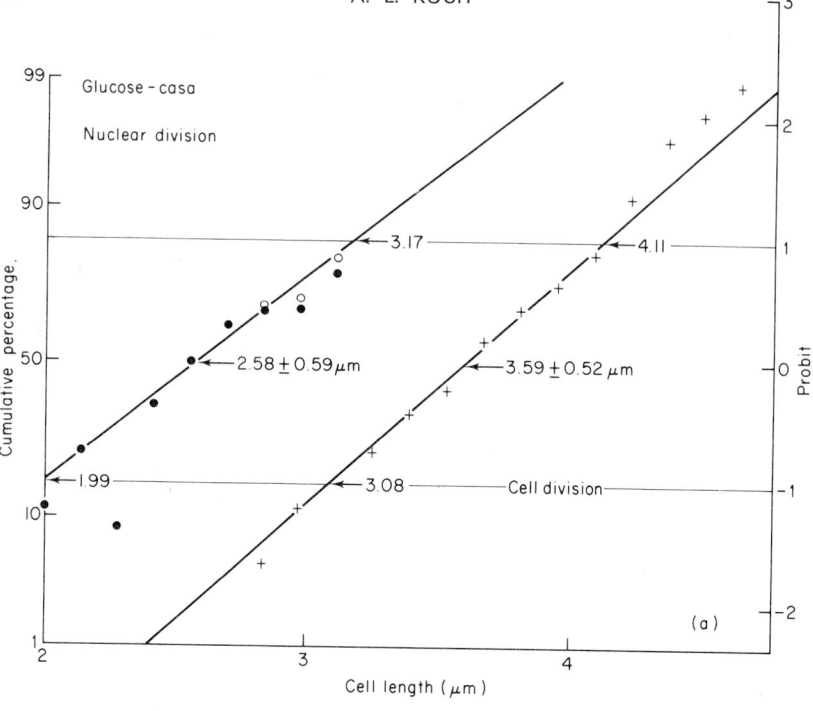

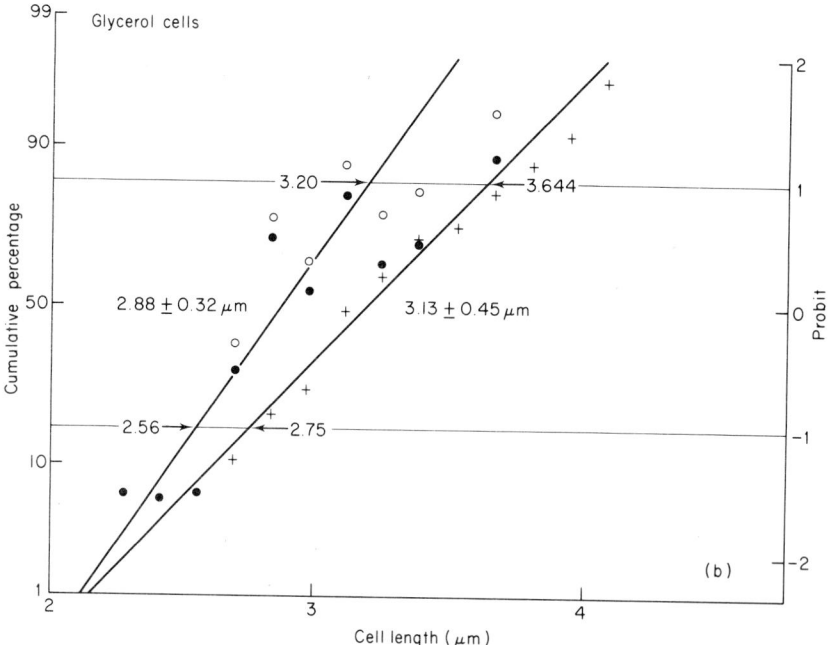

TABLE 5. Variation in cell cycle and nuclear to cell division interval[a]

Organism	Cell cycle time		Nuclear division to cell division	
	mean	s.d.	mean	s.d.
	(min)	(min)	(min)	(min)
Escherichia coli B/4 (expt 23)	26.0	5.6	16.2	6.1
Proteus vulgaris (expt 20 plus 22)	43.0	9.5	20.9	8.9
Salmonella typhimurium (expt 24)	28.5	4.6	13.1	5.3

[a] Taken from unpublished data obtained as part of the same study reported in 1962 by Schaechter, William, Hood and Koch. The nuclear division-cell division interval and the cell division–nuclear division interval are negatively correlated: −0.62 for the *E. coli* sample; −0.88 for *P. vulgaris*; and −0.52 for the *S. typhimurium* sample. Calculations are based on 43–83 cells for each species.

is identified by the separation of DNA containing material, by the thin section technique, using the electron microscope. This phase occurs later than completion of DNA synthesis, as adjudged by comparison with the millipore filtration experiments of the Helmstetter school, or the time stage at which blockage of DNA, RNA, or protein synthesis will not prevent the ensuing cell division.

The second conclusion is that nuclear division is not a stage of the cell cycle through which the cell must pass in order to start the timing of the next stage; that is, it is a consequence of some process of the cell cycle, but does not time a subsequent process, although it may limit it in the sense of a veto power. This conclusion follows from the similarity of the data obtained using electron and phase-contrast microscopy, respectively. The former is less ambiguous than is the latter with respect to adjudging the completeness of the separation of the two nuclear bodies. On the other hand, with the phase-contrast microscope we were able to measure the interval between the nuclear division and cell division of individual cells. Previously unpublished

FIG. 5. Probit plot of the cumulative percentage of nuclear and cell division. Data taken from Waldringh (1974; 1976). Data in open circles is cumulative percentage for nuclear division, corrected for those cells that have not only undergone nuclear division, but have also proceeded through cell division. The corrected value is simply the sum of the observed cumulative percentages for nuclear and cell division divided by 100% plus the cumulative percentage in cell division. A. Cells growing rapidly, in glucose-casamino acids medium, with doubling time of 32 min. B. Cells growing in glycerol containing medium with a doubling time of 60 min.

parts of the study of Schaechter *et al.* (1962) indicate that the standard deviation of the interval is comparable in magnitude to that for the cell division process itself (see Table 5).

X. Temporal Accuracy of DNA Synthesis Cycle

Is there some factor with which the initiation of chromosome replication correlates more closely than with size or age of the cell? I believe there is, based on a perceptive experiment of Nagata and Meselson (1968). These workers pulse-labelled a growing culture for two minutes with tritiated thymidine, removed the label by filtration, and allowed growth to resume in non-radio-active medium. At various times, portions of the culture were transferred to a medium containing heavy isotopes and were allowed to grow for a third of a generation. Then the DNA of the cells was displayed on caesium chloride gradients. Had DNA synthesis been regulated very precisely, it would be anticipated that the hybrid density peaks would alternately rise to a maximum and fall to zero over a time equal to a third of a generation, with a third of a generation rise and fall time. These periodic rises should re-occur with a period equal to the culture doubling time. Although they found periodicity of the radio-active label in the hybrid density peak corresponding with that of the average doubling time, the phasing was not all-or-none. This was not unexpected, because of various kinds of experimental "noise", but instead of a progressive loss of synchrony expected from various experimental and biological sources of "noise", they found the waves of pulse radio-activity in the hybrid density DNA were almost as damped even before the first doubling time as it was after the fifth or sixth generation. The fact that a significant fraction of the labelled DNA is of hybrid density, even at short times after the tritium pulse, must mean that certain cells in the population have chromosomes with multiple forks, even under conditions where most chromosomes in the population do not. As discussed above (p. 72), it is believed that repair synthesis and/or contamination could not have accounted for the initial blurring. On the other hand, if such irregular initiation events are accidental, and uncorrelated with growth generally, then the phasing ought to lessen progressively and ought to be abolished in one or two generations. Obviously the observed persistence of periodic association of the previous pulse label with newly synthesized DNA means that there are control mechanisms that function well in the culture viewed as a *whole*, and followed in real time, and not by time counted from the particular chromosome initiation or cell division event creating a particular cell.

XI. Computation of the Fraction of Cells of a Size Class Engaged in DNA Replication

The computer program functioning as described earlier (p. 58) included features such that it could be augmented by giving it an arbitrary rule to compute for every combination of b, d, and m, whether the cells in that class would be replicating DNA. On the basis of this part of the program, the contribution of cells is to be awarded either to the distribution for cells in replication or to the distribution for cells not engaged in replication. These distributions are then normalized so that the sum of the totals of both is fixed at 100%. The program then prints out the replicating and non-replicating distributions, and the combined distribution. It also yields the statistics of the various distributions and the percent cells engaged in replication.

The simplest assumption concerning the control of DNA replication is to specify an invariant critical size of the cell for DNA initiation and an invariant time for the C period which thus fixes the critical size (f_c) at termination. On this basis, all cells of the intermediate size are assigned to the replication array, and the others to the non-replicating array.

The procedure is straightforward for those cases where the entire process of DNA replication occurs in a size range included between the birth and division size of the cell under consideration. However, when b is larger than c, this procedure is wrong, and this condition causes the program to use a different code because it corresponds to those cells in which the new round of replication started before the previous cell division. Thus an appropriate contribution to cells of twice the size under consideration (i.e. cells of size $2m$) must be made. Actually the program apportions half of the increment to cells of size $2m$, and divides the remainder between the cells of the adjacent size classes. It corrects for this contribution to the distribution of replicating cells by subtracting a like amount from the corresponding memories for cells not replicating DNA. Similarly, a different code is needed for those cases where d is less than f; that is, where a cell is scheduled to divide before there are two completed chromosomes. At present, the program is arranged so that these cells wait until they have chromosome replication before they divide. Thus, the need for the f command overrides the d command. This, of course, distorts the cell size distribution, and will increase the breadth of the cell size distribution, because it effectively increases the coefficient of variation of the cell size at division. At the present time, the computer program does not make a similar change for the distribution of birth sizes; consequently it broadens the cell size distribution. For this reason, only the results of cases where the distribution of cell size has not been appreciably broadened will be presented in this paper.

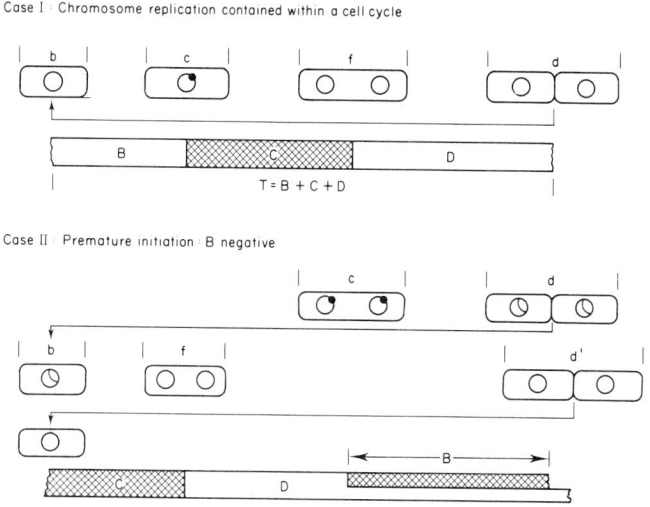

FIG. 6. Models for the relationship of chromosome replication. Two of the four cases considered in the text for slow growing cells are considered.

Some of these concepts are shown in Fig. 6. Case I is the usual one that most bacteriologists think about. Case II is appropriate to a cell cycle where c is less than b. As mentioned, the other two cases are avoided in the present report, but will serve as a basis for future refinements. If every cell in the population had a particular exact value of c (and f) then the distribution of labelled cells on the size scale will be an all-or-none. There will be labelled cells in certain size class such as diagrammed in the bottom lines of the two cases in Fig. 5.

In Fig. 7, are shown the regions of cell size expected to contain only labelled cells. These regions of labelling and non-labelling were computed for the parameters that fit the conditions of the Chai and

Lark (1970) experiment; that is, coefficients of variation of the division distribution of 10%, with 68% of the cells engaged in DNA replication (which corresponds to C period of 80 minutes), and a $\bar{T}$ period of 120 minutes. For the example given c was set at 1.1. Clearly it is impossible to fit the data of Chai and Lark (1970) with any choice of a fixed value of c.

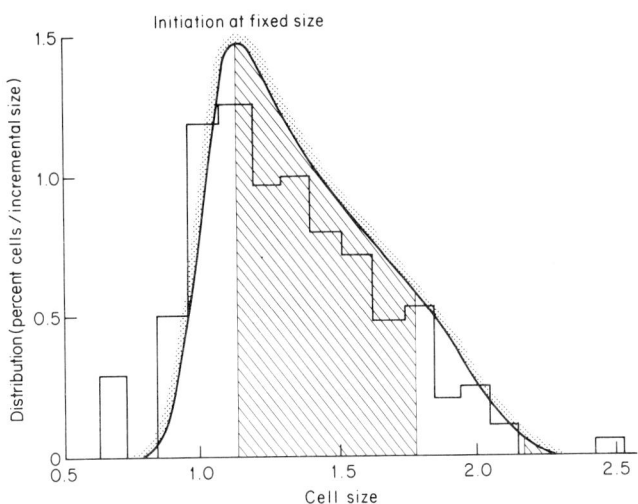

FIG. 7. Distribution of cells engaged in DNA synthesis if there is a precise size for initiation. The outer distribution is a replicate of the theoretical distribution shown in Fig. 1, where the coefficient of variation of the size at division is taken as 0.1. The distribution shown as cross-hatching is that expected if the size at initiation of chromosome replication in every cell's population is exactly 1.1 times the population mean birth size $\bar{b}$. The length of the C period was taken as 80 min, and the other parameters were taken to be consistent with the experiment of Chai and Lark (1970). Note the largest cells in the population are all labelled. These are cells that have initiated a new round before dividing. Also shown again is the data of Chai and Lark (1970).

As a second model, we can fix DNA initiation in time relative to the time of the last cell division (Fig. 8). If a cell divides when too small, and yields small daughters, then initiation takes place when the cells are smaller than the average size at which initiation takes place. It can be seen that there is no positive value of B (the gap between cell division and initiation) that makes predictions in agreement with the data of Chai and Lark (1970). Compare experimental data in Fig. 7 with Fig. 8.

A third model postulates that initiation sizes fluctuate from cell to cell in the population but that once initiation occurs, synthesis finishes exactly C minutes later, and division ensues exactly D minutes after than. Calculations based on this model are shown in Fig. 9, also for the parameters of the Chai and Lark (1970) experiments. In this case as well, no agreement with that experiment can be obtained. Compare Fig. 7 with Fig. 9.

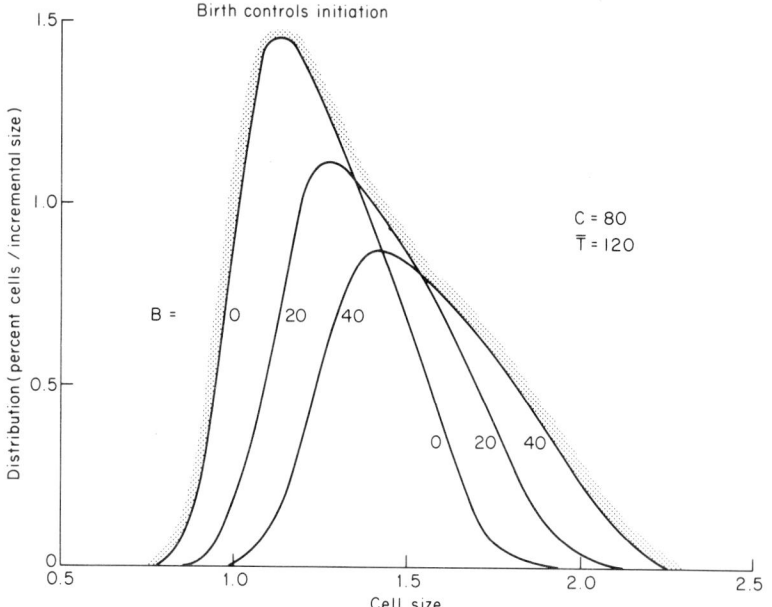

FIG. 8. Theoretical distribution of labelled cells on the assumption that there is a fixed delay after a cell arises by division before initiation starts (constant B). Other parameters used as in Figs. 7 and 9.

Failure of these data to fit any of the first three models derives from the fact that all these models predict certain cell-size classes in which 100%, or 0%, of the cells should be synthesizing DNA; this, simply, does not accord with the data. Any of these models could be modified so as to fit the experimental data by assuming that the real population consists of at least two sub-populations with different values of the relevant parameters. For example, the data of Chai and Lark (1970) could be fitted roughly if two-thirds of the population corresponded in Fig. 8 to $B = 0$ and the remainder to $B = 40$ minutes. Other subdivisions of the population could be proposed, but it is not reasonable to assume that the division and replication of cells are distributed bimodally in the population. So although we will return to the cases embodied by Fig. 8 and Fig. 9, we now consider the case where the size at initiation is independent of the size at the previous or the ensuing cell division, and its variation is randomly distributed within the population.

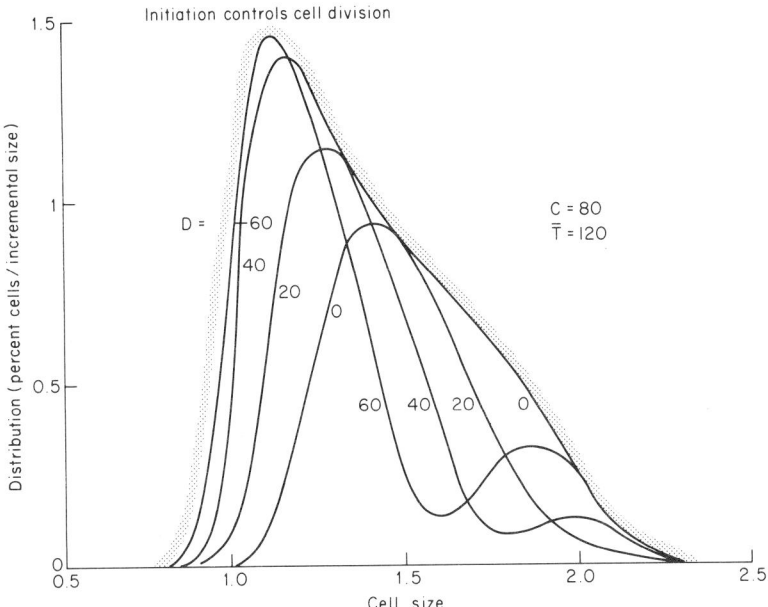

FIG. 9. Theoretical distribution of labelled cells on the assumption that there is a delay after the chromosome is completed before cell division can take place (constant D). Conditions of Figs. 7 and 8.

This fourth model of operation assumes that initiation of DNA synthesis occurs at some critical size in each cell, modified by fluctuations uncorrelated with other events of the cell cycle of that cell. I have assumed in this case that c is normally distributed about some mean value. This computer model uses a great deal of computer time to obtain accuracy, since the program considers 60 discrete values for each of the three variables (b, c, and d), and 300 potential values for m. Consequently there are $60^3 \times 300$, or 6.5×10^7 individual contributions to construct a distribution of labelled and non-labelled cells.

However, with the addition of a third random variable, distributions somewhat in closer agreement with the experiment could be obtained. Attempts (not shown) at computer simulation made clear the point that a reasonable fit to the data of Chai and Lark (1970) could only be obtained if the size distribution at initiation was extremely broad (coefficient of variation of 25% or greater); that is, the data imply that initiation can take place throughout at least most of the cell cycle. So, if

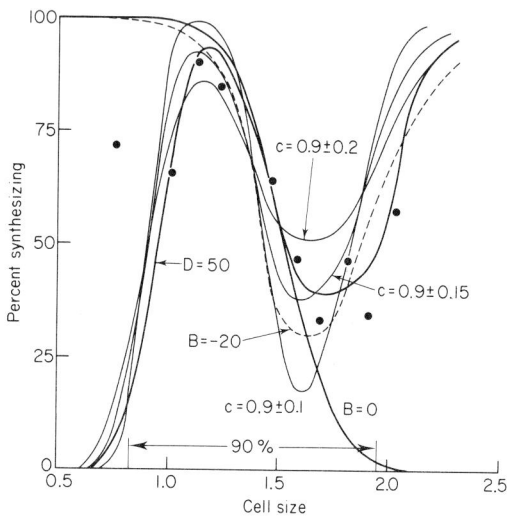

FIG. 10. Labelled cells as a function of cell size for several models. The experimental, and a variety of theoretical models are shown in attempts to fit the data of Chai and Lark (1970). While no acceptable fit to a model in which the initiation size fluctuates independently of the size of the cell at birth and division with a coefficient of variation of 0.1 could be made, almost acceptable fits could be made if certain modifications were introduced. These are: (a) increase the coefficient of variation of the size at division from 0.1 to 0.15; (b) assume that chromosome initiation takes place, on average, prior to cell division so that most dividing cells are engaged in DNA synthesis. The several possibilities shown here are discussed in the text.

the reader is willing to accept the data of Chai and Lark at face value, and the analysis of the cell size distribution presented above as indicating that most of the cells grow and divide with a 10% coefficient of variation for the size at division, and that, in addition, there are a few pathologically small and large cells, then it can be rigorously concluded *that initiation is much less well controlled than cell division, and therefore initiation cannot control or time cell division.*

However, to give every opportunity to those models that mechanistically link chromosome replication with cell division, and not reject them unfairly, I tried to see if I could obtain somewhat more reasonable fits by assuming a broader distribution of cell size at division. So I tried increasing the coefficient of variation of the b and d distributions from 0.10 to 0.15. Figure 1 shows that a value 0.15 fits the cell size distribution of the population of cells almost as well as the 0.1 value. The percent of cells synthesizing DNA are displayed against cell size in Figs. 10 and 11. For the several types of model, a range of size

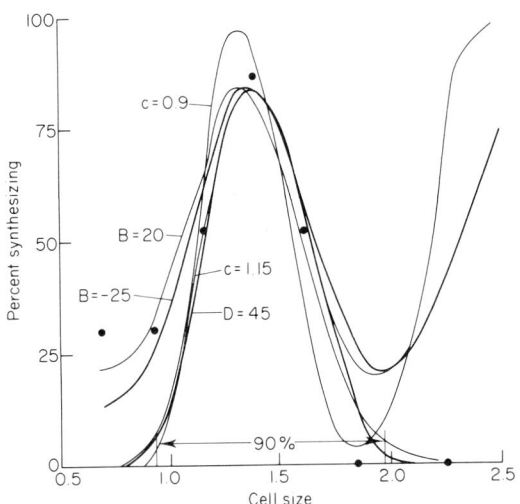

FIG. 11. Labelled cells as a function of cell size. This is a plot similar to Fig. 10. The data shown by solid circles are those from the unpublished experiment of Forro. (See text.)

for the central 90% of the cells in the balanced population is shown, and the experimental results are indicated by closed circles; those of Chai and Lark (1970) in Fig. 10 and those of Forro in Fig. 11.

Figure 10 shows the fit of the fourth model with three independent variables (b, c, d) to the data of Chai and Lark by thin solid lines. The curves shown in the figures were fitted to an overall percent of cells synthesizing DNA of 68 ± 1%, corresponding to their observed result of 68%. The case of $c = 0.9 \pm 0.15$ fits better than 0.9 ± 0.1, or 0.9 ± 0.2 (also shown), and other cases (not shown) where the mean value of c was varied, and the variation in c kept at 0.15. The cases where $\bar{c} < 1.0$ are those where a cell, on the average, initiates a round of DNA synthesis on both of its previously completed chromosomes slightly before the ensuing cell division. Thus a rough fit can be achieved by assuming that, in the experiments of Chai and Lark (1970), most of the cells entering cell division are already engaged in a new round of DNA replication. The best fitting value of the coefficient of variation is 0.15, equal to the 0.15 assumed for the cell division process that creates and removes each cell from the distribution.

Although this fit is better than any of those presented in Figs. 7–9, the suggestion that initiation of the new rounds of DNA synthesis might precede an average cell division convinced me to go back and recalculate first for the case of cell division "controlling" initiation retro-actively (B fixed but assigned negative values), and second for the case where initiation controls not the next, but the succeeding cell division (that is, a model where $C + D$ is fixed at a value greater than $\bar{T}$).

The value that fits the data best for the first case is $B = -20$ (thin dashed line in Fig. 10). The case where $B = 0$ (the Cooper–Helmstetter model) is also shown (thick solid line) for comparison. It clearly fits very poorly. The value that fits the data best for the latter case is $D = 50$ minutes (thick solid line). For the second case, $D = 50$ minutes probably fits the data as well as the $c = 0.9 \pm 0.15$ or $B = -20$ minutes cases shown. Thus, all of the three cases with parameters chosen to optimize the fit correspond to a most typical behaviour of the cells such that a new round of DNA replication occurs shortly before cell division in both of the already completed and separated chromosomes. The random initiation $c = 0.9 \pm 0.15$ fits a little better than scheduled cell division 20 minutes after the initiation of the new round ($D = -20$), while the case where $D = 50$ minutes fits as well. The point that needs to be stressed is that the only model with a mechanism whereby chromosome replication and cell division are co-ordinately controlled is the $D = 50$ minutes model, and this can only be fitted roughly, and with the assumption of a coefficient of variation of 0.15 for the distribution of cell size of those cells in division.

Figure 11 shows the results of a fit to the privately communicated data of Forro, analysed above (p. 65). The cell size distribution in Forro's cell sample was broader than that for the data of Chai and Lark (1970), and corresponded to a coefficient of variation at cell division of 0.15. The doubling time is 20 minutes longer, and the best fitting C period is 70 minutes (that is, 10 minutes shorter than for the data of Chai and Lark). Again, there is no acceptable fit to a fixed size at initiation of chromosome replication. The best fit for the Forro experiment is obtained on the assumption that the new round of DNA replication usually starts somewhat after the cell division that partitions the previously replicated chromosomes. Either a fixed $B = 20$ minutes, $D = 40$ minutes, or $c = 1.15 \pm 0.15$ described the data particularly well in the central part of the distribution, where the data are most accurate. None describe it well on extremes of cell size.

Putting all these three estimates on a time scale, initiation on the average starts 20 minutes (fixed B), 30 minutes (fixed D), or 28 minutes ($\bar{c} = 1.15 \pm 0.15$). Taking 26 minutes as an average B time, since $\bar{C} = 70$ and $\bar{T} = 140$ minutes, the average D time of the three models is 44 minutes for Forro's data.

In conclusion of this section, it can be stated that none of the models really fit the data well. But we can exclude the constant c model, the $B = 0$ model of Cooper–Helmstetter for slowly growing cells, or the $D = 20$ minutes, which is the Kubitschek "self-consistent" model. Rather, initiation is either random and centres toward the end of the previous cell cycle, for the data of Chai and Lark (1970), and shortly after the start of the new cells, in the data of Forro. In either case these data fit the hypothesis that cell division occurs in some of the slowly growing cells while a round of DNA synthesis is taking place. It can be concluded that an initiation of chromosome replication does little to time or control the cell division that occurs after the completion of the DNA replication.

The computer program described here now waits for better autoradiographic experiments; these should combine the electron microscope approach of the Chai and Lark (1970) experiment with the analysis of the grain counts, as done here with the data of Forro. Above all, they should be done in the presence of purine deoxyribosides to shorten the C time of the thymine auxotrophs, as found by Pritchard and Zaritzky (1970). Any shortening of C would increase the resolution of the method. It also would be important to ensure that many generations of balanced growth had occurred before labelling. Today, higher specific activity thymidine is available to increase the time resolution. Of course, one would choose the strain with due regard to the plasmid problem, and purify the thymidine immediately prior to use. With all these improvements, I feel confident that the critical experiment could be executed.

XII. Precision of Initiation of Chromosome Regulations in *Myxococcus xanthus*

Although above I concentrated almost exclusively on work with *E. coli*, there is more extensive and very critical work available for analysis on *Myxococcus xanthus*. Moreover all the necessary measurements have been made in one laboratory (Rosenberg's) on one strain of organisms growing on one medium. *Myxococcus xanthus* is a fruiting myxomycete, but we are here concerned with its vegetative growth.

During balanced growth on defined medium, at 30°C, with a doubling time of 390 minutes, the organism appears to grow and divide in a manner consistent with the model developed above. First, the rate of protein synthesis is directly proportional to cell size. Actually, Zusman *et al.* (1971) found, with an experiment in which a pulse of labelled valine was added to the medium, that the mean grain count per unit length increased about 30% when comparing the smallest class of cells with the largest. If one corrected for rounded ends (new formed cells are about 2.5 times longer than they are wide) and the fact that walls are not as rich in protein as the remainder of the cell, then the rate of protein synthesis would be very nearly proportional to protoplasm content throughout.

Cell division takes place over a very narrow range of cell size. They classified dividing cells as either newly septated, medium septated cells, or mature septated cells. The first two size classes had a coefficient of variation of only 0.083 and 0.094, respectively. Since these phases last only a short time, the ranges of cell size for similar degrees of septation probably are larger, but not much larger than the true precision of the regulation of the cell division process.

The mature septated cells had a coefficient of variation of 0.128. This class, as well as the class of all septated cells, represents a range of stages in the cell cycle, and is therefore somewhat larger than the range for the control process for division. In another publication, Zusman and Rosenberg (1970) found that the dividing population (these three classes combined) had a coefficient of variation of 0.12 or 0.13, depending on whether each septating cell was classified as two small or one large cell. For either classification system, the distributions of the total population of growing cells had a coefficient of variation of 0.23. This value of 0.23, and interpolation from Table 4, leads to the expectation of a coefficient of variation of the size division of 0.14.

Since this agrees well with the observation of the 0.12 and 0.13 estimates, we may further conclude that the simple Koch and Schaechter model works, and that cell division produces equivalent daughter cells; that is, cell division evenly divides the protoplasm. Secondly, both from the shape of the distribution of cell sizes in balanced growth and from the agreement of the theoretical and observed coefficient of variation of the population, it can be concluded that very large and very small cells grow proportional to their size and not abnormally slowly (as for some of the *E. coli* data discussed previously; p. 63). In short, the data obtained with cultures of *M. xanthus* show less deviation from the simple model than do data from *E. coli* cultures. They do, however, initiate the cell division process more precisely than the final separation. It is fair to note that my interpretation of Zusman's data is somewhat different that of Zusman and Rosenberg (1970), who depended heavily on their amplification of the Collins and Richmond (1962) analysis, whereas I have relied on the widths of the two kinds of distributions, and on their valine pulse data.

Zusman and Rosenberg (1970) carried out pulse labelling with thymidine, and classified the number of silver grains over of cells of various size. Thus the analysis developed in Section VI.A (p. 65) is applicable. The DNA synthesis data is simpler than for the cases considered above in that there do appear to be both times in the cell cycle when none (or very few) of the cells are engaged in DNA synthesis, and others when all (or almost all) are engaged in synthesis. These lead to preliminary estimates of $\bar{x}_g = 0.2$ and $\bar{x}_s = 2.8$. From the combined data for 1600 cells (taken from their Table 3), a fit to the computer table led to values of $\bar{x}_g = 0.3$, $\bar{x}_s = 2.9$, and $S = 60\%$. While these parameters gave good fits, when the data was fitted as well as possible to slightly different values of $\bar{x}_g$ and $\bar{x}_s$ the generated value of S did not vary more than 5 percentage points from 60%.

The fact that such a precise fit could be obtained is consistent with the hypothesis that all cells engaged in synthesis lay down DNA at the same rate. This can also be concluded from their published $H(n)$ *vs* n plots. Then it is a simple matter to compute the percentage of all cells engaged in DNA synthesis from the observed mean grain counts and $\bar{x}_s = 2.9$ and $\bar{x}_g = 0.3$. These calculated values are given in Fig. 12. Assuming initiation sizes for chromosome replication are normally distributed in the population, the initial rise portion of the curve is basically the integral normal curve. We can estimate the variation in initiation size by taking one-half of the difference in cell size at one standard deviation below the mean (size corresponding to 84.1%, upper arrow) and one standard deviation below the mean (size corresponding to 15.9%, lower arrow) divided by the mean (median) size (size corresponding to 50%, middle arrow). This calculation yields 0.081 for the coefficient of variation.

Also shown in Fig. 12 (by the smooth line) is the computer simulation for the following model and parameters, $\bar{T} = 390$ minutes, $C = 234$ minutes, $D = 78$ minutes, the coefficient of variation of cell size

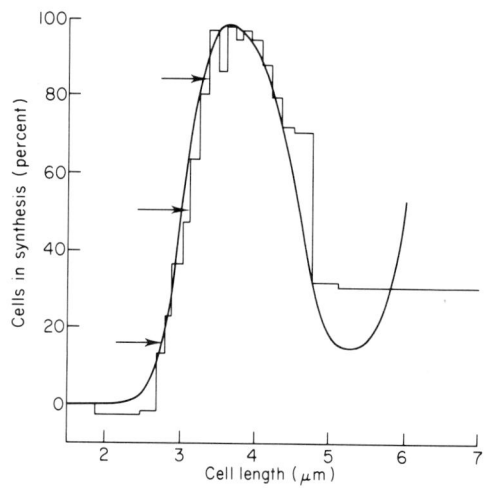

FIG. 12. Labelled cells of *Myxococcus xanthus*. Data taken from the experiments of Zusman and Rosenberg (1970). Type of plot is the same as shown in Figs. 10 and 11. Histogram is the observed data; curved line is fitted on the assumption that the coefficient of variation of the size at initiation of chromosome replication is 0.1, and that C and D are the same for all cells in the population.

at initiation equals 0.1 with no fluctuation allowed for either C or D. This model fits the Zusman's data quite well. I show this fit for an additional reason. This is to point out that the coefficient of variation of the size at initiation, when calculated from this theoretical curve by the approximate method as described at the end of the last paragraph, yields 0.092, quite close to the value of 0.1 used by the computer to generate the curve.

Thus, in the case of *M. xanthus*, I must reach the opposite conclusion to that from the *E. coli* cultures. The control of the sizes at initiation is sufficiently precise that it *could* time the cell division process—if no further sources of variation intervene between initiation and cell division.

XIII. Conclusions

In spite of numerous experimental studies, the mechanism of control of cell division and chromosome replication continues to be obscure. I have faith that ultimately the ability of the organism to produce protoplasm from the resources of the environment must limit growth, and must time the initiation of chromosome replication, nuclear division, and cell division. Here arguments, based on the variability of these processes within cultures of bacteria in balanced growth, are made to try to perceive a hierarchy in the triggering of those processes. The results of this analysis are largely based on computer simulations to fit the autoradiographic data of Chai and Lark (1970) and of Forro (unpublished) of slowly growing cultures of *Escherichia coli* strain $15T^-$. Almost all previous models are excluded for this organism.

The simulations exclude models in which initiation of replication takes place only when a cell has grown and reached a precise size; in fact, the autoradiographic data can only be fitted on the assumption that there is at least 15% coefficient of variation in the size of cells at initiation. Neither set of autoradiographic data fit the model of Cooper and Helmstetter (1968) for slowly growing bacteria in which cell division triggers the immediate initiation of a round of DNA synthesis; since in one set of data the average time of the initiation event precedes cell division, and in the other it follows, although not immediately. The present analysis excludes the Kubitschek and Freedman (1971) "self consistent" model, since the average time of synthesis is almost twice as long as proposed by them, and the average gap between cell division and the ensuing DNA initiation is not a large fraction of the cell cycle in slowly growing cells; in fact, the average gap must be negative for the experiment of Chai and Lark (1970). Although the computer modelling shows that there is considerable variation in cell size at initiation, it is less sure whether there is a comparable, or greater, variation in cell size at initiation than there is at division. If the variation is greater at initiation, then this less precise process cannot control, nor time, the more precise process of cell division. Even if the variation is the same at initiation and division, then either these two processes are unconnected events with accidentally similar variations, or they are connected by a total time (from the initiation of chromosome replication to cell division) that has so little variation from cell to cell that it introduces negligible additional random variation to the size at division. Arguments are presented against the latter hypothesis of constancy, within a population, of the time interval from initiation to cell division. Together these considerations argue against the class of hypotheses in which cell division is timed relative to some aspect of chromosomal synthesis, and for a variation on the Koch and Schaechter (1962) model in which both chromosome initiation and cell division are independently triggered by some aspects of cell growth. They would also be consistent with the model of Donachie et al. (1973), that a single trigger sets in motion chromosome replication and, independently, a process that has cell division as its consequence at some later time. This model should become unpopular because it would require, at least in some cases, that the accuracy with which cell division is controlled is much better than the accuracy with which chromosome initiation is controlled. This requires that the much longer time between this trigger and cell division be more accurately controlled than the much shorter time between the trigger and chromosome initiation.

This conclusion, that there may be multiple triggers, is reinforced by the recent report of Meacock and Pritchard (1975) who found that as the DNA synthesis was slowed by decreased thymine concentration, the $\bar{D}$ became shorter as $\bar{C}$ became longer. On the other hand, the analysis of Zusman's results are fully consistent with a single trigger controlling initiation, which deterministically and precisely controls the rest of the cell cycle, except for some additional variability in the actual size of separation of divided cells. The data do not exclude multiple, and quite precise, triggers but they do not require them.

Evidently, this discussion does not close the problem: I hope it opens up and focuses attention on the core of the problem, "How does a cell know how big it is?" This is the question that needs critical study in addition to the question of how it tells itself to initiate the discontinuous processes of chromosome replication and cell division.

Finally, the diligent reader will have noted that this paper is not a review at all. Instead, I hope this contribution is an *Advance,* but feel that it might be published instead in *Retreats in Microbial Physiology.*

Recent proper reviews of fields relevant to the issue considered here are: Matsushita and Kubitschek (1975) on "DNA replication in bacteria"; Ghuysen (1977) on "Biosyntheses and assembly of bacterial cell walls"; and Daneo-Moore and Shockman (1977) on "The bacterial cell surface in growth and division".

XIV. Acknowledgements

Without the help of congenial friends, this paper could not have been constructed: thank you Sara Bechtold, Jerry Blumberg, Steve Cooper, M. C. Escher, Fred Forro, Robert Harvey, Merijean Kelley, Marty Kraimer, Herb Kubitschek, Gordon Lark, Jerry Marr, Chick Newman, Bob Pritchard, Elio Schaechter, Conrad Woldringh, and Arieh Zaritsky. I especially wish to thank Drs. Forro, Kelley, Lark and Woldringh for making available unpublished data. The work was supported by the U.S. Energy Research and Development Administration. It is also a pleasure to thank the N.I.H. and the N.S.F. who have supported work in my laboratory for the past 20 years. Although the subject discussed here has never been explicitly supported by these agencies, they have never winced at my devotion to the problem of regulation of the cell cycle. Finally, I must thank the Argonne National Laboratory for awarding me the Argonne Universities Association Distinguished Appointment Award, which allowed me to be a scientific bum for a year and to follow where my nose and computer led. Argonnites of great help in this adventure were Herb Kubitschek, Marcia Rosenthal, Shelby Miller, and John Thomson.

REFERENCES

Abbo, F. E. and Pardee, A. B. (1960). *Biochimica et Biophysica Acta* **39**, 478.
Bird, R. E., Louarn, J., Martuscelli, J. and Caro, L. (1972). *Journal of Molecular Biology* **70**, 549.
Chai, N-C. and Lark, K. G. (1970). *Journal of Bacteriology* **104**, 401.
Chandler, M., Bird, R. E. and Caro, L. (1975). *Journal of Molecular Biology* **94**, 127.
Collins, J. F. and Richmond, M. H. (1962). *Journal of General Microbiology* **28**, 15.
Cooper, S. and Helmstetter, C. E. (1968). *Journal of Molecular Biology* **31**, 519.
Cutler, R. G. and Evans, J. E. (1966). *Journal of Bacteriology* **91**, 469..
Dennis, P. P. and Bremer, H. (1974). *Journal of Bacteriology* **119**, 270.
Daneo-Moore, L. and Shockman, G. D. (1977). *In* "Cell Surface Reviews", vol. 4. (G. Post and G. L. Nicolson, eds.). North Holland Publishing Co.
Donachie, W. D. (1968). *Nature, London* **219**, 1077.

Donachie, W. D., Jones, N. C. and Teathes, R. (1973). *Symposia of the Society for General Microbiology* **23**, 9.
Ecker, R. E. and Kokaisl, G. (1969). *Journal of Bacteriology* **98**, 1219.
Errington, F. P., Powell, E. O. and Thompson, N. (1965). *Journal of General Microbiology* **39**, 109.
Forro, F., Jr. (1965). *Biophysical Journal* **5**, 629.
Forro, F., Jr. and Wertheimer, S. A. (1960). *Biochimia et Biophysica Acta* **40**, 9.
Grivell, A. R. and Hanawalt, P. C. (1973). *Biophysical Society Abstract* **37a**.
Gudas, L. I. and Pardee, A. B. (1974). *Journal of Bacteriology* **117**, 1216.
Ghuysen, J.-M. (1977). *In* "Cell Surface Reviews", vol. 4. (G. Post and G. L. Nicolson, eds.). North Holland Publishing Co.
Hanawalt, P. C., Maaløe, O., Cumming, D. J. and Schaechter, M. (1961). *Journal of Molecular Biology* **3**, 156.
Harvey, R. S., Marr, A. G. and Painter, R. P. (1967). *Journal of Bacteriology* **93**, 604.
Helmstetter, C. E. (1967). *Journal of Molecular Biology* **24**, 417.
Helmstetter, C. E. (1968). *Journal of Bacteriology* **95**, 1634.
Helmstetter, C. E. (1974). *Journal of Molecular Biology* **84**, 21.
Helmstetter, C. E. and Cooper, S. (1968). *Journal of Molecular Biology* **31**, 507.
Hoffman, H. and Frank, M. E. (1965). *Journal of Bacteriology* **89**, 513.
Hohlfeld, R. and Vielmetter, W. (1973). *Nature New Biology* **242**, 130.
Kelley, M. S. (1974). Thesis, The University of Texas Health Science Center at Dallas, pp. 1–110.
Kendall, D. G. (1948). *Biometrika* **35**, 316.
Koch, A. L. (1966). *Journal of General Microbiology* **43**, 1.
Koch, A. L. and Blumberg, G. (1976). *Biophysical Journal* **16**, 389.
Koch, A. L. and Pachler, P. F. (1967). *Journal of Molecular Biology* **28**, 531.
Koch, A. L. and Schaechter, M. (1962). *Journal of General Microbiology* **29**, 435.
Kubitschek, H. E. (1962). *Experimental Cell Research* **26**, 439.
Kubitschek, H. E. (1967). "Proceedings of the Fifth Berkeley Symposium on Mathematics, Statistics and Probability", pp. 549–572. University of California Press.
Kubitschek, H. E. (1968a). *Biophysical Journal* **8**, 792.
Kubitschek, H. E. (1968b). *Biophysical Journal* **8**, 1401.
Kubitschek, H. E. (1971). *Journal of Bacteriology* **105**, 472.
Kubitschek, H. E. (1974). *Biophysical Journal* **14**, 119.
Kubitschek, H. E. and Freedman, M. L. *Journal of Bacteriology* **107**, 95.
Kubitschek, H. E., Bendigkeit, H. E. and Loken, M. R. (1967). *Proceedings of the National Academy of Sciences of the United States of America* **57**, 1611.
Lark, C. (1966). *Biochimica et Biophysica Acta* **119**, 517.
Lark, K. G. (1963). *In* "Molecular Genetics" (J. Taylor, ed.), part I, pp. 153–206. Academic Press, New York.
Lark, K. G. (1966). *Bacteriological Review* **30**, 3.
Maaløe, O. (1961). *Cold Spring Harbor Symposia on Quantitative Biology* **26**, 45.
Maaløe, O. and Kjeldgaard, N. O. (1966). *In* "Control of Macromolecular Synthesis", pp. 1–197. W. A. Benjamin, Inc., New York.
Mason, D. J. and Powelson, D. M. (1956). *Journal of Bacteriology* **71**, 474.
Masters, M. and Broda, P. (1971). *Nature New Biology, London* **232**, 137.
Matsushita, T. and Kubitschek, H. E. (1975). *In* "Advances in Microbial Physiology" (A. H. Rose and D. W. Tempest, eds.), vol. 12, pp. 247–327. Academic Press, New York.

Meacock, P. A. and Pritchard, R. H. (1975). *Journal of Bacteriology* **122**, 931.
McFall, E. and Stent, G. S. (1959). *Biochima et Biophysica Acta* **34**, 580.
Mitchison, J. M. (1971). "The Biology of the Cell Cycle", pp. 313. Cambridge University Press, Cambridge.
Mitchison, J. M. and Vincent, W. S. (1965). *Nature London* **205**, 987.
Nagata, T. and Meselson, M. (1968). *Cold Spring Harbor Symposia on Quantitative Biology* **33**, 553.
Pachler, P. F., Koch, A. L. and Schaechter, M. (1965). *Journal of Molecular Biology* **11**, 650.
Painter, P. R. (1974). *Genetics* **76**, 401.
Painter, P. R. and Marr, A. G. (1968). *Annual Review of Microbiology* **22**, 519.
Powell, E. O. (1964). *Journal of General Microbiology* **37**, 231.
Pierucci, O. and Zuchowski, C. (1973). *Journal of Molecular Biology* **80**, 477.
Pritchard, R. H. (1968). *Heredity* **23**, 472.
Pritchard, R. H. (1974). *Philosophical Transactions of the Royal Society of London* Series B. **267**, 303.
Pritchard, R. H. and Zaritzky, A. (1970). *Nature London* **226**, 126.
Pritchard, R. H., Barth, P. T. and Collins, J. (1969). *Symposia of the Society for General Microbiology*
Rahn, O. (1931–1932). *Journal of General Physiology* **15**, 257.
Rodriguez, R. L. and Davern, C. I. (1976). *Journal of Bacteriology* **125**, 346.
Rosenberg, B. H., Cavalieri, L. and Unders, G. (1969). *Proceedings of the National Academy of Sciences, Washington* **63**, 1410.
Schaechter, M., Bentzon, M. W. and Maaløe, O. (1959). *Nature London* **183**, 1207.
Schachter, M., Williamson, J. P., Hood, J. R., Jr. and Koch, A. L. (1962). *Journal of General Microbiology* **29**, 435.
Woldringh, C. L. (1974). Thesis, The University of Amsterdam, pp. 129.
Woldringh, C. L. (1976). *Journal of Bacteriology* **125**, 248.
Young, E. and Fitz-James, P. (1959). *Nature London* **183**, 372.
Zaritsky, A. and Pritchard, R. H. (1971). *Journal of Molecular Biology* **60**, 65.
Zaritsky, A. and Pritchard, R. H. (1973). *Journal of Bacteriology* **114**, 824.
Zusman, D. and Rosenberg, E. (1970). *Journal of Molecular Biology* **49**, 609.
Zusman, D., Gottlieb, P. and Rosenberg, E. (1971). *Journal of Bacteriology* **105**, 811.

ADDITIONAL NOTES

Newman and Kubitschek (Personal Communication) have recently conducted experiments to measure the variability of the time interval from the synthesis of a region of chromosome to the time of synthesis of the same region of the chromosome in the daughter cell. They pulse labelled cells for 2.5 min. with 5-bromo uracil and followed the killing caused by high doses of 313 nm ultra violet. After a generation of growth, abruptly the cells became radiation resistant. This happens when the bromo uracil contain region is replicated to yield a complete genome devoid of bromo uracil. The abruptness of the change from radiation sensitivity to resistance corresponded to a coefficient of variation of 9.3% for the replicating fork interval (RFI). For technical reasons this must be an upper estimate. The true estimate must be an estimate of the coefficient of variation of interval between initiations, $I = C + D + B'$, where B' is the gap between cell division and initiation in the daughter generation.

These new experiments extend the interpretation given on pp. 80–81 of the experiment of Nagata and Meselson (1968). They reinforce the idea that basically the cell cycle of procaryotes is very precise even though the timing of all divisions is quite imprecise. It is difficult to resolve these findings except by assuming postulate 1 and 4 of the Koch–Schachter model: that deterministic growth in the long run times chromosome replication and cell division and that cell division in *E. coli* precisely equiportions cell substance.

Repair of Damaged DNA in Bacteria

B. E. B. MOSELEY and E. WILLIAMS

Department of Microbiology, University of Edinburgh, School of Agriculture, Edinburgh EH9 3JG, Scotland

I. Survival Value of DNA Repair Mechanisms	100
II. Radiation and Chemical Damage to DNA	101
A. Ultraviolet Radiation Damage	101
B. Ionizing Radiation Damage	102
C. Mitomycin-C and Psoralen-plus-Near-Ultraviolet Radiation Damage	105
D. The Extent of Repair of DNA Damage	106
III. The Isolation of Mutants Defective in DNA Repair	109
A. Mutations Leading to Increased Sensitivity of *Escherichia coli* to Radiation and/or Chemical Damage	111
IV. Photoreactivation Repair	121
A. The Substrate for Photoreactivating Enzyme	122
B. The Photoreactivating Enzyme (PRE)	122
C. The Enzyme–Substrate Complex	124
D. Photoreactivation as an Analytical Tool	125
V. Excision Repair	125
A. Methods for Studying Excision Repair	126
B. Incision (*uvrA, uvrB*)	127
C. Pre-excision (*uvrC*)	128
D. Excision (*polA*)	128
E. Re-insertion of Bases	130
VI. Post-Replication Recombination Repair	131
A. Methods for Studying Post-Replication Recombination Repair	133
B. Gap Formation Opposite Pyrimidine Dimers	133
C. Gap Repair	135
D. Recombination-Deficient Repair (Genetic Control of Post-Replication Repair)	135
VII. The Repair of Ionizing Radiation Damage	137
A. Type III Repair (*recA, recBC, exr, dnaB, ror, lig*)	137
B. Type II Repair (*polA, lig*)	138

	C. Type I Repair (*lig?*)	139
	D. The Contribution of Types I, II and III Repair in the Irradiated Wild-Type Cell	140
	E. The Repair of DNA Double-Strand Breaks	141
VIII.	The Repair of Cross-Link Damage	142
IX.	The Mutagenic Consequences of Repair	144
	A. Photoreactivation	145
	B. Excision Repair	145
	C. Recombination	146
	D. The Inducible Nature of Ultraviolet-Radiation-Induced Mutagenesis in *Escherichia coli*	147
	E. Repair-Dependent and Independent Mutagens	148
X.	Summary	149
	References	150

I. Survival Value of DNA Repair Mechanisms

In those bacterial species which have been intensively investigated genetically, the chromosome or genome is a single, circular molecule of DNA containing several million base pairs, and is about a thousand times longer than the bacterial cell. In non-growing populations of bacteria, or those growing with extremely long generation times, there may be only one copy of the genome present per cell and so the future of that bacterium is dependent on the DNA remaining undamaged and retainings its genetic continuity. However, DNA is very sensitive to damage by both environmental radiation and chemicals. For example, a suspension of a double mutant (AB2480, *uvrA6 recA13*) of *Escherichia coli K-12* strain AB1157 which was totally deficient in DNA dark-repair capability, and physiologically unable to photoreactivate ultraviolet (u.v.) damage, lost more than 99% viability as a result of a one-minute exposure to sunlight in Dallas, Texas, one October day, and more than 90% on a December day (Harm, 1969). The evidence that sunlight was causing DNA damage was that 85% of the damage could be photoreactivated. The only known substrate for the photoreactivating enzyme is the cyclobutane-type pyrimidine dimer, a DNA photoproduct of u.v. radiation (Setlow and Setlow, 1963). In the wild-type strain *E. coli* K12S, in which the DNA repair mechanisms were fully operative, the bacteria could repair the usual bactericidal damage caused by sunlight so extensively that at very long exposure times the ultimate inactivation that was expressed arose from non-repairable damage (Harm, 1965). This is an excellent example of the survival value to bacteria of DNA repair mechanisms, and indicates the necessity for the evolution of at least a mechanism for the repair of u.v. radiation damage along with the evolution of a chromosome.

The purpose of this article is to describe the methods by which DNA repair mechanisms are investigated, genetically and biochemically, the nature of the repair modes which have so far been described, and their effect on survival and mutation.

II. Radiation and Chemical Damage to DNA

The structure of DNA is basically open to three kinds of damage, all of which may lead to lethal or mutational events. These are damage or removal of bases, strand breakage (either single or double-stranded), and covalent cross-linking of the DNA. These three kinds of damage pose quite different problems to a repair mechanism in terms of the restoration of the normal DNA structure. However, since ultraviolet and ionizing radiations cause a preponderance of one or the other types of damage, it is perhaps more convenient to describe the lesions in DNA caused by particular treatments.

A. ULTRAVIOLET RADIATION DAMAGE

The range of ultraviolet-irradiation products formed in DNA, either *in vivo* or *in vitro*, has been reviewed previously (Setlow, J. K., 1966, 1967; Moseley, 1968). Briefly, u.v. radiation can cause phosphodiester strand breaks (Marmur *et al.*, 1961), DNA intrastrand cross-links (Marmur and Grossman, 1961), nucleic acid–protein cross-links (Smith, 1962; Smith and Aplin, 1966), hydration products of cytosine (Sinsheimer, 1957), and pyrimidine dimers (Wacker *et al.*, 1960). Strand-breaks and intrastrand cross-links are only formed in very low numbers at excessive doses of u.v. radiation, well beyond the doses required to inactivate bacteria. Pyrimidine hydrates are not formed in double-stranded DNA and are therefore unlikely to contribute to the u.v. inactivation of bacteria. The role of DNA-protein cross-links is not clear but, in the event of protein playing a major role in u.v. inactivation, one might expect that the u.v. action spectrum (the efficiency of cell inactivation as a function of wavelength; Gates, 1930; Wykoff, 1932), would indicate a greater efficiency of killing at 280 nm than it does. The role of pyrimidine dimers in u.v. inactivation is undisputed. The pyrimidine dimer is formed from two pyrimidine bases adjacent to each other in the same DNA strand which become linked to each other by carbon-to-carbon bonds between their respective C-5 and C-6 atoms, forming a cyclobutane ring between the two pyrimidines (Fig. 1). The yield of pyrimidine dimers in the DNA of bacteria is approximately 2.6×10^{-9} dimers per thymine residue per erg/mm² at 254 nm. Thus a u.v. dose of 1 erg/mm² will produce about six pyrimidine dimers in the genome of *E. coli* (Howard-Flanders and Boyce, 1966). In a bacterium effectively lacking DNA repair mechanisms, the recombination-deficient, excision-deficient *E. coli* double mutant AB2480, 1.5 dimers per genome are lethal (Howard-Flanders

FIG. 1. Structural formulae of: I, thymine; II, thymine-thymine dimer; III, conformation of the dimer in DNA.

and Boyce, 1966). Thus for bacteria to survive u.v. irradiation the elimination, or by-passing, of pyrimidine dimers is absolutely necessary.

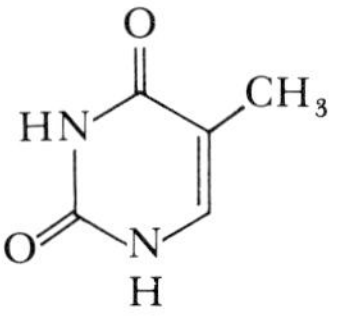

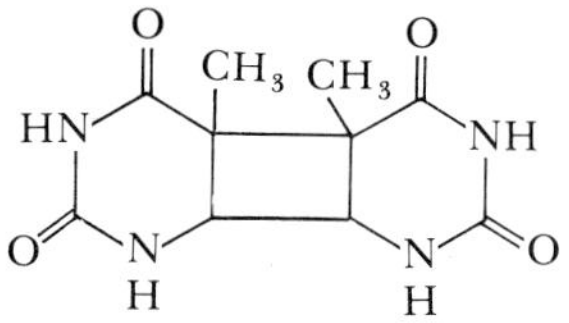

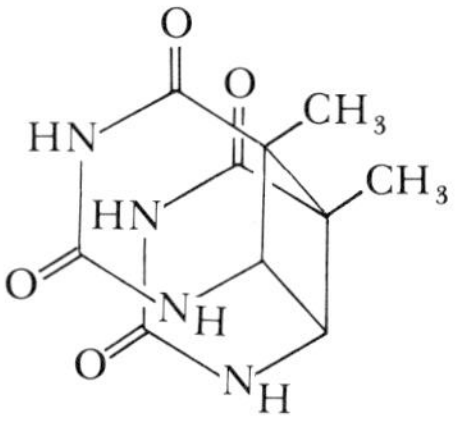

B. IONIZING RADIATION DAMAGE

Of the lesions produced in DNA by ionizing radiation, the one most extensively studied to date, and most closely correlated with cell killing, is the breakage of one or both strands of the sugar-phosphate backbone. Evidence for base damage is accumulating but has been little studied. For reviews on ionizing radiation damage to DNA, and its repair, see Setlow and Setlow (1972) and Town *et al.* (1973b).

One of the easiest parameters to measure, following the ionizing irradiation of DNA, is the breakage of chains by measuring the sedimentation velocity of the irradiated DNA and comparing it with unirradiated DNA. The method used is that developed by McGrath and Williams (1966) who showed that it was possible to isolate and sediment relatively high molecular-weight single-stranded DNA from bacteria, by lysing the cells on top of an alkaline sucrose gradient. For DNA isolated from phage, bacteria or mammalian cells, the energy absorption per single-strand break is about 50 eV (see Setlow and Setlow (1972) for a summary of experimental values). This value holds whether or not the DNA source is irradiated in oxygen or nitrogen, and it follows that the well-known oxygen effect cannot be explained in terms of increased strand breakage in oxygen. An energy absorption of 50 eV/break is equivalent to one single-strand break per krad ionizing radiation per 5×10^8 daltons. The number of double-strand breaks is 10- to 20-fold less than the number of single-strand breaks. Thus, in a repair-proficient *E. coli* strain AB1157, the lethal dose of 18 krad X-rays (Howard-Flanders and Boyce, 1966) would cause about 80 single-strand breaks and approximately four double-strand breaks per genome. In surviving ionizing radiation, therefore, the ability to mend strand gaps is of prime importance.

The evidence for base damage being caused by ionizing radiation comes firstly from an investigation being carried out by R. B. Setlow and his colleagues (Paterson and Setlow, 1972; Setlow and Carrier, 1973; Setlow, 1973) based on the action of an endonuclease which makes single-strand incisions in the region of u.v. photoproducts, the incisions being detected as a reduction in the molecular weight of the DNA fragments. Since ionizing radiation itself makes single-strand breaks, they measured the ability of crude endonuclease preparations to make further breaks in the irradiated DNA. The endonuclease preparation, although having no effect on unirradiated DNA, caused the molecular weight of the irradiated DNA to fall by a factor of five. The data suggest that the enzymes recognize damage other than strand breaks (presumably base damage) in γ-irradiated DNA.

Secondly Hariharan and Cerutti (1971, 1972) have developed a reduction assay for identification and determination of the major radiolysis products of thymine and its derivatives. Radiolysis products of the type 5,6-dihydroxy-5,6-dihydrothymine and 5-hydroperoxy-6-hydroxy-5,6-dihydrothymine are reductively cleaved with sodium borohydride and a four-carbon alcohol is released containing the thymine–methyl group (Fig. 2. If thymine derivatives carry a radio-active label in the methyl group, the amount of four-carbon fragments produced by sodium borohydride (mainly 2-CH_3-glycerol after irradiation in air) can be determined by radio-activity measurements, and is a measure of thymine radiolysis. The amount of the thymine derivatives formed in the DNA of irradiated *Micrococcus radiodurans* is

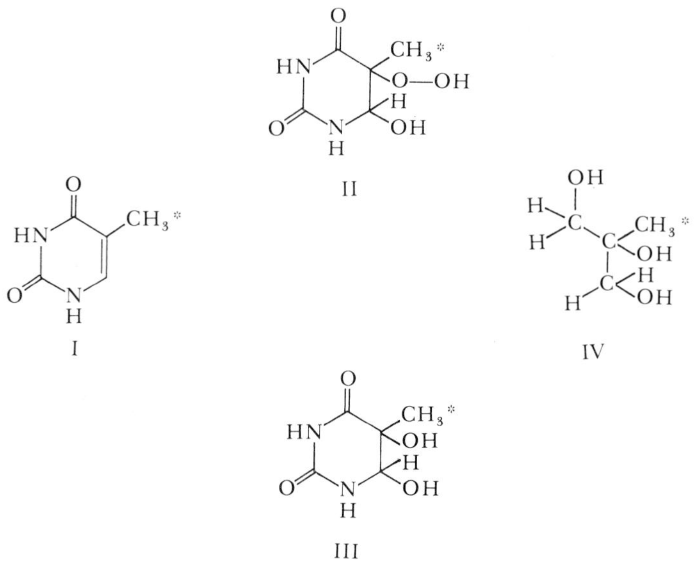

FIG. 2. Structural formulae of: I, thymine; II. 5-hydroperoxy-6-hydroxy-5,6-dihydrothymine; III. 5,6-dihydroxy-5,6-dihydrothymine. II and III are radiolysis products of thymine resulting from γ-irradiation of DNA; IV, 2-methyl glycerol, formed by reductive cleavage of II and III. A tritiated methyl group (*) enables quantitative estimation of the radiolysis products to be made. After Hariharan and Cerutti (1972).

1.2×10^{-6} per rad per 10^6 daltons (which is equivalent to about 8400 modified thymine bases per genome at the lethal dose in air). Only small amounts of these derivatives were found in the medium or acid-soluble fraction immediately after irradiation, but the amount in the medium increased with incubation time. On average, about 300 undamaged thymine residues are removed from the DNA per damaged residue during four hours post-irradiation incubation (Hariharan and Cerutti, 1972). Whether or not this is a DNA repair mechanism which is removing potentially lethal lesions has not been demonstrated, but it is not unreasonable to suppose that degradation of *M. radiodurans* DNA, after irradiation, results from single-strand breaks introduced into the DNA in the vicinity of damaged thymine bases.

C. MITOMYCIN-C AND PSORALEN-PLUS-NEAR-ULTRAVIOLET-RADIATION DAMAGE

An important consequence of the treatment of bacteria with the antitumour drug mitomycin-C, or treatment with 4,5',8-trimethyl-psoralen or 8-methoxy-psoralen followed by exposure to near u.v. radiation (360 nm), is the formation of covalent cross-links. The structure of mitomycin-C and derivatives of psoralen are shown in Fig. 3.

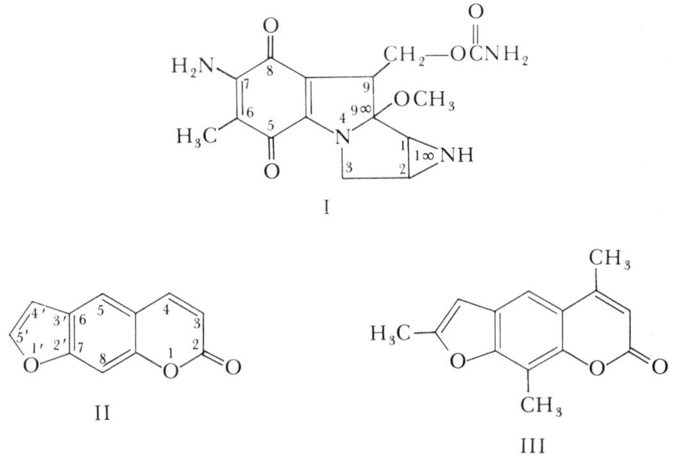

FIG. 3. Structural formulae of some DNA cross-linking agents: I, mitomycin-C; II, psoralen; III, 4,5',8-trimethylpsoralen.

Evidence that treatment of bacteria with mitomycin-C leads to formation of thermostable cross-links between the complementary DNA strands was first reported by Iyer and Szybalski (1963). The mitomycin-induced cross-links are caused by covalent bonding of the antibiotic to DNA as a result of bifunctional alkylation of groups on each of the complementary DNA strands (Iyer and Szybalski, 1964). Mitomycin has no effect on purified DNA *in vitro* but must be activated before any reaction can take place. Activation requires reduction of the mitomycin molecule. This is done enzymically in the cell, or by cell extracts, but can be achieved chemically with agents such as sodium hydrosulphite. The active reduced form of mitomycin is unstable. Szybalski and Iyer (1964) found that only one out of five to ten molecules participates in the formation of cross-links while the others form mono-adducts. For a review see Waring (1966).

Exposure of bacteria to the action of substituted psoralens, even at high concentrations and for quite long periods, does not cause cell inactivation. Similarly the exposure of bacteria to near u.v. radiation of 360 nm does not cause inactivation, e.g. the mean lethal dose for a uvrA mutant of E. coli is 3×10^7 ergs/mm^2 (Webb and Lorenz, 1970). However, the mean lethal dose for a psoralen-treated culture is about 2×10^4 ergs/mm^2 (Cole, 1971). The psoralen-sensitized reaction is independent of oxygen, and the reaction proceeds more rapidly at low temperatures (Oginsky et al., 1959) so that irradiation is best carried out at 0°C. As with mitomycin-C, a considerable proportion of the molecules form mono-adducts rather than cross-links. Cole (1971) has presented evidence that the cross-links are the psoralen photoproducts which cause the biological damage. He calculated, using 4,5',8-trimethyl psoralen, that the yield of cross-links per minute per dalton of E. coli DNA in vivo (with a 360 nm light intensity of 2.4 mW/cm^2) is 4.2×10^{-9}, and with 2.5×10^9 daltons as the estimated molecular weight of the E. coli genome was able to calculate the average number of cross-links per mean lethal dose. The values obtained were, 65 for the wild type E. coli AB1157, 6.7 for the recA13 mutant, 16 for the uvrA6 mutant, and less than one for the double mutant recA13 uvrA6. It has already been pointed out that one pyrimidine dimer is lethal to the same double mutant (Howard-Flanders and Boyce, 1966). The removal of a cross-link appears to be more difficult than either the mending of a single-strand break or a pyrimidine dimer since these forms of damage involve only one strand. However, models have been proposed for their removal while maintaining the genetic continuity of the DNA (Cole, 1973: Howard-Flanders, 1973; see p. 142).

D. THE EXTENT OF REPAIR OF DNA DAMAGE

Because of the nature of radiation and chemical damage to DNA, lethal events are scattered randomly through the population. Thus the average dose required to inactivate a cell is the dose which kills 63% of the population, because of single or multiple inactivating events, and which therefore allows 37% to survive. The dose of radiation or time of exposure to a chemical which allows 37% survival can be obtained from dose–response (survival) curves and can be converted into numbers of pyrimidine dimers, or DNA strand breaks, or DNA cross-links (Fig. 4). In *E. coli,* approximately six dimers per genome are formed per erg/mm² of 254 nm u.v. radiation (Howard-Flanders and Boyce, 1966), and since the 37% survival dose for the wild type AB1157 is 500 ergs/mm² this means about 3000 dimers per genome represents a lethal dose (and, of course, that one less than this can be repaired). It

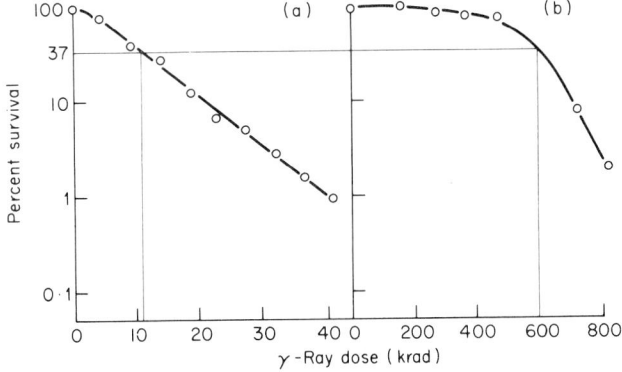

FIG. 4. Gamma-irradiation survival curves of (a) *Escherichia coli* B/r and (b) *Micrococcus radiodurans.* The dose which allows 37% of the initial population to survive (D_{37}), and which is the dose required to inactivate a single viable unit, is indicated. The values for *E. coli* B/r and *M. radiodurans* are 11 and 600 krad, respectively, and can be used to calculate the numbers of single- and double-strand breaks per genome. Data from Sweet and Moseley (1976).

is the last dimer that breaks the back of the DNA! If the number of base pairs in the genome of *E. coli* is taken as 3×10^6, then this represents one pyrimidine dimer per 1000 base pairs, or about one per gene. It has already been stated that the D_{37} for X-rays on the same strain of *E. coli* would cause about 80 single-strand breaks and about four double-strand breaks (Howard-Flanders and Boyce, 1966; Setlow and Setlow, 1972), and that *E. coli* AB1157 can repair up to 64 cross-links induced by 4,5′,8-trimethyl-psoralen and near u.v. radiation (Cole, 1971).

However, even the most resistant strains of *E. coli,* namely *E. coli* B/r (Witkin, 1947) and *E. coli* K-12 AB1157, are relatively sensitive when compared with *Micrococcus radiodurans* (Anderson *et al.*, 1956) which features in the Guiness Book of Records (20th edition, 1973, p. 56) as the "toughest" bacterium, an accurate description only insofar as it refers to radiation resistance. Other vegetative bacteria are also extremely resistant to radiation, e.g. *Micrococcus roseus* ATCC 19172 isolated from irradiated haddock (Davis *et al.*, 1963), *Micrococcus radiophilus* isolated from irradiated Bombay duck (Lewis, 1971), and *Micrococcus radioproteolyticus* isolated from llama faeces (Kobatake *et al.*, 1973).

Driedger (1970) has provided an estimate of the amount of DNA per cell, and of the genome size of *M. radiodurans*. His values are $1.92 \pm 0.12 \times 10^{-14}$ g DNA per cell, and a genome size of 1.7×10^7 base pairs per replicating genome or 1.2×10^7 base pairs per non-replicating genome. Since the molecular weight of an A–T base pair is 617, and a G–C base pair 618, the molecular weight of a replicating genome is 1.0×10^{10}. This is about three times larger than the genome of *E. coli*. However, it does mean that D_{37} values obtained for u.v. radiation, γ-radiation, and other treatments can be related to the genome.

The dose delivered by the u.v. lamp that we have used in our studies on *M. radiodurans* has been overrated. This does not affect the calculation since this results in an underestimate of dimers formed per erg/mm^2. We have estimated the number of thymine-containing pyrimidine dimers (TT and TC) induced per genome as 5.7 per erg/mm^2, and the D_{37} as 12,200 ergs/mm^2 which gives a value of 70,000 dimers/genome, or one per 250 base pairs. The D_{37} for γ-irradiation is 600–700 krads when irradiated in air. Burrell *et al.* (1971) showed that double-strand breaks were introduced into the DNA of *M. radiodurans* by X-rays at an efficiency of 520 ± 50 eV/break (equivalent to one break/krad per 5×10^9 daltons) and that the scission of the DNA twin helix caused by double-strand breaks occurred at a frequency some 13-fold lower than single-strand breaks (40 eV/single-strand break). Using these values, it is possible to calculate that *M. radiodurans* survives about 18,000 single-strand breaks and 1400 double-strand breaks per genome!

For the repair of cross-link damage, it has been calculated from the inactivation of a recombination-deficient mutant of *M. radiodurans* by a single mitomycin-C-induced cross-link per genome that the wild type, which is 300 times more resistant than the mutant, can repair 300 cross-links per genome (Moseley and Copland, 1975).

A recent study has been carried out in which the lethal and mutagenic effects of u.v. and ionizing radiation, and of a number of chemicals, have been compared in *E. coli* B/r and *M. radiodurans* (Sweet and Moseley, 1976). *Micrococcus radiodurans* was 33 times more resistant than *E. coli* B/r to the lethal effects of u.v. radiation, 55 times more resistant to γ-radiation, 15 times more resistant to N-methyl-N'-nitro-N-nitrosoguanidine (NTG), and 62 times more resistant to nitrous acid. *Micrococcus radiodurans* showed an intermediate resistance to mitomycin-C and hydroxylamine, being four and seven times as resistant, respectively, as *E. coli* B/r, but had only twice the resistance to β-propiolactone and the same resistance to ethyl methane sulphonate

TABLE 1. Amount of damage required per genome to inactivate *Escherichia coli* K-12, *Micrococcus radiodurans* and some of their mutants. Data from Howard-Flanders and Boyce (1966) and Moseley and Copland (1975)

	Single-strand breaks	Double-strand breaks	Pyrimidine dimers	Cross-links
Escherichia coli				
AB1157	80	4	3000	65
AB1886 *uvrA6*	65	4	50	16
AB2463 *recA13*	12	1	20	6.7
AB2480 *recA13 uvrA6*			1.5	1
Micrococcus radiodurans				
wild type	18000	1400	70000	300
rec30	150	12	4500	1

as *E. coli* B/r. Why *Micrococcus radiodurans* should have evolved such a superb DNA repair system is not at all clear since it would seem to be at least an order of magnitude more efficient than would be required to survive any environmental hazard. Table 1 shows the relative resistances of *E. coli* and *M. radiodurans*, and those of some of their mutants.

III. The Isolation of Mutants Defective in DNA Repair

The normal procedure for studying mechanisms of DNA repair is to isolate mutants which are either more or less resistant to radiation, for example, than the wild type, in order to compare the mutants and wild type biochemically. Technically, isolation of radiation-resistant mutants is easy; one needs only to apply an appropriate dose of radiation to a suspension of the wild type bacterium so that the viability is decreased to an extremely low level. Radiation-resistant mutants might be expected to be among the survivors. In spite of the technical ease, very few radiation-resistant mutants have been isolated. A notable exception is *E. coli* B/r, a radiation-resistant mutant derived from *E. coli* (Witkin, 1947) and now known to be a *sul* mutant (Donch et al., 1969).

Isolation of sensitive mutants requires a more tedious approach, but has been very successful. Normally a wild-type population is treated with a mutagen, such as nitrous acid or NTG, to increase the probability of finding a mutant. After several hours growth to allow phenotypic expression, the bacteria are plated and incubated. The resulting colonies are replicated onto a fresh plate and the latter exposed to a radiation dose which allows the wild type to survive and produce colonies. Any colonies which do not grow on the irradiated plate are picked from the original plate and tested for radiation sensitivity. Of course, the original master plate could be replicated onto other plates (e.g. ones containing mitomycin-C) and the plates checked for missing colonies in the same way.

When the sensitive mutants are isolated, their phenotype is examined. For example, if the mutant was isolated as sensitive to u.v. radiation, then its resistance to agents such as ionizing radiation, mitomycin-C and ethyl methanesulphonate should be determined. One should also measure its ability to carry out genetic recombination, to excise u.v.-induced dimers, to mend DNA strand breaks, and the rate and extent to which its DNA is degraded following irradiation. On the basis of the phenotype of radiation-sensitive mutants of *E. coli* of known genotype, intelligent guesses can be made of the unknown's genotype.

Not all radiation-sensitive mutants have been obtained this way. Sometimes mutants isolated on some quite different basis have turned out to be radiation sensitive. Examples of such are the recombination-deficient mutants (Clark and Margulies, 1965) and the DNA polymerase-1 deficient mutant (DeLucia and Cairns, 1969) of *E. coli*.

The almost total lack of success in isolating radiation-resistant mutants from wild-type populations infers that the wild types have already evolved and now maintain adequate DNA-repair capacity. This can be contrasted with antibiotic resistance in wild populations of bacteria where the majority are sensitive, and resistant mutants easily selected.

A. MUTATIONS LEADING TO INCREASED SENSITIVITY OF
Escherichia coli TO RADIATION AND/OR CHEMICAL DAMAGE

1. *The phr Gene*

Harm and Hillebrandt (1962) treated a population of *E. coli* B with the mutagenic agent nitrous acid until the number of survivors was about 10^{-7} of the total number of organisms. From among 400 colonies derived from surviving cells, they isolated one *phr* mutant. The isolation was facilitated by using the colonies as plating bacteria for u.v.-irradiated T4 v_1 phage. With exposure of the plates to photoreactivating light, about 10% of the phage survived compared with 0.1% when the plates were held under non-photoreactivating conditions. Harm and Hillebrandt (1962) found a mutant which gave low survival of phage under photoreactivating conditions and which could not photoreactivate u.v. damage to its own genome. The *phr* gene was later shown to be closely linked to *gal* (Van De Putte *et al.*, 1965) and has now been mapped between *gal* and the attachment site for phage λ (*att*λ) which are at 17.0 and 17.3 minutes, respectively, on the *E. coli* chromosome map (Sutherland *et al.*, 1972).

2. *The uvr Gene*

The first *uvr* mutant isolated was *E. coli* B_{s-1} (Hill, 1958). Its isolation was remarkably easy. A spread plate of *E. coli* B was irradiated with u.v. radiation and incubated. Twenty-two colonies grew from surviving bacteria and 12 were tested for u.v. sensitivity. Ten were *E. coli* B, one was resistant and presumed to be *E. coli* B/r and one, *E. coli* B_s was very sensitive to u.v. radiation. This mutant was later called *E. coli* B_{s-1} (Hill and Simson, 1961). It was on this strain that Setlow and Carrier (1964) did their classical work in which they showed that the u.v.-resistant *E. coli* B was able to excise thymine-containing pyrimidine dimers from its DNA while the u.v.-sensitive mutant *E. coli* B_{s-1} could not. Thus the excision of pyrimidine dimers was a DNA repair mechanism leading to enhanced survival. In fact *E. coli* B_{s-1} is a double mutant, being mutant at both the *uvrB* and *exr* (*lex*) loci (Mattern *et al.*, 1966; Greenberg, 1967). The *exr* mutation is responsible for its increased sensitivity to X-rays.

Although unirradiated stocks of the phages T1, T3 and T7 give identical plaque counts on both *E. coli* B and B_{s-1}, the survival of these phages after irradiation is considerably lower on *E. coli* B_{s-1} than on *E.*

coli B (Ellison *et al.*, 1960). The ability to reactivate irradiated phage is known as host-cell reactivation (Hcr), so that *E. coli* B_{s-1} is Hcr$^-$. This observation led Howard-Flanders and Theriot (1962) to isolate the first *uvr* mutants in *E. coli* K-12. They treated *E. coli* AB1157 with the mutagen nitrous acid and reduced the viability of the population by a factor of more than 10^5. The surviving bacteria were allowed to grow through ten divisions before mutants were selected by their ability to survive exposure to u.v.-irradiated phage T1. Fourteen mutants were isolated, uv_1^s, uv_2^s etc. (uv^s = u.v. sensitive). The mutant uv_6^s was eventually called *uvrA6* (AB1886) and was the strain in which Boyce and Howard-Flanders (1964a) demonstrated the lack of an excision mechanism, which was present in AB1157. Howard-Flanders and his colleagues showed that the *uvr* mutations map at three distinct loci, which they called *uvrA*, *uvrB* and *uvrC*, mapping at 81, 18 and 36 minutes, respectively, on the *E. coli* chromosome map (Howard-Flanders (in Setlow, 1964); Howard-Flanders *et al.*, 1966). Gene *uvrA*$^+$ is dominant to *uvrA*$^-$ in zygotes.

The mutants *uvrA*, *uvrB* and *uvrC* are all unable to excise pyrimidine dimers from their DNA or from u.v.-irradiated phage. As a result, *E. coli* AB1886 (*uvrA6*) is about 60 times more sensitive to u.v. radiation than the wild type AB1157 while it remains resistant to ionizing radiation (Howard-Flanders and Boyce, 1966). Although they are similar in their sensitivity to u.v. radiation, and their sensitivity to host-cell reactivation, attempts to demonstrate that the *uvrA*, *uvrB* and *uvrC* genes have sequential functions were unsuccessful (Howard-Flanders *et al.*, 1966). Recent work has demonstrated that *uvrA* and *uvrB* mutants lack an endonuclease specific for u.v.-irradiated DNA, but that *uvrC* mutants have wild-type levels of this enzyme (Braun and Grossman, 1974). The enzyme binds very tightly and specifically to u.v.-irradiated DNA, and the binding can be prevented by prior treatment of the irradiated DNA with photoreactivating enzyme.

Mutations in two other genes in *E. coli* (*uvrD* and *uvrE*) also cause an increase in u.v. sensitivity and a decreased ability to "host-cell reactivate" u.v.-irradiated phage. Compared with *uvrA*, *uvrB* and *uvrC* mutants, *uvrD* mutants, obtained following NTG treatment, are less sensitive to u.v. radiation, are sensitive to γ-radiation, show rapid and extensive breakdown of DNA after low doses of u.v. radiation and *uvrD*$^-$ is dominant over *uvrD*$^+$. The gene *uvrD* maps at 75 minutes on the *E. coli* chromosome map (Ogawa *et al.*, 1968). While *uvrA*, *uvrB* and *uvrC* mutations show no enhanced u.v. sensitivity, when occurring with each other in double mutants (Howard-Flanders *et al.*, 1966), the double mutant *uvrD uvrB* is about three times as sensitive as the *uvrB* mutant. Since the double mutant shows much less post-irradiation DNA degradation than the *uvrD* mutant, it has been suggested that the *uvrD* function occurs after the *uvrB* incision step (Ogawa *et al.*, 1968).

The *uvrE* (*mutU*) mutants, isolated following NTG treatment, map at 75 minutes on the *E. coli* chromosome map, are u.v. sensitive and have an enhanced spontaneous mutation rate (Siegel, 1973a). The *uvrE* gene is recessive to *uvrE*$^+$ both as a mutator and as a u.v. sensitive gene. The mutants show normal recombination and normal host-cell reactivation of u.v.-irradiated phage T1, but a decreased ability to host-cell reactivate u.v.-irradiated phage λvir. Double mutants of *uvrE* with *recA*, *recB* or *recC* are extremely sensitive to u.v. radiation while the *uvrE*, *uvrA6* double mutant is only slightly more sensitive than *uvrA6* alone. The *uvrE polA* double mutant is non-viable (Siegel, 1973b). The mutations *uvrE*, *uvr502* (Smirnov and Skravronskaya, 1971) and mutU are in the same cistron (Siegel, 1973b; Mattern, 1971). All attempts to separate u.v. sensitivity from an increased spontaneous mutation frequency have been unsuccessful.

3. *The rec Gene*

The first recombination-deficient (*rec*) mutants were isolated on the basis of their defect in recombination, and not primarily as radiation-sensitive mutants which *pari passu* had a defect in recombination. Clark and Margulies (1965) isolated two mutants from 2000 survivors of an nitrosoguanidine-treated culture of an F$^-$ *E. coli*. The mutants were unable to form recombinants by conjugation and thus appeared infertile in crosses with Hfr strains. The F′*lac* factor could be transferred into both mutants and the chromosomal transfer of λ prophage led to zygotic induction. However there were no *leu* recombinants, even though *leu* preceded λ in the cross. Thus the defect was not in conjugation but in the recombination process itself. Both mutants were very sensitive to u.v. radiation. Revertants selected for u.v. resistance regained the ability to form recombinants.

Further *rec* mutants were isolated by Howard-Flanders and Theriot (1966) and by Van De Putte *et al.* (1966). Howard-Flanders and Theriot (1966) isolated five X-ray-sensitive mutants from a nitrosoguanidine-treated *E. coli* K-12 strain by replica-plating surviving colonies for radiation sensitivity. Each was defective in the ability to form recombinants, the frequency being less than 1% of a rec^+ cross. The properties of one of these mutants (*rec13*) were reported by Howard-Flanders and Boyce (1966). It was about 170-fold more sensitive to u.v. radiation than its rec^+ parent and six-fold more sensitive to X-rays, but showed normal host-cell reactivation of u.v.-irradiated phage T1. The double mutant, *rec13 uvrA6*, was 3000 times more sensitive to u.v. radiation than the wild type. Howard-Flanders and Boyce (1966) recognized other types of recombination-deficient mutants. For example a mutant *rec21* was intermediate in X-ray sensitivity between *rec13* and the original strain AB1157, and whereas *rec13* showed spontaneous degradation of its DNA and was "reckless" in its degradation of DNA following u.v. radiation, *rec21* showed no spontaneous breakdown and was "cautious" in post-irradiation DNA degradation. Van De Putte *et al.* (1966) found that rec^+ was dominant over rec^- since in crosses with Hfr strains the appearance of stable recombinants was delayed until the entrance of the allelic rec^+ marker of the donor into the rec^- recipient. This allowed mapping to be done. A further mutant (*rec22*) was isolated (Emmerson and Howard-Flanders, 1967). Howard-Flanders, in the discussion of the paper of Clark (1967), indicated that the mutation *rec13* could complement both *rec21* and *rec22*, and that *rec21* and *rec22* could complement each other. He concluded that *rec13*, *rec21* and *rec22* appear to define three complementing groups of mutants and at least two genes *rec13* and *rec21*. This was confirmed by Emmerson (1968). However, Willetts *et al.* (1969) and Willetts and Mount (1969) mapped *rec13* at the *recA* locus, approximately half-way between *cysC* and *pheA*, at 51.7 minutes on the *E. coli* chromosome map, *rec21* at the *recB* locus and *rec22* at the *recC* locus. The order of *recB* and *recC* on the linkage map of *E. coli* is *thyA-recC-recB-argA*, with *recB* at 54.2 and *recC* at 54.5 minutes.

In summary, the *recA* mutation confers u.v. sensitivity and X-ray sensitivity, shows spontaneous degradation of DNA and "reckless" degradation following u.v. radiation, and has very low frequency of recombination (about 0.001 to 0.1% compared with $recA^+$. The *recB* and *recC* mutants are less sensitive to ionizing radiation, show no spontaneous degradation of DNA and are "cautious" in post-irradiation breakdown, and have a higher recombination frequency than *recA* of about 1% to 2% compared with $recB^+$, $recC^+$.

The reason for the identical phenotype of *recB* and *recC* mutants is that these genes code for two non-identical subunits of an ATP-dependant exonuclease which has a molecular weight of 270,000 daltons (Goldmark and Linn, 1972). This enzyme is present in extracts of wild-type *E. coli* but not in *recB* or *recC* mutants (Wright *et al.*, 1971). Two exonuclease activities can be detected. Firstly an exonuclease (exonuclease V) which acts on double-stranded DNA, only in the presence of ATP. The products are 3'-hydroxyl and 5'-phosphoryl-ended oligonucleotides of an average length of 4.5 nucleotides. Exonuclease V attacks DNA from ends generated by double-strand breaks. Secondly a 3'-exonuclease acts on single-stranded DNA in the absence of ATP. The enzyme can also cleave single-stranded DNA endonucleolytically, being stimulated seven-fold by ATP in this reaction.

4. *The pol Gene*

Isolation of the first polymerase-deficient (*polA1*) mutant was reported in 1969 (De Lucia and Cairns, 1969). It had long been suspected that the Kornberg polymerase (*pol1*) was a DNA repair enzyme and not the normal replication polymerase. De Lucia and Cairns (1969) treated a population of *E. coli* W3110 with nitrosoguanidine and tested several thousand surviving colonies for the *in vitro* activity of the Kornberg polymerase. One colony (number 3478!) had less than 1% of the normal level of DNA polymerase activity. The mutant had a normal growth rate, but was found to be sensitive to u.v. radiation and methyl methanesulphonate inactivation. Thus the Kornberg polymerase had a role in DNA repair. Both mutant and parent were equally susceptible to infection with phages T4, T5, T7 and λ and were Hcr$^+$. The mutation, an amber nonsense mutation, was mapped at 75 minutes on the *E. coli* chromosome map and shown to be recessive to the wild-type gene (Gross and Gross, 1969). Genetic recombination was not defective. Kato and Kondo (1970) reported the isolation of X-ray-sensitive mutants from *E. coli* B which turned out to be defective in DNA polymerase activity. They called these *res* mutants. Strains deficient in the Kornberg polymerase are moderately sensitive to u.v. radiation, and about three to five times more sensitive to X-rays than the wild type (Town *et al.*, 1971b). The double mutant, *polA1 recA*, is non-viable (Gross *et al.*, 1971; Monk and Kinross, 1972).

The discovery of mutants which were defective in Kornberg polymerase activity, but which had normal growth, stimulated a search for other DNA polymerases which might be responsible for semi-conservative DNA replication in *E. coli*, and two (DNA polymerases II and III) have been discovered (for a review of their properties see Gefter, 1974). A mutant of *E. coli* defective in DNA polymerase II activity has been isolated and characterized (Campbell *et al.*, 1972). It was isolated in much the same way as the original *polA1* mutant. *Escherichia coli polA1* was treated with NTG and extracts of surviving colonies tested for polymerase activity. The *polA1 polB1* double mutant contains less than 0.5% of the normal levels of DNA polymerase II. The only polymerase activity detected was DNA polymerase III. The mutant was able to grow normally at 25°C and 42°C, and supported the growth of phages T4, T7 and λ. The *polB* mutation did not affect sensitivity to u.v. irradiation or recombination frequencies.

The recognition of DNA polymerase III as the enzyme required for normal DNA replication in *E. coli* was made by Gefter *et al.* (1971). They constructed a series of double mutants carrying one of the mutations temperature sensitive for DNA synthesis (dnaA, B, C, D, E, F and G) and the *polA1* mutation (De Lucia and Cairns, 1969) and assayed polymerase II and III activities in each mutant. Polymerase II activity was normal in each double mutant. Polymerase III activity was temperature sensitive specifically in those strains having temperature-sensitive mutations at the *dnaE* locus. They concluded that polymerase III, the product of the *dnaE* gene, is an enzyme required for DNA replication in *E. coli*.

The role of the DNA polymerases in DNA repair will be dealt with in detail under the respective pathways of repair. Briefly DNA polymerase I plays a major part in the excision of pyrimidine dimers and in Type II repair of ionizing radiation damage. DNA polymerases II and III do not appear to have a major role in DNA repair in *E. coli*, except in mutants defective in polymerase I.

5. *The exr (lex) Gene*

All the u.v.-sensitive mutants isolated from *E. coli* B by Hill (1958) and Hill and Simson (1961) were also sensitive to ionizing radiation, although some of the mutants were Hcr$^+$ (e.g. *E. coli* B$_{s-2}$) while others were Hcr$^-$ (e.g. *E. coli* B$_{s-1}$). Genetic analysis revealed that *E. coli* B$_{s-2}$ contains a mutation closely linked to *malB* (81 minutes on the *E. coli* chromosome map) and a second linked to *lac* (9 minutes). When the first mutation is replaced by the wild-type allele, the resulting strain is as resistant to u.v. radiation as *E. coli* B, and the restoration of the second wild-type allele makes it as resistant as *E. coli* K-12 or *E. coli* B/r (Greenberg, 1964; Donch and Greenberg, 1968). *Escherichia coli* B_{s-1} also contains a mutation linked to *malB* with a second mutation near *gal* (17 minutes) (Mattern *et al.*, 1966; Greenberg, 1967). The latter mutation is in the *uvrB* gene, and is responsible for *E. coli* B_{s-1} being Hcr⁻, while the mutation linked to *malB* is responsible for the X-ray sensitivity of the strain and is designated *exr* (for X-ray sensitivity; Mattern *et al.*, 1966). Thus the genotype of *E. coli* B_{s-1} is $uvrB^- exr^- fil^+$, the tendency to form filaments being suppressed by the mutation in *exr*, and the genotype of *E. coli* B_{s-2} is $uvrB^+ exr^- fil^+$.

Howard-Flanders and Boyce (1966) isolated an X-ray-sensitive mutant of *E. coli* K-12 which had a mutation near *malB*, in the same location as *exr* in *E. coli* B, and which they designated *lex* (locus for X-ray sensitivity). The mutation in *lex* also increased the sensitivity of the mutant to u.v. radiation, and increased the extent of radiation-induced DNA degradation, but its effect was not as extreme as a mutation in *recA*. Perhaps the most interesting feature of *exr* or *lex* mutants is that they are almost non-mutable by u.v. radiation and other repair-dependent mutagenic treatments (Witkin, 1967a). Witkin (1969a, b) reported that *exr* and *lex* mutants have a lower efficiency of recombination commensurate with their two- to three-fold increase in sensitivity to u.v. radiation, but in a very thorough study of three *lex* mutants Mount *et al.* (1972) showed them not to be defective in recombination, although in some crosses they might fail to produce a normal yield of genetic recombinants (58–70%) depending on the time of mating and the marker selected. Stable lex^+/lex^- heterozygotes have the mutant phenotype of the *lex*⁻ haploid strains so that the *lex* mutation is dominant. The *lex* gene maps between *malB* and *uvrA* at 80.9 minutes on the *E. coli* K-12 chromosome map (Mount *et al.*, 1972, 1973).

The original *exr* mutation described by Mattern *et al.* (1966) is now referred to as *exrA* since a second *exr* mutation (*exrB*) has been described which is also linked to *malB* (Greenberg *et al.*, 1974). It differs from *uvrA* in causing sensitivity to NTG and γ-radiation, in being able to host-cell reactivate u.v.-irradiated phage T3 and in forming filaments during normal growth, as well as after irradiation (Donch and Greenberg, 1976).

6. *The lig Gene*

Pauling and Hamm (1968) attempted to isolate a mutant defective in the last step of excision repair, namely the rejoining process catalysed by polynucleotide ligase. They assumed that a mutation leading to the loss of this function would be lethal in that such rejoining processes would probably be involved in the repair of single-strand DNA breaks, generated during normal growth, and they therefore searched for a temperature-sensitive u.v.-sensitive mutant. They isolated such a mutant (ts7) after subjecting *E. coli* $\overline{\text{TAU}}$ to NTG mutagenesis. Sensitivity to u.v. radiation was tested by growing a culture of ts7 at 25°C, irradiating, and holding the irradiated culture at 40°C for two hours before incubating at 25°C. The excision of u.v.-induced pyrimidine dimers, and repolymerization of the resulting single-strand gaps, both proceed at 40°C. Under these conditions, ts7 is two to three times more sensitive than $\overline{\text{TAU}}$ under the same conditions, although rather less sensitive than that when compared with ts7 held at 25°C throughout. DNA ligase assays are consistent with its classification as a *lig* mutant. In some properties it resembles *recA* mutants, e.g. there is spontaneous degradation of DNA during normal growth and this becomes "reckless" after u.v. irradiation. It is also sensitive to ionizing radiation, having a sensitivity intermediate between *recBC* and wild type (Dean and Pauling, 1970).

A method was developed by Gellert and Bullock (1970) to search for both ligase-deficient and overproducing mutants of *E. coli* based on the finding that a functional DNA ligase, supplied either by the host or phage, is required for T4 growth. A mutant which overproduces DNA ligase supports growth of ligase-defective (gene 30 mutant) T4 mutants Double mutants of T4, mutant at the rII locus and in gene 30 and which are able to grow in normal *E. coli,* cannot grow in cells deficient in DNA ligase. An *E. coli* K-12 mutant *lig4* was isolated which made a temperature-sensitive ligase. However, it differed from the $\overline{\text{TAU}}$ ts7 mutant in that, although it became radiation sensitive at high temperature, it also grew normally at the high temperature and showed no obvious defect in DNA replication.

It was thought that the difference in phenotype of the two mutants might be due to the different genetic backgrounds of the two strains in which the mutations occurred, but Konrad *et al.* (1973) transferred the ts7 mutation to a wild type K-12 strain and it became temperature sensitive for growth, and displayed the same phenotype as $\overline{\text{TAU}}$ ts7. These findings suggest that a functional DNA ligase is essential for normal DNA replication and repair in *E. coli*, and that the *lig4* mutant of *E. coli* K-12 may have sufficient residual ligase activity at 42°C to allow growth. The *lig* gene maps at 46.5 minutes on the *E. coli* K-12 chromosome map.

7. *The fil (lon) Gene*

It has been known since the initial isolation of the radiation-resistant mutant *E. coli* B/r that, following irradiation, some strains of *E. coli* (e.g. strain B) show a loss in cell viability even though growth and nucleic acid synthesis are normal, and on microscopic examination can be seen to have formed long, non-septate filaments or "snakes" which eventually lyse. *Escherichia coli* B/r, on the other hand, does not form such filaments (Witkin, 1947; Payne *et al.*, 1956). Rörsch and Van Der Kamp (1961) showed that *E. coli* B/r, B and B_s give a similar response when grown in a medium containing ^{32}P-phosphate; i.e., filament formation is almost non-existent in cultures of *E. coli* B/r, good in cultures of *E. coli* B and even better in cultures of *E. coli* B_s, and thus the tendency to form filaments reflects the differences in radiation sensitivity between the strains. Rörsch *et al.* (1962) then isolated a radiation-resistant mutant (B_{11}) from *E. coli* B which gave radiation survival curves similar to those of *E. coli* B/r and showed that the mutation, designated *fil*, responsible for the radiation resistance also led to the loss of the tendency to form filaments. The mutation was from *fil*$^+$ to *fil*$^-$ and mapped near the *gal* locus (Van De Putte *et al.*, 1963).

Escherichia coli K-12 has about the same resistance to radiation as *E. coli* B/r (Howard-Flanders *et al.*, 1964). Following u.v. irradiation of a culture of *E. coli* K-12, Howard-Flanders *et al.* (1964) isolated some radiation-sensitive mutants by picking surviving colonies which had a mucoid appearance on supplemented minimal medium. These mutants all produced filaments, or long forms, following irradiation and all carried a mutation at a locus designated *lon* (long form) situated close to *tsx* (resistance to phage T6) and at the same locus as *fil* in *E. coli* B. The wild type *E. coli* K-12 AB1157 was *lon*$^+$ and resembled *E. coli* B/r, while the *lon*$^-$ mutants (e.g. AB1899 = AB1157 *lon*-1) were relatively sensitive to both u.v. and ionizing radiation and resembled *E. coli* B. It should be noted, since it could lead to confusion, that lon^+ in the *E. coli* K-12 mutant series is equivalent to fil^- in the *E. coli* B mutant series, although the tendency these days is to refer only to *lon* as describing the locus for filament formation in both. The lon^+ gene is dominant to lon^- regardless of whether the lon^+ allele is on the chromosome or a plasmid (Uretz and Markovitz, 1969).

It may not be clear why mucoid colonies of *E. coli* K-12 should be lon^-. Markovitz (1964) showed that a gene designated R_I, and redesignated *capR* (Markovitz and Baker, 1967), controls the synthesis

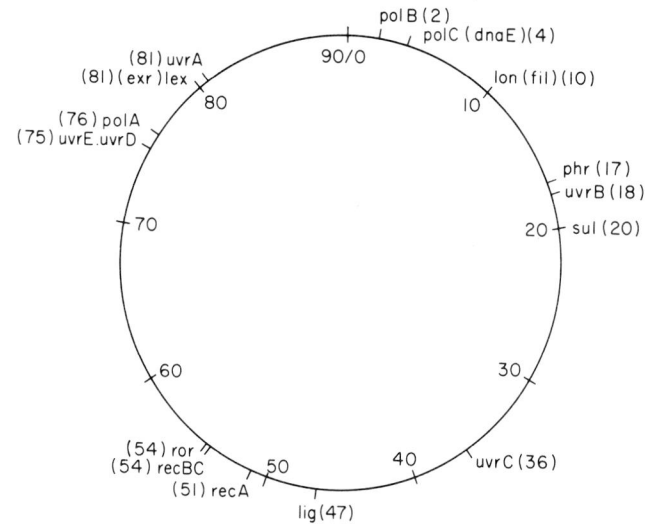

FIG. 5. Chromosomal location of genes associated with repair of damaged DNA in *Escherichia coli* K-12. After Taylor and Trotter (1972).

of capsular polysaccharide, and strains mutant at this locus have a mucoid colony appearance. It has been established that in *E. coli* K-12 the *capR* and *lon* mutant phenotypes are identical, i.e. are mucoid, radiation sensitive (Adler *et al.*, 1966; Adler and Hardigree, 1964; Donch and Greenberg, 1968) and able to host-cell reactivate u.v.-irradiated phages T1 and T7 to the same extent as the wild type (Howard-Flanders *et al.*, 1964; Uretz and Markovitz, 1969). *Escherichia coli* B, however, is non-mucoid because of other genes present which specifically suppress mucopolysaccharide formation but not u.v. sensitivity (Donch and Greenberg, 1970).

The tendency of *E. coli* B and *lon* mutants of *E. coli* K-12 to form filaments after u.v. irradiation can be suppressed physiologically by incubating the irradiated bacteria on agar plates containing pantoyl lactone at an optimum concentration of 0.08 M (about 12 g/l). Under these conditions, not only is filament formation inhibited but the radiation-sensitive strains appears to be resistant (Van De Putte et al., 1963; Adler and Hardigree, 1964). The mechanism of action of pantoyl lactone is not known.

Filament-formation can also be suppressed genetically, by mutation at *exr* (Donch et al., 1968), *recA* (Green et al., 1969) or by a suppressor *sul* (suppressor of *lon*) (Donch et al., 1969) which maps at about 20 minutes on the *E. coli* K-12 chromosome map (Johnson and Greenberg, 1975). In fact the original B/r mutant (Witkin, 1947) is not mutant at the *fil* locus but contains the suppressor *sul* (Donch et al., 1969). The mechanism by which *lon* controls radiation resistance is not known, but is obviously connected with the control of cell division following DNA damage.

The distribution of the various loci described, which influence the capacity of *E. coli* to repair damage to its DNA, is shown in Fig. 5. The genes involved in repair do not appear to be associated in any particular pattern.

IV. Photoreactivation Repair

Photo-enzymatic repair was the first DNA repair mechanism to be recognized both *in vivo* and *in vitro* and is possibly the most primitive one in an evolutionary sense. The biological consequences of photoreactivation were first recognized by Kelner (1949a, b) who found that, when u.v. irradiation of *Streptomyces griseus* conidia, *Escherichia coli, Pencillium notatum* or yeast cells was followed by exposure to visible light, the organisms had an enhanced ability to survive and form colonies. An understanding of the molecular processes involved in photoreactivation began when Goodgal et al. (1957) demonstrated that transforming DNA, which had been inactivated by u.v. radiation, could recover some of its biological activity when illuminated in the presence of an extract of *E. coli*. However, because of the presence of nucleases in extracts of *E. coli* which degrade transforming DNA, in competition with photoreactivation, most *in vitro* studies have been done with extracts of yeast (Rupert, 1960). For reviews of photoreactivation see J. K. Setlow (1966) and Harm et al. (1971).

A. THE SUBSTRATE FOR PHOTOREACTIVATING ENZYME

Photoreactivation removes u.v.-induced pyrimidine dimers from DNA by monomerization of the dimer *in situ*. Proof of the involvement of thymine-thymine dimers in photoreactivation was provided by Setlow and Setlow (1963) who found that there was an overlap in the biological effect of short-wavelength reversal and photoreactivation. Irradiated transforming DNA which has been photoreactivated is not further reactivated by exposure to short-wavelength u.v. radiation (240 nm) which itself favours monomerization of thymine-thymine dimers, and its ability to be photoreactivated is decreased by previous short-wave irradiation. From such results it was concluded that thymine-thymine dimers are involved in the biological effect of photoreactivation.

The involvement of the other pyrimidine dimers formed in DNA by u.v. irradiation (namely cytosine-thymine and cytosine-cytosine dimers) was determined using competition experiments (Rupert, 1961, 1964) in which irradiated polynucleotides are allowed to compete with irradiated transforming DNA for photoreactivating enzyme. If a polynucleotide contains a substrate for the enzyme, then it will compete with substrates in the transforming DNA resulting in less reactivation of biological activity of the transforming DNA. Furthermore if the radiation damage in the synthetic polymer is removed by the photoreactivation enzyme during illumination, its capacity to compete is removed and the damage is considered a substrate for the photoreactivating enzyme. Experiments on a variety of irradiated synthetic polymers have shown that only those which contain adjacent pyrimidine bases can be photoreactivated (Setlow *et al.*, 1965). This explains why photoreactivation repair applies only to u.v. irradiation damage and not to ionizing radiation or chemical damage to DNA. The thymine dimer itself (TpT) and the irradiated (pT)8 do not compete for enzyme (Rupert, 1964), which suggests that the pyrimidine dimer is a substrate for photoreactivating enzyme only if it is part of a particular structure to which the enzyme can bind.

B. THE PHOTOREACTIVATING ENZYME (PRE)

An interesting and ingenious method has been developed for investigating the kinetics of photoreactivation repair—the intense, single-flash illumination (Harm and Rupert, 1968; Harm *et al.*, 1968).

The number of u.v.-induced pyrimidine dimers formed in the genome of $E.\ coli$ is six per erg per mm^2 of 254 nm u.v. radiation. Thus it is possible to calculate the number of pyrimidine dimers per genome for any dose in the biological dose range of u.v. radiation, and one can measure the decreased viability of such irradiated populations of $E.\ coli$. The irradiated bacteria are then subjected to a single intense flash of light (duration of the order of one millisecond) from four electronic photographic flash units, and the bacteria assayed for increased survival. After an interval to allow the released PRE molecules to bind to further dimers, another intense flash is given and survival assayed. This is continued. When no further biological survival is produced as a result of exposure to intense illumination, it is assumed that all of the pyrimidine dimers have been photoreactivated and the number pyrimidine dimers induced per cell, divided by the number of flashes of illumination required to photoreactivate them is a measure of the number of molecules of photoreactivation enzyme per cell of $E.\ coli$. Such methods suggest a value of 20 molecules per cell in $E.\ coli$ B_{s-1} and B, and rather less in $E.\ coli$ K-12. Using the same technique, a value of 272 ± 27 molecules PRE per haploid $Sacharomyces$ cell has been obtained (Yasui and Laskowski, 1975). In this method, use is made of the fact that the PRE will bind to u.v.-irradiated, but not to unirradiated, DNA in the dark, and dissociate only when illuminated (Rupert, 1960, 1961, 1962a, b). The single intense flash of light provides the photons necessary for the repair of pyrimidine dimers $in\ situ$, so that the PRE molecules are free to bind to other dimers. A second light flash immediately after the first does not give more survival than the single flash because the enzyme molecules have not yet bound to further dimers. Thus this method, by measuring increased survival between two timed flashes, gives information on the kinetics of enzyme–substrate complex formation.

Since absorption of the radiation that produces photorepair is due to the presence of the enzyme, the latter must be responsible for the absorbing structure or "chromophore". However, nothing is known about the nature of the sensitizing chromophore. Pyrimidine dimers are split by wavelengths of u.v. radiation which they can absorb, mostly below 280 nm (Setlow, R. B., 1966) but the wavelengths effective in photoenzymic repair are somewhat longer, from 300 to 550 nm (Jagger and Latarjet, 1956) with a maximum for dimers in DNA at 360–370 nm (Rupert $et\ al.$, 1973).

Attempts to identify the nature of the chromophore have involved purification of the photoreactivating enzyme, but the constraining factor in the purification has been the low initial activity present in the cells. The first attempt was by Muhammed (1966) who, working with yeast cells, purified the enzyme 3500-fold, yet his final preparation contained less than 1% PRE (Harm and Rupert, 1968). Minato and Werbin (1971) purified the yeast enzyme an additional 20-fold using columns of cellulose to which irradiated DNA had been adsorbed. Sutherland and Chamberlin (1973) purified the *E. coli* PRE 10,000-fold, but even so the yield was very low because of the small numbers of PRE molecules per cell. To improve the starting material, Sutherland and her colleagues mapped the *phr* gene of *E. coli* between the attachment site for phage λ (*attλ*) and *gal*. They then constructed a *phr* transducing phage which carries the *gal* gene for easy selection, the *phr* gene for production of PRE, has the C1857 mutation which causes it to be induced at 42°C, a mutation (S7) to prevent cell lysis and thus to allow a long time for production of enzyme, and a deletion of the structural genes for phage proteins to prevent the production of unwanted proteins (Sutherland *et al.*, 1972; Sutherland, 1973). When induced, this strain produces levels of PRE 4000 times higher than *E. coli* B. The improved yield of enzyme in the cell lysate has led to a greater purification of the *E. coli* PRE (20-fold over the starting material) but still no chromophore has been identified. It is suggested either that the chromphore is the enzyme–substrate complex, which creates a new absorption band, or a separate molecule other than protein or DNA exists which is required for photoreactivation (Sutherland, 1973).

C. THE ENZYME–SUBSTRATE COMPLEX

Studies *in vitro* using u.v.-inactivated transforming DNA and photoreactivating enzyme (PRE) from yeast have shown that photoreactivation involves a light-independent binding of PRE (E) to its substrate, the pyrimidine dimer (S), followed by a photochemical reaction in the complex (ES) which yields the monomerized dimer as its product (P) (Rupert, 1964):

$$E + S \rightleftharpoons ES \rightarrow E + P$$

Except for the light dependence of the last step, this is the familiar Michaelis–Menten reaction scheme characteristic of other enzymic processes. Using timed flashes after mixing u.v.-irradiated transforming DNA and yeast PRE, Harm and Rupert (1968) showed that the formation of enzyme–substrate complexes takes minutes, mainly because of the extreme dilution of the reactants. They also showed that the rate of complex formation was temperature dependent, confirming the observation of Kelner (1950) that the photoreactivation process is also temperature dependent. However, the photochemical step is temperature independent between 2° and 37°C. *Escherichia coli* B_{s-1}, irradiated with 4.8 ergs/mm^2 u.v. radiation (= 30 dimers per genome) forms a maximum number of complexes in five minutes at room temperature, 50% being formed within 10–15 seconds (Harm *et al.*, 1968). Only one photon is required per repair event, since the amount of photoreactivation increases linearly with the total number of photons and is independent of the light intensity. This is true for photoreactivation *in vivo* in *E. coli* (Jagger and Latarjet, 1956) and photoreactivation *in vitro* of transforming DNA (Setlow and Boling, 1963).

D. PHOTOREACTIVATION AS AN ANALYTICAL TOOL

Photoreactivation is frequently used as a "probe" in other modes of DNA repair. Because the only substrate for photoreactivating enzyme is the pyrimidine dimer, the removal of some biological effect (e.g. cell death or mutation) by photoreactivation means that dimers are involved. Recently Hart and Setlow (1973), for example, used photoreactivation to examine the role of dimers in producing cancerous lesions in the non-sexually reproducing fish, *Poecelia formosa* (Setlow, 1973).

V. Excision Repair

The excision mechanism for the removal of u.v.-induced pyrimidine dimers is a multienzyme process involving single-strand incision of the DNA in the region of the dimer, the excision of the dimer, removal of some adjacent bases by exonuclease activity, repolymerization of the single-strand gap and finally the joining of the single-strand break by a polynucleotide ligase. (For reviews of the excision process see Setlow, 1968; Howard-Flanders, 1968, 1973; Grossman *et al.*, 1973; Grossman *et al.*, 1975.)

A. METHODS FOR STUDYING EXCISION REPAIR

Two methods are available for studying the excision of pyrimidine dimers from the DNA of irradiated bacteria.

The first method is based on the paper chromatographic separation of pyrimidine dimers from mononucleotides (Boyce and Howard-Flanders, 1964a; Setlow and Carrier, 1964). The DNA is radio-actively labelled by incubating the bacteria in media containing tritiated thymidine for about three to four generations. A sample is retained and held at 0°C and the rest of the culture is irradiated. An irradiated, but non-incubated, sample is held at 0°C while the remainder of the irradiated culture is incubated in a growth medium. At various times, samples are removed and held at 0°C. Each sample is then washed with ice-cold 5% trichloroacetic acid (TCA) to remove all TCA-soluble material, and the TCA-insoluble material is hydrolysed with 98% formic acid at 175°C in sealed ampoules. The hydrolysed material is run on a paper chromatogram to separate the pyrimidine dimers from thymine. The amount of radio-activity in the dimer region, relative to that in the thymine region, gives a measure of the rate of specific removal of thymine-containing dimers from the DNA. The dimers can be recovered in short oligonucleotides from the TCA-soluble fraction of the bacterial cells (e.g. in *E. coli*; Boyce and Howard-Flanders, 1964a; Setlow and Carrier, 1964), or from the incubation medium (e.g. *Micrococcus radiodurans*; Boling and Setlow, 1966) since in the latter case the dimers are excreted from the cell. Details for this method are contained in Carrier and Setlow (1971).

The second method is based on the fact that the first step in the removal of a dimer is formation of a single-strand break in the DNA as a result of endonuclease action. The single-strand break can be detected by centrifuging the DNA in an alkaline sucrose density gradient (McGrath and Williams, 1966). Bacterial DNA is radio-actively labelled and the bacteria irradiated, incubated and samples removed as for the first method. The bacteria in each sample are made into sphaeroplasts, using appropriate methods, and layered into 0.1 M-NaOH on the top of an alkaline (pH 12) sucrose gradient (5–20%). The sphaeroplasts lyse immediately and DNA is released onto the gradient. The gradients are centrifuged for about 1.5 hours at 20,000 to 30,000 r.p.m. After centrifugation, a hole is made in the bottom of each tube and the tube contents collected dropwise. The amount of radio-activity in each drop (or equal number of drops) is plotted against drop or fraction number. The distance that the single-stranded DNA sediments is a function of its molecular weight. If the tube is calibrated for molecular weight by using labelled phage DNA, the average molecular weight can be calculated. The average molecular weight of DNA from unirradiated bacteria is about 2×10^8 daltons. Thus, if the average molecular weight of the DNA from an irradiated culture falls to 5×10^7, three single-strand breaks per 2×10^8 daltons were present at the time of cell lysis. If the molecular weight of the double-stranded genome is 3×10^9 daltons, then there were 45 single-strand breaks in the genome, half in each strand, at the time of cell lysis. Of course, this method can be used to measure the rate of strand-mending in bacteria irradiated with ionizing radiation, and in fact was developed for just this purpose (McGrath and Williams, 1966). Double-strand breaks can be assayed by centrifuging the DNA in a neutral sucrose gradient, although great care has to be exercised in interpreting the results (Burrell et al., 1971; Setlow and Setlow, 1972).

B. INCISION ($uvrA$, $uvrB$)

A whole battery of endonucleases capable of making incisions in DNA which contains damaged bases has been identified in bacteria (for review see Grossman et al., 1975). In the review of Grossman et al. (1975), endonucleases which are involved in the repair of DNA are called correndonucleases (for correctional endonucleases) and are classified as I or II on the basis of their specificity for damage involving a single nitrogenous base (e.g. the removal of a deaminated base) or for damage involving more than one nitrogenous base (e.g. a pyrimidine dimer or a cross-link), respectively. Thus an endonuclease specific for the removal of a pyrimidine dimer is referred to as a correndonuclease II. All known correndonuclease II enzymes will incise u.v.-irradiated DNA. The *E. coli* correndonuclease II specified by the $uvrA$, $uvrB$ genes, which makes the first incision in u.v.-irradiated DNA, has been isolated (Braun and Grossman, 1974). Although absent from extracts of $uvrA$ and $uvrB$ mutants of *E. coli*, it is present in normal amounts in $uvrC$ which are nevertheless excision-defective mutants. The enzyme binds very tightly and specifically to u.v.-irradiated DNA, but the binding can be prevented by prior treatment of the irradiated DNA with photoreactivating enzyme indicating that pyrimidine dimers are responsible for the binding. It is a small enzyme having a molecular weight of less than 14,000 daltons, and is active in the presence of mM EDTA. The incision occurs 5' to a pyrimidine dimer, and leaves a 3'-hydroxyl terminus which renders the break amenable to sealing by polynucleotide ligase (Braun and Grossman, 1974). The uvrA and uvrB mutants of E. coli are more sensitive than the wild type to cross-linking agents such as mitomycin-C (Boyce and Howard-Flanders, 1964b) or psoralen-plus-360 nm light (Cole, 1971), and to the action of nitrous acid (Howard-Flanders and Boyce, 1966), indicating that the endonuclease responsible for incision of u.v.-irradiated DNA plays a part also in the removal of these forms of DNA damage.

C. PRE-EXCISION (uvrC)

The uvrC mutant of E. coli possesses wild-type levels of the uvrA, uvrB endonuclease, and yet is an excision-defective mutant. If the endonuclease incises the irradiated DNA, why does excision not follow? When the changes in molecular weight of single-stranded DNA are followed in irradiated uvrC mutants, the extent of incision does not follow that of the wild type (Seeberg and Rupp, 1975). There is a rapid, but only partial, decrease in molecular weight which is quickly restored to control length DNA. When an irradiated double mutant, constructed from a uvrC mutant and a temperature-sensitive polynucleotide ligase mutant, is held at the restrictive temperature, the molecular weight changes simulate those of the wild type with the accumulation of low molecular-weight DNA. At the permissive temperature, the partial decrease in molecular weight is quickly restored. The evidence suggests that the incision made by the endonuclease 5' to a pyrimidine dimer and with a 3'-hydroxyl group can be resealed by polynucleotide ligase and, in the absence of a uvrC gene product, this leaves the dimer in the DNA. So the uvrC gene product prevents the endonuclease incision from being resealed.

D. EXCISION (polA)

Since incision leaves a single-stranded free end, excision is carried out by exonuclease action. Two types of exonuclease action are available, one associated with DNA polymerase and one not associated with DNA polymerase (unassociated exonuclease).

DNA polymerase I of *E. coli*, for long mistakenly regarded as the DNA replicating enzyme, is a single polypeptide chain with a molecular weight of 109,000 daltons which is capable of carrying out several functions (Kelly *et al.*, 1969; Kornberg, 1969). In addition to its polymerizing ability, it has two associated exonucleolytic activities, one a 3'-5' nuclease which, at 37°C, prefers single-stranded DNA possessing a 3'-hydroxyl terminus in which 5' nucleotides are the exclusive product of digestion. An important role for this nuclease activity, in conjunction with the enzyme's polymerizing activity, is its editing function in which non-complementary nucleotides are removed (Brutlag and Kornberg, 1972). However, DNA polymerase I excises dimers because of its 5'-3' exonuclease activity (Cozzarelli *et al.*, 1969). Irradiated poly(dA:dT), incised non-specifically with pancreatic DNase, is hydrolysed by polymerase I with the production of dimer-containing fragments of from two to seven nucleotides. Similar results are obtained using irradiated DNA (Kelly *et al.*, 1969).

The assignment of any *in vitro* role for a polymerase in DNA repair is dependent on experiments carried out on mutants defective in the polymerase. Thus mutants with very low levels of DNA polymerase I (*polA*), although sensitive to u.v. radiation, are considerably more resistant than either *uvr* or *rec* mutants (De Lucia and Cairns, 1969; Gross and Gross, 1969) and yet excise pyrimidine dimers at the same rate, and to the same extent, as the wild type (Boyle *et al.*, 1970). At first sight these two results seem contradictory in that the excision and the polymerizing ability are functions of the same enzyme. However, DNA polymerase I has been isolated from a variety of *polA* mutants, and although the polymerase activity in these mutants is only 0.5 to 3% of that of the wild type, the associated 5'-3' exonuclease activity is normal (Lehman and Chien, 1973). The reciprocal mutant has been isolated in which the polymerase I activity is normal but the 5'-3' exonucleolytic activity is substantially decreased. This mutant (*ror*) is u.v. sensitive and has a decreased ability to excise pyrimidine dimers (Glickman *et al.*, 1973). *PolA* mutations result in a preponderance of "long patch" repair which could be consistent with the lack of polymerizing ability but normal 5'-3' exonuclease activity (see Section on re-insertion; p. 131).

Two unassociated exonucleases capable of excising pyrimidine dimers in irradiated bacteria have been identified (Kaplan *et al.*, 1971; Chase and Richardson, 1974). Both types of exonuclease (a u.v.

exonuclease and exonuclease VII) are specific for denatured DNA and are capable of hydrolysing u.v.-damaged and undamaged substrates at comparable rates, and to the same extent. In this way, they differ from *E. coli* exonuclease I since the hydrolytic properties of this enzyme are inhibited by u.v.-induced photoproducts (Grossman *et al.*, 1968).

E. RE-INSERTION OF BASES

Once the region of DNA containing the photoproduct has been excised, it is necessary for the resulting single-strand gap to be repaired by the re-insertion of nucleotides catalysed by DNA polymerase, using the complementary strand as a template. The process can be studied using the density label 5-bromodeoxyuridine in place of thymidine during DNA repair of u.v.-irradiated *E. coli* and analysing the extent of incorporation (Pettijohn and Hanawalt, 1964). Two extremes have been identified, namely "short patch" repair, in which regions of 10–30 nucleotides are replaced for each dimer removed, and "long patch" repair, in which nucleotide lengths of 1000–3000 are replaced per dimer excised (Cooper and Hanawalt, 1972).

1. *Short Patch Repair (uvr, polA, lig)*

The data of Cooper and Hanawalt (1972) suggest that the majority of repaired DNA in u.v.-irradiated *E. coli* is in short patches synthesized *via* the DNA polymerse I pathway. Only in *polA* mutants does the "long patch" repair predominate.

"Short patch" repair has been mimicked *in vitro* by the repair of u.v.-irradiated *Bacillus subtilis* transforming DNA (Heijneker *et al.*, 1971). They used an endonuclease from *Micrococcus luteus* to make incisions, and then incubated the DNA with a mixture of *E. coli* polymerase I and polynucleotide ligase. Approximately 20 nucleotides were removed per pyrimidine dimer excised, and the transforming activity of the DNA was partially restored. When transforming DNA was 50% or less inactivated, there was a quantitative restoration of biological activity by these three enzymes. Thus, *in vitro*, the combination of an endonuclease, DNA polymerase I and polynucleotide ligase is adequate to complete repair of u.v.-irradiated DNA and restore biological activity.

2. Long Patch Repair
(uvr, recA, recBC, unwinding protein, polB, dnaE, lig)

Cooper and Hanawalt (1972) found that the amount of repair replication, in terms of the extent of incorporation of 5-bromodeoxyuridine, was much greater in *polA* mutants than in *polA*$^+$ strains. The *uvrB* gene product is required, as in the case of "short patch" repair, together with the *recA* and *recBC* gene products (Masker and Hanawalt, 1973). One reason for the *recA*, *recBC* gene products (Grossman *et al.*, 1975) is that the size of the single-strand gap, following excision of a dimer by an unassociated exonuclease, is 6–8 nucleotides long and, whereas DNA polymerase I will bind to such a site, the binding properties of DNA polymerases II and III require much larger regions. Thus, before re-insertion of nucleotides can occur *via* these polymerases, considerable expansion of the single-stranded gap is required (for a review of the properties of DNA polymerase II and III, see Gefter, 1974). The necessary gap expansion is carried out by the ATP-dependent double-stranded activity associated with the exonuclease V enzyme coded for by the *recBC* genes (Goldmark and Linn, 1972) and controlled by the *recA* gene. This activity occurs in both a 5'-3' and 3'-5' manner (Mackay and Linn, 1974). As the gap expands, of course, a larger and larger region of the complementary strand of DNA becomes exposed therefore susceptible to attack by the ATP-dependent single-stranded endonucleolytic activity of the *recBC* gene product, and must be protected. This protection can be provided either by the unwinding protein (Molyneux and Gefter, 1974) or by the *recA* gene product which must play a similar role during post-replication recombination repair.

Once the nucleotides have been re-inserted into the single strand gap by polymerases II and III, using the complementary strand as a template, the formation of the final phosphodiester bond is catalysed by the enzyme polynucleotide ligase, to give a completely repaired region. The whole process of excision repair is summarized in Fig. 6.

VI. Post-Replication Recombination Repair

The current view of recombination repair of u.v.-induced pyrimidine dimers is that DNA replication proceeds normally until it reaches an unexcised dimer. Since the dimer is a non-coding lesion, a

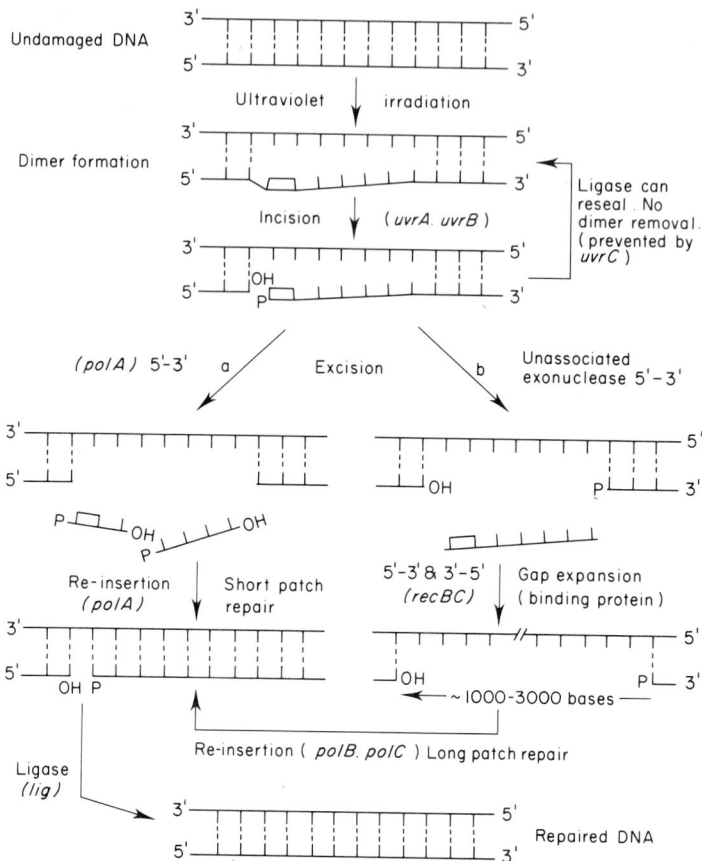

FIG. 6. Excision repair. As a result of ultraviolet irradiation, a pyrimidine dimer is formed which distorts the DNA helix and breaks the hydrogen bonding between the strands. The *uvrA* and *uvrB* endonucleases make a single-strand nick, 5' to the dimer. In a *uvrC* mutant, this break can be resealed without removal of the dimer. In a *uvrC* + cell, excision can proceed by one of two routes: (a) In a *polA*⁺ cell "short patch" repair predominates with the *polA* associated exonuclease releasing the dimer and about 10 adjacent bases. Resynthesis of this region is catalysed by polymerase I, and sealed by polynucleotide ligase. (b) "Long patch" repair occurs in *polA*⁻ cells when the dimer and between 1000–3000 bases are removed by unassociated exonuclease activity. The large single-strand gap is repolymerized by DNA polymerases II and III sealed by ligase.

gap is left opposite the dimer and replication begins again at a new initiation site some distance from it. Of course, dimers are likely to be in both parental strands of the DNA and so the gaps left opposite them during replication will be in both daughter strands, and thus none of the four strands of DNA will have full coding potential. In recombination-proficient strains, the gaps in the daughter strands are repaired and the pyrimidine dimers are effectively by-passed. For reviews of this process see Howard-Flanders (1968, 1973) Smith (1971) and Clark (1971).

A. METHOD FOR STUDYING POST-REPLICATION RECOMBINATION REPAIR

In this method, the DNA which needs to be radio-actively labelled is not that present at the time of irradiation, as in the study of dimer excision, but that synthesized after irradiation. So instead of prelabelling the bacteria, they are first irradiated and then incubated in growth medium containing tritiated thymidine. An unirradiated culture must be added as a control, since DNA in an unirradiated cell is normally first synthesized as low molecular-weight fragments (Okazaki fragments: Okazaki et al., 1968) and is then joined to high molecular weight DNA. Because the molecular weight of the DNA synthesized at short time intervals after irradiation is of prime interest, it is necessary to use a thymine-requiring mutant in order to get enough radio-active label into the DNA to be analysed on a gradient. The bacteria (both irradiated and unirradiated) are converted to sphaeroplasts, after increasing times of incubation in growth medium containing tritiated thymidine, lysed on alkaline sucrose gradients, the gradients centrifuged and the molecular weights of the DNA synthesized, post-irradiation, calculated (Rupp and Howard-Flanders, 1968; Rupp et al., 1971). The original observations were made using a *uvr* mutant of *E. coli* so that excision repair was excluded, but, though convenient, it is not necessary (Smith, 1971). For all of the experimental details see Rupp and Howard-Flanders (1971).

B. GAP FORMATION OPPOSITE PYRIMIDINE DIMERS

The original experiments by Rupp and Howard-Flanders (1968) were carried out on *E. coli* K-12 *uvrA6*, which is defective in the excision of pyrimidine dimers but can nevertheless survive the formation of about 50 u.v.-induced dimers in its genome. When the irradiated bacteria are incubated in a medium containing tritiated thymidine, the presence of the dimers inhibits DNA synthesis, but not permanently.

Sedimentation of the labelled single-stranded DNA in alkaline sucrose gradients showed that, whereas that from unirradiated bacteria has an average molecular weight in excess of 10^8 daltons, that in cells given 60 ergs/mm^2 u.v. radiation has an initial molecular weight of 1.4×10^7 daltons, which indicates that the number of strand gaps is similar to the number of pyrimidine dimers in an equivalent length of parental DNA. From the delay caused to DNA synthesis, it was calculated that DNA synthesis past a dimer takes 10 seconds. Since *E. coli* replicates its chromosome of 3×10^9 base pairs in 40 minutes at 37°C, this is about 10^7 times as long as replicating two pyrimidine bases during normal replication. The DNA that was labelled prior to irradiation had the same sedimentation characteristics as the DNA from unirradiated bacteria, even if allowed only 10 minutes in the labelled growth medium. Thus strand breakage does not occur in the parental DNA during the post-irradiation incubation period.

If daughter strand DNA is synthesized only between dimers on the parental strand, then the molecular weight should decrease as the dose, and therefore the numbers of dimers, increases. Six pyrimidine dimers are induced per erg/mm^2 of u.v. radiation per *E. coli* genome (Rupp and Howard-Flanders, 1968), and since the molecular weight of the *E. coli* genome is known (Cairns, 1963) the average distance between dimers can be calculated for a given dose. The average molecular weight of the DNA synthesized after irradiation was close to the average molecular weight of the parental DNA between dimers. Using photoreactivation as a probe for the involvement of pyrimidine dimers, it was found, not unexpectedly, that u.v.-irradiated cells which were exposed to photoreactivating light before being incubated in radio-active medium synthesized larger molecular-weight daughter DNA than cells not subjected to photoreactivation (Smith and Meun, 1970).

What size are the single-strand gaps left opposite the dimers? To examine sister exchanges of DNA fragments, bacteria were grown for several generations in a medium containing ^{13}C and ^{15}N so that their DNA would be heavy. They were then u.v. irradiated at varying doses and grown for less than one generation in a light medium containing tritiated thymidine. Thus, the double-stranded DNA made following the irradiation would be hybrid with tritium in the light strand. Under these conditions, tritium-labelled chains would become associated with heavy strands if exchanges occurred between sister duplexes. DNA extracted from unirradiated bacteria was denatured by heating and separated into light and heavy strands by centrifugation in neutral caesium chloride gradients. That from irradiated cells did not separate completely, but contained some single-stranded DNA of intermediate density. This separated into light and heavy fractions only after shearing to a molecular weight of 5×10^5 daltons. Therefore segments greater than 5×10^5 daltons molecular weight had been exchanged. There was approximately one exchange per dimer (Rupp et al., 1971). A molecular weight of 5×10^5 daltons is equal to about 800 nucleotides. Another estimate for the gap size was obtained by using calibrated benzoylated naphthoylated DEAE cellulose chromatography columns that are selective for single-stranded DNA (Iyer and Rupp, 1971). This gave the value of 1000 nucleotides, or about the size of an Okazaki fragment (Okazaki et al., 1968).

C. GAP REPAIR

In their initial description of the nature of the recombinational repair of u.v.-irradiated *E. coli*, Rupp and Howard-Flanders (1968) observed that, although the DNA synthesized immediately after u.v.-irradiation had a low molecular weight, on subsequent incubation the molecular weight increased and approached control values. At 90 minutes after irradiation, the molecular weight covers a wide range and extends fully to control size (Rupp et al., 1971). This suggests that during incubation the lower molecular-weight material is joined together to form high molecular-weight DNA. From a consideration of the experiments of Rupp et al. (1971), there is an exchange of sister DNA which replaces the genetic information missing from the gap left opposite the dimer. After the post-replication gaps have been filled, the resulting high molecular-weight DNA can serve as a template during the next round of replication for the direct synthesis of high molecular-weight DNA (Ganesan and Smith, 1971). A summary of post-replication recombination repair is shown in Fig. 7.

D. RECOMBINATION-DEFICIENT MUTANTS: (GENETIC CONTROL OF POST-REPLICATION REPAIR)

In terms of the molecular events described for post-replication recombination repair, how do the recombination-deficient mutants differ from recombination-proficient strains? In *recA* mutants, the

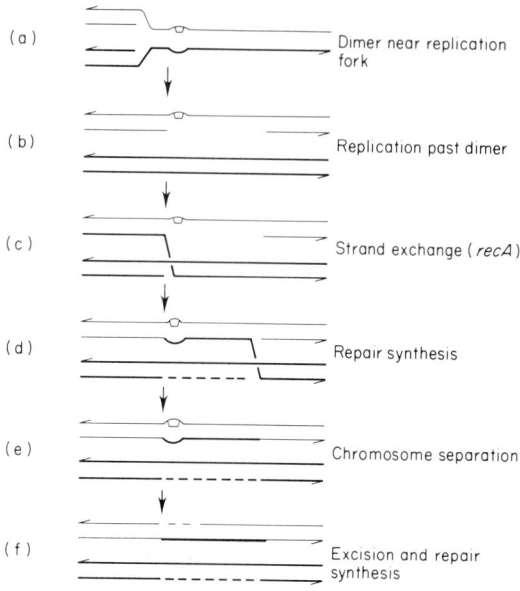

FIG. 7. Post-replication recombination repair. (a) An ultraviolet-induced pyrimidine dimer is formed just ahead of the advancing replication fork. (b) Since a pyrimidine dimer is a non-coding lesion, a gap is left opposite the dimer and replication begins again at a new initiation site, leaving a gap of about 1000 nucleotides. The other parental strand is replicated normally in this region. (c) Strand breakage and re-union between sister strands occurs, dependent on *recA*. (d) Strand exchange replaces the genetic information missing from the gap while the sister strand replicates this region (---) from the intact complementary strand. Although the correct sequence is now opposite the dimer, the hydrogen bonding will not be re-established in this region. (e) The two chromosomes separate. (f) The dimer can now be excised in the normal way and the gap repaired because the complementary strand is intact. After Rupp and Howard-Flanders (1968).

DNA synthesized after u.v. irradiation sediments more slowly than that made in unirradiated cells, indicating that gaps are left opposite pyrimidine dimers during replication. However, incubation of the irradiated cells does not result in an increase in molecular weight. In the absence of the *recA* gene product, the gaps opposite the dimers are not repaired (Smith and Meun, 1970).

In *recB* and *recC* mutants, the situation is different. In both, the DNA synthesized after u.v. irradiation sediments more slowly than that from unirradiated cells, but after incubation the molecular weight begins to increase and reaches control size (Smith and Meun, 1970; Howard-Flanders *et al.*, 1971). Since the *recB* and *recC* mutants appear able to repair all of the post replication gaps in the daughter DNA, it suggests that the ATP-dependent exonuclease V coded for by the *recBC* genes, while not necessary for gap-filling, may be necessary for some subsequent step in post-replication repair (Howard-Flanders *et al.*, 1971).

VII. The Repair of Ionizing Radiation Damage

The initial observation that X-ray-induced single-strand breaks in the DNA of *E. coli* could be repaired in the radiation-resistant *E. coli* B/r, but not in the radiation-sensitive *E. coli* B_{s-1} (McGrath and Williams, 1966), was a great stimulus for research on the chemical nature of the single-strand break and its repair. As mentioned previously (p. 117), *E. coli* B_{s-1} is a multiple mutant, differing in three genes from *E. coli* B/r (namely *uvr, exr* and *fil*) and it is the mutation at *exr* which is responsible for its inability to repair single-strand breaks (Youngs and Smith, 1973a). In fact, the repair demonstrated by McGrath and Williams (1966) is only one type of repair of single-strand breaks, and two other types have now been observed. The different repair systems have been called Types I, II and III, based on their rate of completion of repair (for a review see Town *et al.*, 1973b) and are described below in the chronological order of their discovery.

A. TYPE III REPAIR
(*recA, recBC, exr, dnaB, ror, lig*)

The observation of McGrath and Williams (1966) was that X-ray induced single-strand breaks in the DNA of *E. coli* B/r were rejoined when the bacteria, which had been irradiated in buffer, were incubated in a growth medium. A similar type of repair has been demonstrated in wild type *E. coli* K-12 (Kaplan, 1966), and shown to be absent from *recA* mutants of *E. coli* K-12 (Morimyo *et al.*, 1968; Kapp and Smith, 1970). Type III repair is defined as the slow rejoining of single-strand breaks which occurs only when bacteria are incubated under conditions which allow growth. At doses of 15–20 krad, it takes *E. coli* K-12 40–60 minutes to complete repair in a minimal medium at 37°C, although less time is required after lower doses (Ganesan and Smith, 1972). Type III repair requires DNA replication of the semiconservative type, or at least the *dnaB* function (Fangman and Russel, 1971), but is completed in less time than is necessary for one round of chromosome replication (Ganesan and Smith, 1972). This observation excludes a mechanism of recombination between homologous daughter chromosomes, in a manner analogous to the recombination repair of u.v.-induced damage, while recombination between homologous parent chromosomes is excluded by the result of Bridges (1971) that acetate-grown cells of *E. coli*, which are mononucleate, are only slightly more sensitive than glucose-grown cells, which are binucleate. Using a quite different technique to measure the rate and time of completion of recombination repair of ionizing radiation damage in *Micrococcus radiodurans*, Moseley and Copland (1976) found that for doses up to 120 krad, recombination repair was complete before DNA synthesis resumed at normal rates.

In *E. coli, recA, recB* and *recC* gene products are all required for Type III repair; *recA* mutants are completely deficient in Type III repair whereas *recB* and *recC* mutants have considerably less Type III rejoining ability than wild type (Kapp and Smith, 1970; Youngs and Smith, 1973b). Other genes involved in this type of repair are *exr* (Youngs and Smith, 1973a), *dnaB* (Fangman and Russel, 1971), *ror* (Glickman et al., 1971) and presumably *lig* (Dean and Pauling, 1970).

Since Type III repair occurs only under conditions which allow growth, and it has been known for many years that different growth conditions can alter the radiation sensitivity of a culture, Town et al. (1971a) studied the effect of different growth conditions on both the survival of *E. coli* B/r and on the Type III single-strand mending process. They found that *E. coli* B/r, grown to stationary phase in a peptone-containing medium, was 3.4 times more resistant to radiation than when grown without glucose, and that exponential phase cells, growing in a complex medium, were 1.7 times more sensitive to radiation than stationary phase cells. In each case the more sensitive bacteria had a decreased ability to repair single-strand breaks, and there was greater DNA degradation during post-irradiation incubation.

B. TYPE II REPAIR (*polA, lig*)

Type II repair was discovered when the response of *polA* mutants of *E. coli* to ionizing radiation was investigated. Town et al. (1971b) found that the *polA1* mutant (De Lucia and Cairns, 1969) was three to five times more sensitive to X-rays than the wild type. The ability of the mutant to repair single-strand breaks was examined and an unexpected result found, namely, that there were many more single-strand breaks in the DNA of the *polA1* mutant than in the wild type, after identical doses of radiation and before incubation in growth medium. By irradiating the bacteria at 0°C, and lysing them immediately on alkaline sucrose gradients, they found that the same number of breaks were produced in the DNA of pol^- and pol^+ bacteria but that in pol^- bacteria about 90% of them were repaired in two to five minutes, when incubated in a buffer at room temperature.

Type II repair is therefore defined to include all of the repair processes which rejoin single-strand breaks in buffer after the completion of Type 1 repair. In wild-type bacteria, the Type II rejoining process at temperatures between 20 and 37°C has a half-time of one to two minutes, and at 0°C proceeds with a half time of about 10 minutes (Town et al., 1973a). Type II repair can be inhibited by 0.1 M EDTA (Town et al., 1971b).

DNA polymerase I (the *polA* gene product) plays a major role in Type II repair, and DNA ligase also would be expected to be necessary. This is supported by the work of Laipis and Ganesan (1972) who were able to increase both the biological activity and the molecular weight of transforming DNA which had been extracted from an X-irradiated *polA* mutant of *B. subtilis*, by incubating it with purified *E. coli* DNA polymerase I plus ligase.

C. TYPE I REPAIR (*lig?*)

To distinguish between Type I and II repair, Type I is defined as those rejoining processes which occur immediately after irradiation, do not require growth conditions and do not require DNA polymerase I. Whether Type I repair requires ligase activity or is a chemically, as opposed to enzymically, modified change has not been adequately demonstrated. Repair is complete in buffer at 0°C in one to two minutes.

The existence of this type of strand mending was demonstrated by comparing the yields of X-ray-induced single-strand breaks under aerobic and anoxic conditions, i.e. in the absence of oxygen, in untreated *polA* bacteria and in pol^- and pol^+ bacteria which had been inactivated either by heat treatment or cold shock (Town et al., 1972). It was assumed that these treatments would inactivate any enzymes involved in repair, and would give the most accurate measure of X-ray-induced single-strand breaks. Town *et al.* (1972) found that the number of single-strand breaks was 2.8 per single-strand genome of *E. coli* per krad, and was independent of the presence or absence of oxygen during irradiation. A similar conclusion had been reached previously by Dean *et al.* (1969) who inhibited strand mending in *Micrococcus radiodurans* with EDTA. However the number of single-strand breaks in untreated bacteria, after anoxic irradiation, was only 25% of that in the inactivated bacteria (i.e. 75% had been rejoined), while the number of breaks in bacteria irradiated in the presence of oxygen was 75% of that in inactivated bacteria (i.e. only 25% had been rejoined). Thus, after Type I repair had occurred, there were three times as many single-strand breaks present in aerobically irradiated as in anoxically irradiated bacteria.

D. THE CONTRIBUTION OF TYPES I, II AND III REPAIR IN THE IRRADIATED WILD-TYPE CELL

An extremely good picture of the contributions of Types I, II and III repair to the overall repair of X-ray damage in *E. coli* has been built up by Town *et al.* (1973a, b). Values for the number of single-strand breaks actually induced in irradiated *E. coli* as a function of the dose were obtained from irradiated heat- or cold-inactivated bacteria. The number of single-strand breaks remaining after Type I repair was completed was measured in *polA* mutants of *E. coli* held in buffer at 0°C. The number of breaks remaining after completion of Types I and II repair was measured in *pol*$^+$ strains after incubation in buffer at room temperature for 15 minutes, and the number of breaks remaining after completion of Types I, II and III repair by using *pol*$^+$ strains incubated under growth conditions in minimal medium at 37°C for 60 minutes. This was done for a large number of X-ray doses given both in the presence and absence of oxygen.

Essentially the same number of single-strand breaks are made in the DNA of *E. coli* whether irradiated in the presence or absence of oxygen, the value being about 2.8 breaks per single-strand genome (1.4×10^9 daltons) per krad. After anoxic irradiation, Type I repair mends 75% of the breaks but only 25% are mended after irradiation in the presence of oxygen. Thus, after the completion of Type I repair, there are three times as many single-strand breaks in the aerobically-irradiated bacteria as in the anoxically-irradiated ones, and this may account for the well-known dose-modifying oxygen effect in which the inactivation of irradiated biological material in oxygen appears about three times more efficient as compared with anoxic irradiation. At anoxic doses of less than about 8.3 krad, all of the breaks remaining after the completion of Type I repair can be mended by Type II repair, but at higher doses a constant proportion of about 90% of the single-strand breaks remaining after Type I repair are rejoined by the Type II repair. Of the breaks remaining after Types I and II repair are complete, a maximum of approximately two can be rejoined by Type III repair. At anoxic doses above about 35 krad, single-strand breaks accumulate because the total repair capacity of the three types of repair systems is exceeded.

For aerobic irradiation, after the mending of 25% of the original single-strand breaks by Type I repair, the remaining breaks can all be rejoined up to a dose of about 3.7 krad by Type II repair. At doses exceeding this, 90% of the breaks remaining after Type I repair are mended by Type II repair, and a maximum of two breaks are mended by Type III repair. Thus, single-strand breaks begin accumulating at about 12 krad, when the total repair capacity is exceeded.

It should be noted that the D_{37} values (i.e. the average dose of radiation required to kill a bacterium) are considerably less than 35 krad and 12 krad for anoxic and aerobic irradiation, respectively, the doses up to which all single-strand breaks can be repaired (see Town et al., 1973a). The actual values are close to 10 krad for anoxic, and 3 krad for aerobic, irradiation. Thus the bacteria are beginning to be inactivated at doses at which repair of single-strand breaks is still very efficient, and it must be assumed that the eventual inability to repair single-strand breaks is irrelevant since the cells have already received lethal damage of some other form (e.g. double-strand DNA breaks or base damage which is not excisable).

E. THE REPAIR OF DNA DOUBLE-STRAND BREAKS

The mechanism by which bacteria repair DNA double-strand breaks is not known. Even the most resistant strains of *E. coli* do not repair more than four double-strand breaks per genome (Setlow and Setlow, 1972) so that the possession of large numbers of defined mutants defective in repair processes is not helpful. On the other hand, *Micrococcus radiodurans*, which can repair about 1400 double-strand breaks per genome, has poorly developed genetics! At the physiological level, Burrell et al. (1971) have shown that DNA isolated from unirradiated *M. radiodurans* is present in a rapidly-sedimenting membrane complex, whereas random shearing of the DNA by exposure of the bacteria to X-rays releases free DNA molecules. If the bacteria are incubated under growth conditions following X-irradiation, the DNA becomes re-associated with the membrane complex. This was interpreted to mean that a membrane complex has multiple attachment sites to the chromosome, and the damaged chromosome is thus stabilized while repair of DNA double-strand breaks occurs. This must remain only a theoretical possibility until it is demonstrated that the membrane-chromosome association is a unique one which does not occur in radiation-sensitive bacteria.

VIII. The Repair of Cross-link Damage

It has already been pointed out that mutants isolated on the basis of defects in their ability to repair radiation damage are also more sensitive to the lethal effects of cross-linking agents, such as mitomycin-C and psoralen-plus-near u.v. light, than the wild types from which they were derived (Boyce and Howard-Flanders, 1964b; Moseley, 1967; Cole, 1971). It follows that cross-link damage is subject to repair mechanisms which operate also in the repair of radiation damage. Cole (1971) showed that the excision and recombination-proficient *E. coli* AB1157 was able to repair 64 cross-links per genome, an excision-deficient mutant (*uvrA6*) only 15, a recombination-deficient mutant (*recA13*) six, and an excision and recombination-deficient double mutant (*uvrA6 recA13*) none. Cole (1973) went on to show that, during incubation of *E. coli* AB1157, which had been treated with psoralen-plus-near u.v. light, cross-links were excised and the cellular DNA cut into discrete pieces. The average molecular weights of these pieces corresponded to about twice the average single-strand distance between cross-links. This excision did not occur in excision-deficient strains carrying mutations at *uvrA* or *uvrB*. On further incubation, the DNA fragments were covalently joined into high molecular-weight DNA. This joining did not occur in *recA* mutants. On the basis of these results, Cole (1973) proposed a model for the removal of DNA interstrand cross-links which involved the sequential action of excision and genetic recombination. The model is shown in Fig. 8.

Obviously, if an interstrand cross-link is completely excised in a single step, a double-strand gap would be produced with lethal

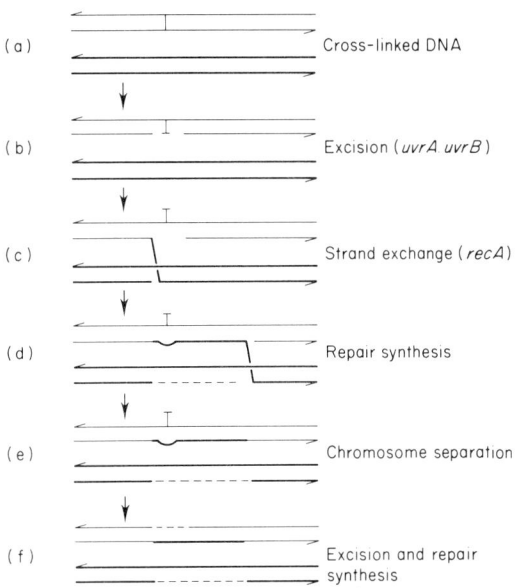

FIG. 8. Removal of cross-links from the DNA of *Escherichia coli*, based on the model of Cole (1973). (a) This model assumes the existence of an homologous chromosome adjacent to the chromosome containing the cross-link. (b) The *uvrA* and *uvrB* endonucleases and *polA*-associated exonuclease make single-strand breaks, thus freeing one arm of the cross-link and leaving a single-stranded gap opposite a non-coding mono-adduct. (c) Strand breakage and re-union between sister strands occurs, dependent on *recA*. (d) Strand exchange replaces the genetic information missing from the gap opposite the mono-adduct although hydrogen bonding in that region may be defective. The gap left in the homologous chromosome is repolymerized by repair replication (---). (e) The two chromosomes separate. (f) The mono-adduct can now be excised and the resulting gap filled by repair replication.

consequences in *E. coli*. It was therefore proposed (Cole, 1973) that only one side of the cross-link is excised and that this step is controlled by the *uvrA* or *uvrB* genes. If only one side of each cross-link is excised, this would leave single-strand DNA fragments of average molecular weight equal to twice the distance between cross-links, and each fragment would, on average, be covalently linked to one molecule of mitomycin-C which had been excised from the complementary strand. A situation would now exist, at the site of a cross-link, of one DNA strand containing an attached mitomycin-C molecule and a gap in the complementary strand opposite the mitomycin-C molecule. Cole (1973) suggests that this is analogous to the situation where DNA replication has occurred past an unexcised pyrimidine dimer, such that a gap is left opposite a non-coding region and repair replication is excluded. This means that, for recovery of the missing genetic information in the region of a cross-link, an homologous parental chromosome must participate in the recombination repair. This step is controlled by the *recA* gene. The mitomycin molecule which is still attached as a mono-adduct is then removed by excision and repair replication.

One major flaw in this model is that it does not explain why *recA* and *uvrA* mutants are still able to repair DNA cross-links, albeit fewer than the wild type. If there were obligatory requirements for the participation of *uvrA* and *recA* gene products in a sequential manner, as described, then one might expect that the loss of either function would prevent cross-link removal, thus making the bacterium as sensitive to cross-linking agents as one which had lost both functions, and this is clearly not the case. Moseley and Copland (1975) have isolated a recombination-deficient mutant of *Micrococcus radiodurans* which is probably inactivated by a single mitomycin-C-induced cross-link. Since the wild type is 300 times more resistant to mitomycin-C than the recombination-deficient mutant, and is assumed therefore to repair about 300 cross-links per genome, and the excision of pyrimidine dimers in the *rec* mutant is normal, there must be an obligatory requirement for recombination in the repair of cross-link damage in *M. radiodurans*.

IX. The Mutagenic Consequences of Repair

In both excision and post-replication recombination repair, nucleotides are removed from the damaged DNA and the genetic information recovered either by repolymerization of the missing bases, using the complementary DNA strand as a template, or by a recombination mechanism, or both. If these processes are carried out with absolute fidelity, then potentially lethal lesions are repaired and the surviving clone is genetically identical to the original strain. However, if the repair is less than perfect (for example, by the insertion of an incorrect base), it could result in mutation. It is well known that both u.v. and ionizing irradiation are mutagenic treatments, and the former is used routinely in microbiological laboratories for isolation of mutants. Thus the frequency of induced mutations following DNA damage by irradiation or chemicals can be used as a measure of the fidelity of repair processes, and by comparing such frequencies in both wild-type and radiation-sensitive mutants it is possible to identify those repair pathways which are more, or less, accurate in restoring the original base sequence. For reviews of repair-mediated induced mutation in bacteria see Bridges (1969), Witkin (1969a) and Clarke (1975).

A. PHOTOREACTIVATION

Photoreactivation is essentially a non-mutagenic process, which is not surprising since it does not remove the u.v.-induced pyrimidine dimers from DNA. The dimers are monomerized *in situ*, and therefore the base sequence remains unchanged. The studies of mutagenesis in phr^+ and phr^- strains of *E. coli* merely indicate that a dimer which remains in the DNA is much more likely to lead to mutation, as a result of passing through other repair pathways, than one which is repaired by photoreactivation (Witkin *et al.*, 1963; Witkin, 1964; Kondo and Jagger, 1966; Kondo and Kato, 1966).

B. EXCISION REPAIR

Comparison of induced mutagenesis in strains of *E. coli* possessing (uvr^+) and lacking (uvr^-) the ability to excise u.v.-induced pyrimidine dimers has shown that excision is essentially a high-fidelity repair mechanism. Hill (1965) showed that mutations are induced with a much higher frequency at low doses of u.v. radiation in a *uvr* mutant than in the wild type, an observation which has been confirmed by Witkin (1966, 1967a), Bridges and Munson (1966) and Kondo and Kato (1966). The relative mutagenicity of an unexcised pyrimidine dimer and one that is excised can be illustrated by an example from Witkin (1969a). At a u.v. dose of 20 ergs/mm^2, about 120 dimers are induced per genome of *E. coli*. In a *uvr* mutant all, or most, of these dimers remain in the DNA to be by-passed by recombination repair, while in a strain capable of excision all, or most, of them are removed by excision. At 20 ergs/mm², the *uvr* strain yields about 200 u.v.-induced mutants (to streptomycin resistance) while the wild type produces 0.4 induced mutants, per 10^7 survivors (Witkin, 1966; 1969b). Since all of the mutations produced in a *uvr* mutant must be due to unexcised dimers, an unexcised dimer must be at least 500 times more likely (200/0.4) to cause a mutation than an excised dimer, as a result of its passing along a different repair pathway. Making certain assumptions Witkin (1969a) calculated that, in a *uvr* strain of *E. coli*, one induced mutant arises per 2×10^3 unexcised dimers, while in the wild type the maximum probability that an error occurs, resulting in an induced mutant, is about 1 per 10^6 dimers excised.

C. RECOMBINATION

While photoreactivation and excision repair are essentially high-fidelity repair systems, and are referred to as error-proof (or mutation-proof) modes of survival, the recombination repair pathway appears to be associated with u.v.-induced mutation and was thought to be error (or mutation) prone (Witkin, 1967a). Witkin compared u.v.-induced mutation in *exr*⁻ and *exr*⁺ strains of *E. coli* and found that u.v.-induced mutability was completely lacking in *exr* mutants, irrespective of the presence or absence of an excision mechanism. The decreased recombination ability of *exr* mutants (Witkin, 1969a), but normal excision ability (Witkin, 1967a; Mattern *et al.*, 1966), suggested that u.v.-induced mutation is due to errors in recombination repair of daughter-strand gaps opposite unexcised pyrimidine dimers caused by the presence of the *exr* gene product (Witkin, 1969a; Bridges *et al.*, 1967; Kondo, 1968). In the absence of the *exr* gene product, the error-prone nature of recombination repair is eliminated, but at the expense of survival since *E. coli* strains which carry an *exr* mutation are two to three times more sensitive to u.v. radiation than the wild type. Studies on u.v.-induced mutagenesis, in recombination-deficient mutants of *E. coli*, have shown that the mutagenic pathway requires recombination. The yield of u.v.-induced mutations is decreased substantially in *recB* or *recC* mutants (Witkin, 1969b; 1972), and completely in *recA* mutants (Miura and Tomizawa, 1968; Witkin, 1969b). Thus, the induction of mutations in *E. coli* by u.v. radiation is dependent on both $recA^+$ and *exr*⁺.

D. THE INDUCIBLE NATURE OF ULTRAVIOLET-RADIATION-INDUCED MUTAGENESIS IN *Escherichia coli*

The dependence of u.v.-induced mutation on *recA* and *exr* genes led Witkin and George (1973) to compare u.v.-induced mutagenesis with other u.v.-associated phenomena in *E. coli* which were also dependent on *recA*$^+$ and *exr*$^+$. They included the u.v. induction of λ-prophage (Monk *et al.*, 1971), u.v. reactivation and u.v. mutagenesis of irradiated phage (Miura and Tomizawa, 1968; Donch *et al.*, 1970; Defais *et al.*, 1971; Fauquet and Defais, 1972) and u.v.-induced filamentous growth in *lon* mutants (Witkin, 1967b; Green *et al.*, 1969; Donch *et al.*, 1968). The induction of these functions requires not only the correct genotype, but also the presence of photoproducts, such as dimers, which block or slow down DNA replication. The dose of u.v. radiation required for induction is much lower in *uvr* mutants defective in excision repair, and to explain an apparent anomaly in *polA* mutants which excise pyrimidine dimers at the wild-type rate (Boyle *et al.*, 1970), but require a lower u.v. dose than *polA*$^+$ strains for induction, Witkin and George (1973) postulated that in *polA* mutants the gaps resulting from excision remain open much longer than in the wild type, and that these are, in effect, replication-blocking lesions.

Further evidence that prophage induction, u.v. reactivation, u.v. mutagenesis of phage and filamentous growth are all inducible, with the same set of genetic and physiological requirements for their induction, came from studies on a temperature-sensitive mutant (*tif*) in which both prophage and filament formation are induced when the temperature is raised, although DNA synthesis is not inhibited (Kirby *et al.*, 1967; Castellazi *et al.*, 1972a). The *recA* and *exr* mutants derived from this strain lose the ability to be induced for these functions (Castellazi *et al.*, 1972b). To test whether u.v.-induced mutagenesis in *E. coli* is controlled by the same mechanism as these other u.v.-associated phenomena, Witkin (1973, 1974) studied induced mutagenesis in a *tif-1* mutant. She found that raising the temperature of the wild-type strain, after extremely low doses of u.v. irradiation, did not alter the frequency of u.v.-induced mutations, whereas in the *tif-1* mutant it resulted in a ten-fold increase in the induced mutation yield. This result supported the hypothesis (Defais *et al.*, 1971) that u.v.-induced mutability in *E. coli* depends upon an inducible function that is normally expressed only when DNA replication is inhibited.

Sedgwick (1975) has recently isolated, as a band on polyacrylamide gel, what appears to be the u.v.-induced protein. He labelled bacteria, which had been u.v. irradiated, with ^{35}S-methionine for 20 minutes following irradiation, and identified a band on gels which was only formed in u.v.-irradiated bacteria possessing the $recA^+$ and the exr^+ genotype. The same protein band appeared when *tif* mutants were analysed after growth at an inducing temperature of 42°C. No biochemical studies on this protein have yet been reported.

The most recent view on the nature of u.v.-induced mutation is that it is caused by the repair of pyrimidine dimers when the post-replication recombination repair mechanism attempts to by-pass two pyrimidine dimers which lie relatively close to each other on opposite strands of the irradiated parental DNA. In replicating past these dimers, a strand gap of about 1000 base pairs is left opposite each dimer and the two gaps overlap. It is the attempt to reconstitute this region by recombination in the presence of the induced *exr* gene product which leads to mutation (Green *et al.*, 1976). At present no methodology exists to test such an hypothesis.

E. REPAIR-DEPENDENT AND INDEPENDENT MUTAGENS

Since u.v.-induced mutagenesis is dependent on the *recA* and *exr* gene products, it is possible to classify mutagenic treatments as repair-dependent or independent by comparing their ability to induce mutation in strains carrying mutations at these loci. Treatments which induce mutation in both mutants and wild-type organisms are considered repair-independent, and treatments which induce mutation in the wild type but not, or at a much lower frequency, in *recA* or *exr* mutants are considered repair-dependent. Thus Bridges *et al.* (1968) showed that mutability by ionizing radiation and thymine deprivation is greatly decreased in *exr* mutants of *E. coli*, and they proposed a common mutagenic pathway for u.v. and ionizing radiations and thymine deprivation. Other repair-dependent mutagens are 4-nitroquinoline-1-oxide and mitomycin-C (Kondo *et al.*, 1970). Repair-independent mutagens are hydroxylamine, ethyl methanesulphonate and nitrosoguanidine (Ishii and Kondo, 1975; Witkin, 1967a, 1969a).

This classification depends upon the reaction of *E. coli* mutants to these mutagens. In *Micrococcus radiodurans*, the classification does not hold. Although *M. radiodurans* is as sensitive as *E. coli* B/r to mutation by nitrosoguanidine, it is non-mutable by u.v. and ionizing radiation, mitomycin-C and hydroxylamine, and only very slightly sensitive to mutation by nitrous acid, ethyl methanesulphonate and β-propriolactone (Sweet and Moseley, 1974, 1976). *Micrococcus radiodurans* has a very efficient recombination repair mechanism (Moseley *et al.*, 1972; Moseley and Copland, 1975) and it must be concluded that the recombination repair is an accurate process, and that there is no error-prone pathway present in this bacterium. Thus, it should not be assumed by microbiologists that any compound which is mutagenic in *E. coli* is a mutagen for all bacteria, although nitrosoguanidine may be an exception.

XI. Summary

The bacterial chromosome is vulnerable to damage by both environmental radiation and chemicals, and bacteria have evolved DNA-repair mechanisms to maintain the genetic continuity of the chromosome in the presence of such damaging agents. Damage to DNA may take the form of purine or pyrimidine base damage, single- or double-strand breakage of the phosphodiester backbone, or covalent cross-linking of the twin strands of the DNA helix.

Pyrimidine dimers, induced by u.v. radiation, may be repaired by a photoreactivation mechanism which monomerizes the dimer *in situ*, or by excision of the single-stranded region containing the dimer and its repolymerization using the complementary strand as a template, or they may be by-passed by a post-replication recombination mechanism. Single-strand breaks may also be repaired in three different ways. In *E. coli*, Type I repair occurs within seconds of the single-strand breaks being formed, even in buffer at 0°C, and in the absence of DNA polymerase I activity. This type of repair accounts for the restitution of 75% of the single-strand breaks formed during anoxic irradiation, and 25% of those formed during aerobic irradiation. Type II repair, which takes place within a few minutes at 37°C in buffer, mends 90% of the remainder and requires the activity of DNA polymerase I. Type III repair takes about 40–60 minutes at 37°C, in a growth medium, and mends about two single-strand breaks per genome of *E. coli*. The mechanism of repair of double-strand breaks remains unknown. Covalent cross-links are mended by a combination of excision, which releases one "arm" of the cross-link leaving a single-stranded gap opposite a mono-adduct, and a recombination mechanism which restores the genetic information missing from the gap.

Some treatments which damage DNA may induce mutations in the bacterial population, not as a result of the primary damage, but because the repair of such DNA damage goes through an error-prone repair pathway. Such induced mutation is abolished in bacteria which are mutant at a *recA* or *lex* (*exr*) locus, but at the expense of cell survival since such mutants are more sensitive to the damaging treatments. Some bacteria, (e.g. *Micrococcus radiodurans*) do not have an error-prone pathway and are not therefore mutated by such treatments.

REFERENCES

Adler, H. I. and Hardigree, A. A. (1964). *Journal of Bacteriology* **87**, 720.
Adler, H. I., Fisher, W. D., Hardigree, A. A. and Stapleton, G. E. (1966). *Journal of Bacteriology* **91**, 737.
Anderson, A. W., Nordan, H. C., Cain, R. F., Parrish, G. and Duggan, D. (1956). *Food Technology* **10**, 575.
Boling, M. E. and Setlow, J. K. (1966). *Biochimica et Biophysica Acta* **123**, 26.
Boyce, R. P. and Howard-Flanders, P. (1964a). *Proceedings of the National Academy of Sciences of the United States of America* **51**, 293.
Boyce, R. P. and Howard-Flanders, P. (1964b). *Zeitschrift für Vererbungslehre* **94**, 345.
Boyle, J. M., Paterson, M. C. and Setlow, R. B. (1970). *Nature, London* **226**, 708.
Braun, A. and Grossman, L. (1974). *Proceedings of the National Academy of Sciences of the United States of America* **71**, 1838.
Bridges, B. A. (1969). *Annual Review of Nuclear Science* **19**, 139.
Bridges, B. A. (1971). *Journal of Bacteriology* **108**, 944.
Bridges, B. A. and Munson, R. J. (1966). *Biochemical and Biophysical Research Communications* **22**, 268.
Bridges, B. A., Dennis, R. E. and Munson, R. J. (1967). *Genetics* **57**, 897.
Bridges, B. A., Law, J. and Munson, R. J. (1968). *Molecular and General Genetics* **103**, 266.
Brutlag, D. and Kornberg, A. (1972). *Journal of Biological Chemistry* **247**, 241.
Burrell, A. D., Feldschreiber, P. and Dean, C. J. (1971). *Biochimica et Biophysica Acta* **247**, 38.
Cairns, J. (1963). *Journal of Molecular Biology* **6**, 208.
Campbell, J. L., Soll, L. and Richardson, C. C. (1972). *Proceedings of the National Academy of Sciences of the United States of America* **69**, 2090.
Carrier, W. L. and Setlow, R. B. (1971). *In* "Methods in Enzymology" (S. P. Colowick and N. O. Kaplan, eds.), vol. 21, p. 230.
Castellazi, M., George, J. and Buttin, G. (1972a). *Molecular and General Genetics* **119**, 139.
Castellazi, M., George, J. and Buttin, G. (1972b). *Molecular and General Genetics* **119**, 153.
Chase, J. and Richardson, C. C. (1974). *Journal of Biological Chemistry* **249**, 4553.
Clark, A. J. (1967). *Journal of Cellular Physiology, Supplement 1*, **70**, 165.
Clark, A. J. (1971). *Annual Review of Microbiology* **25**, 437.

Clark, A. J. and Margulies, A. D. (1965). *Proceedings of the National Academy of Sciences of the United States of America* **53**, 451.
Clarke, C. H. (1975). *Science Progress, Oxford* **62**, 559.
Cole, R. S. (1971). *Journal of Bacteriology* **107**, 846.
Cole, R. S. (1973). *Proceedings of the National Academy of Sciences of the United States of America* **70**, 1064.
Cooper, P. K. and Hanawalt, P. C. (1972). *Proceedings of the National Academy of Sciences of the United States of America* **69**, 1156.
Cozzarelli, N. R., Kelly, R. B. and Kornberg, A. (1969). *Journal of Molecular Biology* **45**, 513.
Davis, N. S., Silverman, G. J. and Masurovsky, E. B. (1963). *Journal of Bacteriology* **86**, 294.
Dean, C. J. and Pauling, C. (1970). *Journal of Bacteriology* **102**, 588.
Dean, C. J., Ormerod, M. G., Serianni, R. W. and Alexander, P. (1969). *Nature, London* **222**, 1042.
Defais, M., Fauquet, P., Radman, M. and Errera, M. (1971). *Virology* **43**, 495.
De Lucia, P. and Cairns, J. (1969). *Nature, London* **224**, 1164.
Donch, J. J. and Greenberg, J. (1968). *Molecular and General Genetics* **103**, 105.
Donch, J. J. and Greenberg, J. (1970). *Mutation Research* **10**, 153.
Donch, J. J. and Greenberg, J. (1976). *Mutation Research* **34**, 533.
Donch, J. J., Green, M. H. L. and Greenberg, J. (1968). *Journal of Bacteriology* **96**, 1704.
Donch, J. J., Chung, Y. S. and Greenberg, J. (1969). *Genetics* **61**, 363.
Donch, J. J., Greenberg, J. and Green, M. H. L. (1970). *Genetical Research* **15**, 87.
Driedger, A. A. (1970). *Canadian Journal of Microbiology* **16**, 1136.
Ellison, S. A., Feiner, R. R. and Hill, R. F. (1960). *Virology* **11**, 294.
Emmerson, P. T. (1968). *Genetics* **60**, 19.
Emmerson, P. T. and Howard-Flanders, P. (1967). *Journal of Bacteriology* **93**, 1729.
Fangman, W. L. and Russel, M. (1971). *Molecular and General Genetics* **110**, 332.
Fauquet, P. and Defais, M. (1972). *Mutation Research* **15**, 353.
Ganesan, A. K. and Smith, K. C. (1971). *Molecular and General Genetics* **113**, 285.
Ganesan, A. K. and Smith, K. C. (1972). *Journal of Bacteriology* **111**, 575.
Gates, F. L. (1930). *Journal of General Physiology* **14**, 31.
Gefter, M. L. (1974). *Progress in Nucleic Acid Research and Molecular Biology* **14**, 105.
Gefter, M. L., Hirota, Y., Kornberg, T., Wechsler, J. A. and Barnoux, C. (1971). *Proceedings of the National Academy of Sciences of the United States of America* **68**, 3150.
Gellert, M. and Bullock, M. L. (1970). *Proceedings of the National Academy of Sciences of the United States of America* **67**, 1580.
Glickman, B. W., Zwenk, H., Van Sluis, C. A. and Rörsch, A. (1971). *Biochimica et Biophysica Acta* **254**, 144.
Glickman, B. W., Van Sluis, C. A., Heijneker, H. L. and Rörsch, A. (1973). *Molecular and General Genetics* **124**, 69.
Goldmark, P. J. and Linn, S. (1972). *Journal of Biological Chemistry* **247**, 1849.
Goodgal, S. H., Rupert, C. S. and Herriot, R. M. (1957). *In* "The Chemical Basis of Heredity" (W. D. McElroy and B. Glass, eds.), p. 341. Johns Hopkins Press, Baltimore.
Green, M. H. L., Greenberg, J. and Donch, J. (1969). *Genetical Research* **14**, 159.
Green, M. H. L., Bridges, B. A., Eyfjörd, J. E. and Muriel, W. J. (1976). *Abstracts of the Genetical Society, Heredity*, in the press.
Greenberg, J. (1964). *Genetics* **50**, 639.
Greenberg, J. (1967). *Genetics* **55**, 193.
Greenberg, J., Berends, L. J., Donch, J. and Green, M. H. L. (1974). *Genetical Research* **23**, 175.

Gross, J. D. and Gross, M. (1969). *Nature, London* **224**, 1166.
Gross, J. D., Grunstein, J. and Witkin, E. M. (1971). *Journal of Molecular Biology* **58**, 631.
Grossman, L., Kaplan, J. C., Kuchner, S. R. and Mahler, I. (1968). *Cold Spring Harbor Symposia in Quantitative Biology* **33**, 229.
Grossman, L., Braun, A., Garvik, B., Hamilton, L. and Mahler, I. (1973). *Anais da Academia Brasileira de Ciências* **45**, Suplemento, 167.
Grossman, L., Braun, A., Feldberg, R. and Mahler, I. (1975). *Annual Review of Biochemistry* **44**, 19.
Hariharan, P. V. and Cerutti, P. A. (1971). *Nature New Biology* **229**, 247.
Hariharan, P. V. and Cerutti, P. A. (1972). *Journal of Molecular Biology* **66**, 65.
Harm, H. and Rupert, C. S. (1968). *Mutation Research* **6**, 355.
Harm, W. (1965). *In* "Symposium on the Mutation Process" (M. Kohoutova and J. Hubacek, eds.), p. 51. Academia, Prague.
Harm, W. (1969). *Radiation Research* **40**, 63.
Harm, W. and Hillebrandt, B. (1962). *Photochemistry and Photobiology* **1**, 271.
Harm, W., Harm, H. and Rupert, C. S. (1968). *Mutation Research* **6**, 371.
Harm, W., Rupert, C. S. and Harm, H. (1971). *In* "Photophysiology" (A. C. Giese, ed.), vol. 6, p. 279. Current Topics in Photobiology and Photochemistry. Academic Press, New York.
Hart, R. W. and Setlow, R. B. (1973). Abstract 1st Annual Meeting of the American Society for Photobiology, Sarasota, Florida, p. 120.
Heijneker, H. L., Pannekoek, H., Oosterbaan, R. A., Pouwels, P. H., Bron, S., Arwert, F. and Venema, G. (1971). *Proceedings of the National Academy of Sciences of the United States of America* **68**, 2967.
Hill, R. F. (1958). *Biochemica et Biophysica Acta* **30**, 636.
Hill, R. F. (1965). *Photochemistry and Photobiology* **4**, 563.
Hill, R. F. and Simson, E. (1961). *Journal of General Microbiology* **24**, 1.
Howard-Flanders, P. (1968). *Annual Review of Biochemistry* **37**, 175.
Howard-Flanders, P. (1973). *British Medical Bulletin* **29**, 226.
Howard-Flanders, P. and Boyce, R. P. (1966). *Radiation Research Supplement* **6**, 156.
HowardFlanders, P. and Theriot, L. (1962). *Genetics* **47**, 1219.
Howard-Flanders, P. and Theriot, L. (1966). *Genetics* **53**, 1137.
Howard-Flanders, P., Simson, E. and Theriot, L. (1964). *Genetics* **49**, 237.
Howard-Flanders, P., Boyce, R. P. and Theriot, L. (1966). *Genetics* **53**, 1119.
Howard-Flanders, P., Rupp, W. D., Wilde, C. E. and Reno, D. (1971). Proceedings of the 10th International Congress of Microbiology, 1970.
Ishii, Y. and Kondo, S. (1975). *Mutation Research* **27**, 27.
Iyer, V. N. and Rupp, W. D. (1971). *Biochimica et Biophysica Acta* **228**, 117.
Iyer, V. N. and Szybalski, W. (1963). *Proceedings of the National Academy of Sciences of the United States of America* **50**, 355.
Iyer, V. N. and Szybalski, W. (1964). *Science, New York* **145**, 55.
Jagger, J. and Latarjet, R. (1956). *Annales de l'Institut Pasteur* **91**, 858.
Johnson, B. F. and Greenberg, J. (1975). *Journal of Bacteriology* **122**, 570.
Kaplan, H. S. (1966). *Proceedings of the National Academy of Sciences of the United States of America* **55**, 1442.
Kaplan, J. C., Kushner, S. R. and Grossman, L. (1971). *Biochemistry, New York* **10**, 3315.
Kapp, D. S. and Smith, K. C. (1970). *Journal of Bacteriology* **103**, 49.
Kato, T. and Kondo, S. (1970). *Journal of Bacteriology* **104**, 871.
Kelly, R. B., Atkinson, M. R., Huberman, J. A. and Kornberg, A. (1969). *Nature, London* **224**, 495.
Kelner, A. (1949a). *Proceedings of the National Academy of Science, Washington* **35**, 73.

Kelner, A. (1949b). *Journal of Bacteriology* **58**, 511.
Kelner, A. (1950). *Bulletin of the New York Academy of Medicine* **26**, Ser. 2, 189.
Kirby, E. P., Jacob, F. and Goldthwait, D. A. (1967). *Proceedings of the National Academy of Sciences of the United States of America* **58**, 1903.
Kobatake, M., Tanabe, S. and Hasegawa, S. (1973). *Comptes Rendu de la Société de Biologie, Paris* **167**, 1506.
Kondo, S. (1968). *Proceedings of the 12th International Congress of Genetics* **2**, 126.
Kondo, S. and Jagger, J. (1966). *Photochemistry and Photobiology* **5**, 189.
Kondo, S. and Kato, T. (1966). *Photochemistry and Photobiology* **5**, 827.
Kondo, S., Ichikawa, H., Iwo, K. and Kato, T. (1970). *Genetics* **66**, 187.
Konrad, E. B., Mondrich, P. and Lehman, I. R. (1973). *Journal of Molecular Biology* **77**, 519.
Kornberg, A. (1969). *Science, New York* **163**, 1410.
Laipis, P. J. and Ganesan, A. T. (1972). *Proceedings of the National Academy of Sciences of the United States of America* **69**, 3211.
Lehman, I. R. and Chien, J. R. (1973). *Journal of Biological Chemistry* **248**, 7717.
Lewis, N. F. (1971). *Journal of General Microbiology* **66**, 29.
Mackay, V. and Linn, S. (1974). *Journal of Biological Chemistry* **249**, 4286.
Markovitz, A. (1964). *Proceedings of the National Academy of Sciences of the United States of America* **51**, 239.
Markovitz, A. and Baker, B. (1967). *Journal of Bacteriology* **94**, 388.
Marmur, J. and Grossman, L. (1961). *Proceedings of the National Academy of Sciences of the United States of America* **47**, 778.
Marmur, J., Anderson, W. F., Matthews, L., Berns, K., Gajewska, E., Lane, D. and Doty, P. (1961). *Journal of Cellular and Comparative Physiology* **58**, Suppl. 1, 33.
Masker, W. E. and Hanawalt, P. C. (1973). *Proceedings of the National Academy of Sciences of the United States of America* **70**, 129.
Mattern, I. (1971). In "First European Biophysical Congress" (E. Broda, A. Locker and H. Springer Lederer, eds.), vol. 2, p. 237. Verlag der Wiener Medizinischen Akademie.
Mattern, I. E., Zwenk, H. and Rörsch, A. (1966). *Mutation Research* **3**, 374.
McGrath, R. A. and Williams, R. W. (1966). *Nature, London* **212**, 534.
Minato, S. and Werbin, H. (1971). *Biochemistry, New York* **10**, 4503.
Miura, A. and Tomizawa, J. (1968). *Molecular and General Genetics* **103**, 1.
Molyneux, I. J. and Gefter, M. L. (1974). *Proceedings of the National Academy of Sciences of the United States of America* **71**, 3858.
Monk, M. and Kinross, J. (1972). *Journal of Bacteriology* **109**, 971.
Monk, M., Peacey, M. and Gross, J. D. (1971). *Journal of Molecular Biology* **58**, 623.
Morimyo, M., Horii, Z. and Suzuki, K. (1968). *Journal of Radiation Research, Japan* **9**, 19.
Moseley, B. E. B. (1967). *Journal of General Microbiology* **49**, 293.
Moseley, B. E. B. (1968). In "Advances in Microbial Physiology" (A. H. Rose and J. F. Wilkinson, eds.), vol. 2, p. 173. Academic Press, London.
Moseley, B. E. B. and Copland, H. J. R. (1975). *Journal of Bacteriology* **121**, 422.
Moseley, B. E. B. and Copland, H. J. R. (1976). *Journal of General Microbiology* **93**, 251.
Moseley, B. E. B., Mattingly, A. and Copland, H. J. R. (1972). *Journal of General Microbiology* **72**, 329.
Mount, D. W., Brooks-Low, K. and Edmiston, S. J. (1972). *Journal of Bacteriology* **112**, 886.
Mount, D. W., Walker, A. C. and Kosel, C. (1973). *Journal of Bacteriology* **116**, 950.
Muhammed, A. (1966). *Journal of Biological Chemistry* **241**, 516.
Ogawa, H., Shimada, K. and Tomizawa, J. (1968). *Molecular and General Genetics* **101**, 227.

Oginsky, E. L., Green, G. S., Griffith, D. G. and Fowlks, W. L. (1959). *Journal of Bacteriology* **78**, 821.
Okazaki, R., Okazaki, T., Sakabe, K., Sugimoto, K. and Sugino, A. (1968). *Proceedings of the National Academy of Sciences of the United States of America* **59**, 598.
Paterson, M. C. and Setlow, R. B. (1972). *Proceedings of the National Academy of Sciences of the United States of America* **69**, 2927.
Pauling, C. and Hamm, L. (1968). *Proceedings of the National Academy of Sciences of the United States of America* **60**, 1495.
Payne, J. I., Hartman, P. E., Mudd, S. and Phillips, A. W. (1956). *Journal of Bacteriology* **72**, 461.
Pettijohn, D. and Hanawalt, P. (1964). *Journal of Molecular Biology* **9**, 395.
Rörsch, A. and Van Der Kamp, C. (1961). *Biochimica et Biophysica Acta* **46**, 401.
Rörsch, A., Edelman, A., Van Der Kamp, C. and Cohen, J. A. (1962). *Biochimica et Biophysica Acta* **61**, 278.
Rupert, C. S. (1960). *Journal of General Physiology* **43**, 573.
Rupert, C. S. (1961). *Journal of Cellular and Comparative Physiology* **58**, Suppl. 1, 57.
Rupert, C. S. (1962a). *Journal of General Physiology* **45**, 703.
Rupert, C. S. (1962b). *Journal of General Physiology* **45**, 725.
Rupert, C. S. (1964). *Photochemistry and Photobiology* **3**, 399.
Rupert, C. S., Harm, H. and To, K. (1973). *Anais da Academia Brasileira de Ciências* **45**, Suplemento 1973, p. 151.
Rupp, W. D. and Howard-Flanders, P. (1968). *Journal of Molecular Biology* **31**, 291.
Rupp, W. D. and Howard-Flanders, P. (1971). In "Methods in Enzymology" (S. P. Colowick and N. O. Kaplan, eds.), vol. 21, p. 237.
Rupp, W. D., Wilde, C. E., Reno, D. L. and Howard-Flanders, P. (1971). *Journal of Molecular Biology* **61**, 25.
Seeberg, E. and Rupp, W. D. (1975). In "Molecular Mechanisms in DNA Repair" (P. C. Hanawalt and R. B. Setlow, eds.), p. 439. Plenum Press, New York.
Sedgwick, S. G. (1975). *Nature, London* **255**, 349.
Setlow, J. K. (1966). In "Current Topics in Radiation Research" (M. Ebert and A. Howard, eds.), vol. 2, p. 195. North-Holland Publishing Company, Amsterdam.
Setlow, J. K. (1967). In "Comprehensive Biochemistry" (M. Florkin and E. H. Stotz, eds.), vol. 27, p. 157. Elsevier Publishing Company, Amsterdam.
Setlow, J. K. and Boling, M. E. (1963). *Photochemistry and Photobiology* **2**, 471.
Setlow, J. K. and Setlow, R. B. (1963). *Nature, London* **197**, 560.
Setlow, J. K., Boling, M. E. and Bollum, F. J. (1965). *Proceedings of the National Academy of Sciences of the United States of America* **53**, 1430.
Setlow, R. B. (1964). *Journal of Cellular and Comparative Physiology* **64**, Suppl. 1, 51.
Setlow, R. B. (1966). *Science, New York* **153**, 379.
Setlow, R. B. (1968). *Progress in Nucleic Acid Research and Molecular Biology* **8**, 257.
Setlow, R. B. (1973). *Anais da Academia Brasileira de Ciências* **45**, Suplemento 1973, p. 215.
Setlow, R. B. and Carrier, W. L. (1964). *Proceedings of the National Academy of Sciences of the United States of America* **51**, 226.
Setlow, R. B. and Carrier, W. L. (1973). *Nature New Biology* **241**, 170.
Setlow, R. B. and Setlow, J. K. (1972). *Annual Review of Biophysics and Bioengineering* **1**, 293.
Siegel, E. C. (1973a). *Journal of Bacteriology* **113**, 145.
Siegel, E. C. (1973b). *Journal of Bacteriology* **113**, 161.
Sinsheimer, R. L. (1957). *Radiation Research* **6**, 121.
Smith, K. C. (1962). *Biochemical and Biophysical Research Communications* **8**, 157.

Smith, K. C. (1971). *In* "Photophysiology" (A. C. Giese, ed.), vol. 6, p. 209. Current topics in photobiology and photochemistry, Academic Press, New York.
Smith, K. C. and Aplin, R. T. (1966). *Biochemistry, New York* **5**, 2125.
Smith, K. C. and Meun, D. H. C. (1970). *Journal of Molecular Biology* **51**, 459.
Smirnov, G. B. and Skavronskaya, A. G. (1971). *Molecular and General Genetics* **113**, 217.
Sutherland, B. M. (1973). *Anais da Academia Brasileira de Ciências* **45**, Suplemento 1973, p. 161.
Sutherland, B. M. and Chamberlin, M. J. (1973). *Analytical Biochemistry* **53**, 168.
Sutherland, B. M., Court, D. and Chamberlin, M. J. (1972). *Virology* **48**, 87.
Sweet, D. M. and Moseley, B. E. B. (1974). *Mutation Research* **23**, 311.
Sweet, D. M. and Moseley, B. E. B. (1976). *Mutation Research* **34**, 175.
Szybalski, W. and Iyer, V. N. (1964). *Microbial Genetics Bulletin* **21**, 10.
Taylor, A. L. and Trotter, C. D. (1972). *Bacteriological Reviews* **36**, 504.
Town, C. D., Smith, K. C. and Kaplan, H. S. (1971a). *Journal of Bacteriology* **105**, 127.
Town, C. D., Smith, K. C. and Kaplan, H. S. (1971b). *Science, New York* **172**, 851.
Town, C. D., Smith, K. C. and Kaplan, H. S. (1972). *Radiation Research* **52**, 99.
Town, C. D., Smith, K. C. and Kaplan, H. S. (1973a). *Radiation Research* **55**, 334.
Town, C. D., Smith, K. C. and Kaplan, H. S. (1973b). *In* "Current Topics in Radiation Research Quarterly" (M. Ebert and A. Howard, eds.), vol. 8, p. 351. North-Holland Publishing Company, Amsterdam.
Uretz, R. B. and Markovitz, A. (1969). *Journal of Bacteriology* **100**, 1118.
Van De Putte, P., Westenbroek, C. and Rörsch, A. (1963). *Biochimica et Biophysica Acta* **76**, 247.
Van De Putte, P., Van Sluis, C. A., Van Dillewijn, J. and Rörsch, A. (1965). *Mutation Research* **2**, 97.
Van De Putte, P., Zwenk, H. and Rörsch, A. (1966). *Mutation Research* **3**, 381.
Wacker, A., Dellweg, H. and Weinblum, D. (1960). *Naturwissenschaften* **47**, 477.
Waring, M. J. (1966). *Symposium of the Society for General Microbiology* **16**, 235.
Webb, R. B. and Lorenz, J. R. (1970). *Photochemistry and Photobiology* **12**, 283.
Willetts, N. S. and Mount, D. W. (1969). *Journal of Bacteriology* **100**, 923.
Willetts, N. S., Clark, A. J. and Low, B. (1969). *Journal of Bacteriology* **97**, 244.
Witkin, E. M. (1947). *Genetics* **32**, 221.
Witkin, E. M. (1964). *Mutation Research* **1**, 22.
Witkin, E. M. (1966). *Science, New York*, **152**, 1345.
Witkin, E. M. (1967a). *Brookhaven Symposia in Biology* **20**, 17.
Witkin, E. M. (1967b). *Proceedings of the National Academy of Sciences of the United States of America* **57**, 1275.
Witkin, E. M. (1969a). *Annual Review of Microbiology* **23**, 487.
Witkin, E. M. (1969b). *Mutation Research* **8**, 9.
Witkin, E. M. (1972). *Mutation Research* **16**, 235.
Witkin, E. M. (1973). *Anais da Academia Brasileira de Ciências* **45**, Suplemento 1973, p. 185.
Witkin, E. M. (1974). *Proceedings of the National Academy of Sciences of the United States of America* **71**, 1930.
Witkin, E. M. and George, D. L. (1973). *Genetics Supplement* **73**, 91.
Witkin, E. M., Sicurella, N. A. and Bennett, G. M. (1963). *Proceedings of the National Academy of Sciences of the United States of America* **50**, 1055.
Wright, M., Buttin, G. and Hurwitz, J. (1971). *Journal of Biological Chemistry* **246**, 6543.
Wyckoff, R. W. G. (1932). *Journal of General Physiology* **15**, 351.
Yasui, A. and Laskowski, W. (1975). *International Journal of Radiation Biology* **28**, 511.
Youngs, D. A. and Smith, K. C. (1973a). *Journal of Bacteriology* **114**, 121.
Youngs, D. A. and Smith, K. C. (1973b). *Journal of Bacteriology* **116**, 175.

Note Added in Proof

The error-prone DNA repair mechanism induced by ultraviolet radiation of $recA^+exr^-$ strains of *Escherichia coli* or by raising a tif^- mutant of *E. coli* to its restrictive temperature is now known colloquially as "SOS" repair. This term was introduced by Radman (1974). A recent review (Witkin, 1977) discusses all the evidence for, and the mutational and other consequences of, SOS repair.

A second inducible DNA repair process in *Escherichia coli* has been reported by Samson and Cairns (1977). The repair mechanism is induced by growing the bacteria in a low concentration (1 μg/ml) of N-methyl-N'-nitrosoguanidine (NTG). The bacteria gain an increased resistance to both the lethal and mutational effects of exposure for five minutes to a high concentration (100 μg/ml) of NTG. This contrasts with SOS repair when increased survival results in increased mutation. The mechanism involves a different set of gene products from those used in SOS repair.

Finally, the linkage maps of *Escherichia coli* K-12 in which the total genome is represented by 90 min (Taylor and Trotter, 1972) has been superseded by a recalibrated linkage map in which the genome is represented by 100 min (Bachmann *et al.*, 1976). Thus, although the respective positions of the genes in Fig. 5 are correct, the reader is referred to Bachmann, *et al.* (1976) for their time location on the new map.

Additional References

Bachmann, B. J., Brooks-Low, K. and Taylor (1976). Recalibrated linkage map of *Escherichia coli* K-12. *Bacteriological Reviews* **40**, 116.

Radman, M. (1974). *In* "Molecular and environmental aspects of mutagenesis", (L. Prakash, F. Sherman, M. W. Miller, C. W. Lawrence and H. W. Tabor, eds.) p. 128. Charles C. Thomas, Springfield, Illinois.

Sampson, L. and Cairns, J. (1977). A new pathway for DNA repair in *Escherichia coli*. *Nature, London* **267**, 281.

Witkin, E. M. (1977). Ultraviolet mutagenesis and inducible DNA repair in *Escherichia coli*. *Bacteriological Reviews* **40**, 869.

Structure, Synthesis and Genetics of Yeast Mitochondrial DNA

PHILLIP NAGLEY, K. S SRIPRAKASH and ANTHONY W. LINNANE

Department of Biochemistry, Monash University, Clayton, Victoria 3168, Australia

I. Introduction	158
II. Structure and Physical Properties of mtDNA in Respiratory-Competent Yeast	158
A. *Saccharomyces cerevisiae* and *Saccharomyces carlsbergensis*	159
B. Other Yeast Species (*Petite*-Negative Yeasts)	170
III. Synthesis of mtDNA	173
A. mtDNA Synthesis *in vivo*	174
B. mtDNA Synthesis *in vitro*	175
C. mtDNA Replication and the Mitochondrial Membrane	178
D. Cellular Origin of Components Involved in mtDNA Synthesis	180
E. Level of mtDNA and Number of mtDNA Molecules in Cells of Different Strains	184
F. Synthesis of mtDNA During the Cell Cycle	187
G. Effects of Changes in Cell Physiology on Synthesis of mtDNA	189
IV. Petite Mutants of *Saccharomyces cerevisiae*—Molecular Aspects	193
A. Structure and Organization of mtDNA in *Petite* Cells	193
B. General Aspects of *Petite* Induction	203
C. Mechanism of *Petite* Induction by Ethidium Bromide	205
D. *Petite* Negativity of Yeasts other than *Saccharomyces cerevisiae* and Closely Related Species	214
V. Mitochondrial Genetics in Yeast	216
A. Criteria of Mitochondrial Inheritance	217
B. Genetic Determinants on mtDNA	217
C. Transmission and Recombination of Mitochondrial Genes in Genetic Crosses	230
D. Genetic and Physical Map of Yeast mtDNA	245
E. Suppressiveness in *Petite* Mutants	256
References	263

I. Introduction

The study of mitochondrial DNA (mtDNA) as a molecular entity has now passed through its first decade. A great deal of knowledge has accumulated over this period during which the biogenesis of mitochondria has attained the status of a distinct research area. The study of mitochondrial biogenesis embraces the complete spectrum of the molecular biology of development—the structure, synthesis and assembly of two functional membrane systems, the intracellular transport of RNA, proteins, lipids and a plethora of compounds of low molecular-weight, and finally the control and biosynthesis of macromolecules. The organisms of choice for investigation have ranged through the whole spectrum of eukaryotes to include mammals, plants and moulds. However, yeasts have been by far the most studied; being unicellular eukaryotes they have the great advantage over more complex organisms of the ready amenability to physiological and genetic manipulations that micro-organisms in general possess. The multitude of approaches used in the investigations of biogenesis of mitochondria and the complexity of the field have resulted in the growth of a voluminous and in some areas a difficult literature. Previous reviews covering the biogenesis of mitochondria and the role of mtDNA include those by Borst *et al.* (1967), Borst and Kroon (1969), Roodyn and Wilkie (1968), Granick and Gibor (1967), Rabinowitz (1968), Swift and Wolstenholme (1969), Nass (1969), Rabinowitz and Swift (1970), Ashwell and Work (1970), Schatz (1970), Kuntzel (1971), Linnane and Haslam (1970), Linnane *et al.* (1972b), Borst (1972), Borst and Grivell (1971), Sager (1972), Mahler (1973a), Schatz and Mason (1974), Linnane and Crowfoot (1975) and Hall and Linnane (1977).

We devote this review specifically to the function and structure of yeast mtDNA. By restricting the topic of this article to yeast mtDNA, we hope to do justice to this subject by treating the data in sufficient detail so as to permit a critical appraisal of the literature.

II. Structure and Physical Properties of mtDNA in Respiratory-Competent Yeast

Species of yeast can be broadly divided into two classes on the basis of whether or not respiratory-deficient *petite* mutants can be obtained. The so-called *petite*-positive yeast species consist of *Saccharomyces cerevisiae* and closely related interbreeding species such as *Sacch. carlsbergensis*. Sherman and Lawrence (1974) point out that, because of the interbreeding amongst the *Saccharomyces* group, many strains designated *Sacch. cerevisiae* were derived from crosses with *Sacch. carlsbergensis* and other related species. The mtDNA of *Sacch. cerevisiae* and *Sacch. carlsbergensis* are considered together throughout this review.

The *petite*-negative yeasts (Bulder, 1964) are generally obligate aerobes, and the mtDNA shows considerable size and sequence differences among themselves as well as when compared with mtDNA of *Sacch. cerevisiae*. These *petite*-positive and *petite*-negative organisms may be regarded as separate classes and are considered under their own headings.

A. *Saccharomyces cerevisiae* AND *Saccharomyces carlsbergensis*

The cytological work on mtDNA in yeast and other organisms has been discussed by Swift *et al.* (1968) and by Swift and Wolstenholme (1969) and will not be considered here. The positive biochemical identification of yeast mtDNA took place during 1964 (Schatz *et al.*, 1964; Tewari *et al.*, 1965). Genetic experiments indicative of an extranuclear genetic system affecting mitochondria in yeast had been made some 15 years earlier (for a review see Ephrussi, 1953), and these studies distinguished between the respiratory-competent *grande* strains (wild type) and the extranuclearly inherited respiratory-deficient *petite* mutants. The study of *petites* has become a major aspect of both physicochemical analysis of mtDNA, and mitochondrial genetics.

1. *General Properties and Preparative Techniques*

From the early studies on characterization of mtDNA in *grande* strains of *Sacch. cerevisiae* and *Sacch. carlsbergensis* there is general agreement that the buoyant density in caesium chloride (CsCl) of mtDNA from these two species is 1.683 g/ml and the T_m (measured in the presence of 0.165 M Na^+) is 74–75°C. In contrast, the nuclear DNA has a buoyant density of 1.698 g/ml (also 1.706 g/ml, heavy nuclear satellite) and a T_m of 85°C. These differences allow preparation and analysis of mtDNA in yeast using CsCl centrifugation. Other methods employed to purify mtDNA include use of hydroxyapatite (Bernardi *et al.*, 1968, 1970, 1972) and poly-(L-lysine)-coated kieselguhr (Finkelstein *et al.*, 1972; Blamire *et al.*, 1972b).

2. Physical Size

Many laboratories have attempted to determine the physical size of *Sacch. cerevisiae* mtDNA, but until 1969, only a heterogenous array of both linear and circular molecules had been observed. This applied to both purified preparations of mtDNA as well as to electron microscopic examination of mitochondria lysed to spread DNA by the Kleinschmidt procedure. However, in 1969, Borst and his colleagues examined preparations of osmotically shocked mitochondria of *Sacch. cerevisiae* and *Sacch. carlsbergensis* and observed under the electron microscope a number of highly twisted circular DNA molecules of 25 μm in circumference, that is a size of about 50×10^6 daltons (Hollenberg et al., 1969, 1970). The mass per unit length of DNA spread according to the Kleinschmidt procedure upon a cytochrome *c* film is taken to be 2×10^6 daltons per μm, which is close to the values determined by MacHattie and Thomas (1964) and Lang (1970). It was not, however, possible at that time to isolate and collect purified mtDNA molecules of this size. More recently a gentle technique of layering detergent-treated mitochondria on top of preformed sucrose gradients and immediately centrifuging has indicated that a large portion of the mtDNA sediments at a rate consistent with a molecular weight (for linear molecules) of $45-55 \times 10^6$ daltons (Blamire et al., 1972a). A similar approach was used by Petes et al. (1973), who layered sphaeroplasts directly on a sucrose gradient. The electron microscopic examination of gradient fractions enriched for mtDNA revealed a population of DNA molecules about 21 μm (42×10^6 daltons) in length. With this procedure, mtDNA specifically labelled in the presence of cycloheximide yielded DNA molecules sedimenting at rates corresponding to 45×10^6 daltons (Michels et al., 1974).

Measurements of the genome size of *grande* mtDNA of about 50×10^6 daltons as suggested from the electron microscopic and sedimentation rate studies described above are supported by two independent techniques, namely renaturation kinetics and restriction endonuclease analysis. Hollenberg et al. (1970) calculated an approximate kinetic complexity of $50-100 \times 10^6$ daltons from studying the rate of renaturation in 0.15 M Na$^+$ of sonicated denatured mtDNA (about 2×10^5 dalton fragments). Christiansen et al. (1971) found that the fragment size used for renaturation studies of *Sacch. carlsbergensis* mtDNA strongly influenced the reaction rate. A detailed analysis was carried out by Christiansen et al. (1974) of the effects of both the fragment size and the temperature of renaturation, on the kinetics of renaturation of *Sacch. carlsbergensis* mtDNA. The results showed that, under the conditions usually used, the rate of renaturation *decreased* with increasing fragment size. An analysis of the effect of varying pH value on the sedimentation properties of single stranded fragments of mtDNA of various sizes studied showed that the longer fragments ($> 2 \times 10^5$ daltons) adopted considerable secondary structure. This would explain the unexpectedly slow renaturation rate of the long fragments, because of the steric hindrance of correct base pairing (nucleation) necessary for re-association. The adoption of a folded secondary structure by longer molecules is ascribed to the presence of partially homologous A + T-rich sequences scattered through the yeast mitochondrial genome. When the renaturation kinetic analysis was carried out under conditions which minimized these steric effects, an overall genome size of close to 50×10^6 daltons was calculated (Christiansen *et al.*, 1974).

The application to yeast mtDNA of the restriction endonuclease techniques developed for the analysis of various genomes (see Nathans and Smith, 1975) has been recently made. Bernardi *et al.* (1975) described the products of digestion with HpaII endonuclease (specific for the sequence CCGG) of mtDNA isolated from different *grande* strains. Following agarose-acrylamide slab-gel electrophoresis of the HpaII fragments, about 100 bands were obtained from each *grande* ranging in size from about 4×10^4 to 4×10^6 daltons. Significantly, the sum of the molecular weights of all fragments ranged from 47 to 54×10^6 daltons, for the various *grandes*. Bernardi and his colleagues concluded that the mtDNAs of different *grandes* are not identical. The genetic studies from our laboratory also support this view (Molloy *et al.*, 1976). It is therefore conceivable that small differences in mtDNA sizes reported in the literature are due to the strain differences. Analysis of *grande* mtDNA by other restriction enzymes has also been reported (Morimoto *et al.*, 1975; Sanders *et al.*, 1975a; Bernardi *et al.*, 1976).

3. Circularity

As mentioned above, Hollenberg *et al.* (1969, 1970) reported electron microscopic observation of 25 μm supercoiled circular DNA molecules released from yeast mitochondria by osmotic shock.

Although attempts have been made to isolate circular DNA molecules of density 1.683 g/ml by caesium chloride–ethidium bromide (CsCl–EthBr) equilibrium centrifugation (Radloff et al., 1967), little or no success was achieved (Hollenberg et al., 1970; Clark-Walker, 1972). Nevertheless, other lines of evidence suggest that the *grande* yeast mitochondrial genome is functionally circular. The mapping of mitochondrial genetic markers by *petite* deletion analysis (Molloy et al., 1975), and the complementation between newly arisen *petite* mutants (Clark-Walker and Miklos, 1975) provide two pieces of genetic evidence (discussed on p. 248) in favour of circularity. In addition, the cleavage patterns of *Sacch. carlsbergensis* mtDNA with restriction endonucleases EcoRI and Hind II and III (Sanders et al., 1975a), as well as the denaturation maps of *Sacch. carlsbergensis* mtDNA made by Christiansen et al. (1975), can be interpreted in terms of a circular genome. Nevertheless, functional circularity does not necessarily imply physical circularity, as shown by the circularly permuted set of linear T-even phage DNA molecules which behave, at a population level, as circular genomes.

4. Sequence Organization

A good deal of knowledge of the unusual distribution of bases in mtDNA of *Sacch. cerevisiae* is available mainly due to the studies of Bernardi and colleagues. Direct chemical analysis of the base composition of this DNA yielded 17% G+C, with no unusual bases found (Bernardi et al., 1970; Mehrotra and Mahler, 1968; cf. Grossman et al., 1971). From considerations of the buoyant density in CsCl (1.683 g/ml) and T_m (75°C) one would infer using well established relationships (Schildkraut et al., 1972; Marmur and Doty, 1962) a base composition of 24% G+C from the density or 13% G+C from the melting data. Both of these values seriously disagree with the chemical data. Bak et al. (1969) showed that DNA of relatively low density was not *per se* excluded from the empirical relationships, since mycoplasma DNAs (about 1.685 g/ml) showed good correlation between the density in CsCl and the T_m value. Hence an unusual secondary structure for yeast mtDNA was inferred. This interpretation receives support from the ability of hydroxyapatite columns to separate yeast mtDNA from yeast nuclear DNA under conditions where secondary structure of nucleic acids determines binding to the columns (Bernardi et al., 1968, 1970;

Fukuhara, 1969). Bernardi et al. (1970, 1972) proposed that the unusual physical properties result from an imbalance in base distribution along yeast mtDNA molecules, such that clustered regions rich in adenine and thymine (A + T-rich regions) are present. Careful analysis of the shape of the T_m curve by making a differential plot (absorbance change per degree) revealed a major component melting at 71–73°C, ascribed to base sequences containing both (dA.dT) and (d(AT).d(AT)), followed by a series of components of decreasing importance up to 85°C. If the contributions of all of these components are weighted appropriately in determining the base composition from the absorbance-temperature profile, the value calculated is close to 17% (Bernardi et al., 1970). Further evidence for the proposed A + T-rich regions has come from optical rotary dispersion studies (Bernardi and Timasheff, 1970) where shifts in the spectra of mtDNA in the direction of peaks and troughs characteristic of double-stranded polymers of adenine and thymine both alternating and non-alternating in A and T were observed.

A more refined analysis of Sacch. cerevisiae mtDNA has been made by resolution on hydroxyapatite columns and Ag^+–Cs_2SO_4 centrifugation of spleen DNase fragments of mtDNA (Bernardi et al., 1972; Piperno et al., 1972). The percent G + C of the fragments thus generated ranged from 26 at one extreme to 10 at the other; the majority of the DNA fragments eluted broadly off the column in a heterogeneous array between the extremes. Details of these results, together with pyrimidine-tract analysis on yeast mtDNA (Ehrlich et al., 1972), support conclusions that the mitochondrial genome of yeast consists of relatively G + C-rich sequences of sizes of the order of 10^5–10^6 daltons. Carnevali and Leoni (1972) obtained independent evidence for the intramolecular heterogeneity of Sacch. cerevisiae mtDNA. In their experiments, sonicated mtDNA (0.5–1 × 10^6 daltons) was banded preparatively in CsCl gradients, and various fractions (lower, middle and upper) were rebanded in analytical CsCl gradients; these fractions exhibited discrete density modes of 1.696, 1.683 and 1.681 g/ml, respectively. Comparable data of Bernardi et al. (1972) are that various hydroxyapatite fractions yielded fragments of mtDNA banding at densities of 1.693 down to 1.683 g/ml. The role of A + T-rich sequences in the anomalous CsCl density of bulk yeast mtDNA is emphasized by the finding that even fragments of a low G + C content of 14% still show a density in CsCl of 1.683 g/ml (Bernardi et al., 1972).

Prunell and Bernardi (1974) have undertaken a further analysis of the scattering of A + T-rich regions between G + C-rich regions in *grande Sacch. cerevisiae* mtDNA. These authors used micrococcal nuclease, an enzyme preferring single-stranded DNA over duplex DNA as substrate. Conditions were chosen for digestion of the mtDNA such that the A + T-rich sequences would be preferentially attacked because these are more readily denatured than G +C rich sequences. The fragmented A + T-rich segments were separated from the less degraded G + C-rich segments using gel-filtration procedures. Analysis of the base composition of the variously sized fragments showed that the A + T regions are very homogeneous in base composition, with G + C less than 5%. The G + C regions are very heterogenous in base composition, ranging from G + C contents of 25% to 50%, at fragment sizes of about 1×10^5 daltons, the average G + C content being 32%; but one population of small fragments (4×10^4 daltons) had a G + C of 65%. These authors propose that the A + T-rich segments may be considered as spacer regions separating the G + C-rich segments which probably constitute the genes. Overall, the genes and spacers are of similar average lengths and appear in equal amounts in the mitochondrial genome (Prunell and Bernardi, 1974). A refinement to this scheme has been proposed by Bernardi *et al.* (1976), based on the number (70–100) of double-stranded fragments produced from mtDNA by the restriction enzymes Hpa II (cleavage sequence CCGG) and Hae III (GGCC), and the appearance of G + C-rich single-stranded oligonucleotides in the digestion mixtures. The proposal is that in between the genes and spacers are regions very rich in G + C pairs which contain clusters of sites for these restriction enzymes. A possible regulatory role for these site clusters was suggested, and the existence of about 70 site cluster-gene-spacer units proposed (Bernardi *et al.,* 1976).

An attempt has been made to visualize the positioning of A + T-rich regions in the mtDNA of *Sacch. carlsbergensis* by electron microscopic examination of partially denatured molecules under conditions where both single-stranded and double-stranded regions of DNA can be observed (Christiansen *et al.,* 1975). The spacers would be first to denature, and are seen as the single-stranded loops, whilst the genes remain double stranded. Analysis is complicated by the inability to obtain intact mtDNA molecules in high yield. Christiansen *et al.* (1975) examined 130 molecules with an average of 21% of the length in the denatured form. Thus only the very highly A + T-rich spacers would be expected to be denatured. Using a full length (28 μm) molecule as a master molecule, the other fragments (mean length 10 μm) were compared with this master, considering the location and the length of the denatured regions. The general conclusion possible is that there are at least 30 broadly defined easily denaturable sites scattered throughout the *Sacch. carlsbergensis* mitochondrial genome at intervals of about 0.5 to 3 μm (1–6 × 10^6 daltons). In individual molecules, however, denatured regions ranged in size from 4 × 10^4 to 2 × 10^5 daltons separated by duplex regions ranging in size from 2 × 10^5 to 10 × 10^6 daltons. Some uncertainties in these results are firstly that, in individual molecules, only a fraction of the spacers are visualized because of the stochastic nature of the denaturation process. Secondly, in construction of the composite map containing the 33 denaturable regions already referred to, lack of precision in length measurement may diminish resolution between different A + T-rich spacers separated by relatively short G + C-rich regions. Nevertheless, the order of magnitude of the size of spacers and genes (namely 10^6 daltons) agrees broadly with that found by Bernardi and his colleagues. Finally, comparing the denaturation maps of three full-length molecules, it was possible to conclude that the molecules arose by random scission of a circular genome.

5. *Informational Sequences*

Thus far, the discussion of the sequence organization of yeast mtDNA has been concerned with the number and gross distribution of nucleotides within the mitochondrial genome. By RNA–DNA hybridization studies using defined mitochondrial RNA molecular species, it is possible to study the presence of specific base sequences within the mtDNA. In this section, we consider mainly the ribosomal and tRNA genes in yeast mtDNA. Other information which may be encoded in mtDNA is considered later.

In common with mtDNA of other organisms (see Borst, 1972; Borst and Flavell, 1972, for reviews), each mitochondrial genome of *Sacch cerevisiae* and *Sacch. carlsbergensis* contains one cistron of both the small and large RNA species of the mitochondrial ribosome. This was established in yeast by hybridizing rRNA, extracted from carefully purified mitochondrial ribosomes, to *grande* mtDNA. The mitochondrial rRNA molecules are designated as 15S and 21S (Forrester et al., 1970; Grivell et al., 1971). The 15S rRNA is found to be complementary to 0.65–0.85% of mtDNA, the 21S rRNA is complementary to 1.40–1.55% of mtDNA, and together the rRNA molecules account for 2.1–2.5% mtDNA at saturation (Morimoto et al., 1971; Reijnders et al., 1972; Nagley et al., 1974b). These figures are not quite as large as expected for one cistron each. The sizes of the 15S and 21S rRNAs are 0.70 and 1.30 × 10^6 daltons, respectively (Reijnders et al., 1973). The rRNAs together account for 2 × 10^6 daltons, which is about 4% of the mtDNA. The reason for this discrepancy is not understood. Reijnders et al. (1972) showed that it was not due to incomplete saturation of mtDNA with rRNA. Nevertheless, it will be assumed for the rest of this review that one cistron for each rRNA is present in each mtDNA molecule. Reijnders et al. (1972) compared the hybridization properties of rRNA isolated from whole mitochondria with rRNAs extracted from the purified mitochondrial ribosome subunits. In contrast to subunit-extracted rRNA, the rRNA from fractionated mitochondrial RNA preparations hybridized to greater than 6% for each of the 15S and 21S fractions and, even at high input RNA concentrations in the hybridization reactions, plateaux were not reached. This suggests the presence of transcripts of mtDNA representing a relatively large proportion of mtDNA, but accounting for a small mass compared to that of the rRNA molecules. These extra transcripts may be in part mRNA molecules. A similar phenomenon was reported by Schafer and Kuntzel (1972) for RNA from mitochondria of *Neurospora crassa*, comparing hybridizations of rRNA from purified ribosomes with mtDNA, with hybridizations of rRNA separated from whole mitochondria RNA.

Two independent approaches developed for determining the relative location of the two rRNA genes in yeast have shown that they are separated by approximately 35% of the genome length. Sanders et al. (1975b) hybridized labelled rRNAs to restriction endonuclease cleavage products of mtDNA to establish the separation while Sriprakash et al. (1976a) used a collection of physically and genetically defined *petite* mutants to obtain similar separation and to map the rRNA cistrons relative to other mitochondrial markers (see Section V.D.3, p. 252). The wide separation of the two genes is the only case known so far in nature. In mtDNA from HeLa cells

(Robberson et al., 1972) and *Neurospora crassa* (Kuriyama and Luck, 1973), the pair of rRNA cistrons are adjacent to each other.

Hybridization to mtDNA of bulk 4S RNA (assumed to be tRNA) from *Sacch. carlsbergensis* mitochondria has been reported by Reijnders and Borst (1972) to result in saturation of 0.9% of mtDNA. This would provide enough information for about 20 genes for tRNA (if each tRNA is 25,000 daltons in size). Details of specific tRNA genes have been investigated by Rabinowitz and his colleagues who charged bulk tRNA preparations with particular radio-active amino acids, and examined the hybridization of radio-activity to mtRNA. The amino acids corresponding to the 16 aminoacyl-tRNAs able to hybridize mtDNA that to date have been detected in this way are listed by Rabinowitz et al. (1976). The list (see also Baldacci et al., 1975; Schneller et al., 1975) is probably not yet complete and it is likely that a complete set will eventually be found (Rabinowitz et al., 1976).

As already discussed, mitochondria of yeast and other organisms contain transcripts of mtDNA apart from rRNA and tRNA, and it is thought that some of this may be mRNA. Although several discrete size classes of RNA transcripts of mtDNA with polyadenylic acid chains attached have been found in HeLa cell mitochondria (Ojala and Attardi, 1974), polyadenylic acid-containing RNA was not found in yeast mitochondria (Groot et al., 1974). However, Padmanabam et al. (1975) report the existence of a yeast mitochondrial RNA fraction containing polyadenylic acid residues, which possibly codes for polypeptides of cytochrome oxidase.

6. *Sequence Differences Between mtDNAs of Different* Grande *Strains*

One of the important findings made by Bernardi et al. (1975) in their restriction-enzyme analysis of mtDNA from *grande* yeast was that different strains of *Sacch. cerevisiae* and *Sacch. carlsbergensis* gave different band patterns in gel electrophoresis. Sanders et al. (1976) have mapped the cleavage sites for the enzymes EcoRI and Hind II and III in the mtDNA of three strains, and showed that there are differences in the number and position of the cleavage sites in the different mtDNAs.

It is important to note that, when DNA–DNA hybridization is used as a sequence homology test between unfractionated mtDNA of different *grande* strains of *Sacch. cerevisiae* (K. S. Sriprakash, P. Nagley and A. W. Linnane, unpublished data) as well as comparing *Sacch. cerevisiae* and *Sacch. carlsbergensis* (Groot et al., 1975), significant differences are not apparent. This technique is not capable of resolving differences of less than 10% sequence divergence. It is a truism that different *grande* strains differ in their mtDNAs at the level of point mutations and perhaps small deletions and insertions. This is shown by the various allelic forms of antibiotic resistance determinants in mtDNA, and other markers, including the polarity determinant *omega* (see Section V.B.1, p. 218). The question is: how different are *grande* strains from one another? Bernardi's data on restriction-enzyme cleavage would suggest substantial differences in specific sequences (deletions, amplifications, perhaps also re-arrangements). Two other approaches studying the differences between *grande* genomes in restricted segments of the genome support the conclusion that the variations among *grandes* are more than point mutations. Differences between the mtDNAs of various *grande* strains were detected by Michaelis *et al.* (1972) studying complementary RNA synthesized *in vitro* using *Escherichia coli* RNA polymerase and a *grande* mtDNA as template. These authors studied the ability of complementary RNA from one strain to hybridize with its template mtDNA and to the mtDNA of other *grande* strains of *Sacch. cerevisiae*. It was found that the (*grande*) mtDNAs tested differed using this test, with some DNAs binding even more radio-actively labelled RNA than the original template DNA. Because it is highly unlikely that all sequences present in the template *grande* mtDNA were represented equally in the *in vitro* transcripts, these results suggest that some of the sequences represented in the transcripts are present in the various *grande* mtDNAs tested to different extents. This method, although it does point to the existence of detectable differences, does not readily indicate the nature of sequences of difference between *grande* strains. A similar but more definite approach to this question has been adopted in our laboratory (K. S. Sriprakash, P. Nagley and A. W. Linnane, unpublished data). We examined the ability of different *grande* mtDNAs to hybridize to labelled mtDNA from a *petite* retaining a defined segment of the mitochondrial genome. The conditions used for hybridization were such that the *grande* mtDNA was in vast excess over the labelled *petite* mtDNA. The ability of the *grande* mtDNA bound to a filter to hybridize quantitatively the labelled *petite* sequences out of solution was measured. Two *grandes* could hybridize nearly 100% of the sequences of a particular *petite,* but two other *grandes* could only hybridize 50–70% of the *petite* sequences, even when the amount of *grande* mtDNA present was three times that normally used. Taken as a whole, the hybridization results suggest that not only may various sequences in *grandes* be amplified to different extents in different *grande* genomes, but that some sequences may be missing altogether from some *grandes* (*cf.* Sanders *et al.*, 1976).

The quantitative importance of the differences in sequences is probably no more than 10% of total sequences in most *grandes*, since the DNA–DNA hybridization between unfractionated *grandes* would have picked up levels of sequence divergence greater than 10%. However, these DNA–DNA hybridizations would not, under normal circumstances, detect the amplifications (and perhaps rearrangements) suggested by the results of Bernardi *et al.* (1975) and Michaelis *et al.* (1972), because such DNA–DNA hybridizations are designed to measure an overall sequence homology between two genomes, that is, the total representation of the sequences present.

7. *Omicron DNA-2 μm Circles Associated with Mitochondria*

A population of covalently closed circular DNA molecules 2 μm in length, termed oDNA (*omicron*) by Clark-Walker and Miklos (1974a), as well as oligomeric forms, has been isolated from the yeast mitochondrial fraction (Sinclair *et al.*, 1967; Avers *et al.*, 1968; Guerineau *et al.*, 1971; Clark-Walker, 1972, 1973). These molecules appear to be membrane bound *in vivo* as judged by their apparent insensitivity to DNAse digestion *in situ* (Billheimer and Avers, 1969) and observed attachment to cytoplasmic membranes in electron micrographs (Guerineau *et al.*, 1971). The density of these circles, which can be isolated from whole-cell DNA preparations by separation in CsCl—EthBr gradients, is close to that of main-band nuclear DNA, namely 1.698–1.701 g/ml (Billheimer and Avers, 1969; Hollenberg *et al.*, 1970; Guerineau *et al.*, 1971; Stevens and Moustacchi, 1971; Clark-Walker, 1972). The best evidence that the circles are not of mitochondrial origin is that they occur in rho^0 cells lacking mtDNA and devoid of mitochondrial genetic information (Clark-Walker, 1972), and have also been detected in unchanged quantities in rho^- cells (Avers *et al.*, 1968). In one particular *petite* mutant, it was determined by Clark-Walker and Miklos (1974a) that there were 62 molecules of *omicron* DNA per cell. Moreover, these authors showed that the *omicron* DNA was not localized in the nucleus. Synthesis of *omicron* DNA is inhibited by cycloheximide (Zeman and Lusena, 1974a), exactly as is nuclear DNA synthesis, but in contrast to the behaviour of mtDNA (Grossman et al., 1969). Double-stranded replicating forms of *omicron* DNA have been observed (Petes and Williamson, 1975). It can be considered that, if all *omicron* DNA molecules are identical as suggested by renaturation kinetic analysis (Bak et al., 1972), this population of DNA molecules has restricted coding capacity. The biological role of *omicron* DNA is unknown, but the unusual behaviour of certain antibiotic resistance determinants known not to be encoded in mtDNA (Guerineau et al., 1974; Griffiths et al., 1975a) may lead to their localization on *omicron* DNA, perhaps in some sort of episomal role with nuclear or mtDNA.

B. OTHER YEAST SPECIES (*petite*-NEGATIVE YEASTS)

There are many species in the general group of organisms called yeasts, and for the purposes of this review *Sacch. cerevisiae* and *Sacch. carlsbergensis* are considered in one group, and other yeast species in another group. This should not be taken to imply that any particular similarities necessarily exist within this second group, except that they fall under the broad heading of *petite*-negative yeasts, that is to say they do not give rise to the extranuclearly inherited *petite* mutants characteristic of *Sacch. cerevisiae* and related species. A detailed survey of the properties of the various yeasts can be found in Rose and Harrison (1969). The genetic (Mortimer and Hawthorne, 1966; Sherman and Lawrence, 1974) and biochemical aspects (Hartwell, 1970), as well as an analysis of the nucleic acid homologies (Bicknell and Douglas, 1970) of different yeast species have been considered. The DNA–DNA hybridization studies of Bicknell and Douglas (1970) showed very little sequence homology between any of the nuclear DNAs of *Sacch. cerevisiae, Kluyveromyces lactis* and *K. fragilis,* confirming that there is considerable degree of evolutionary divergence between these species.

1. *Properties of mtDNA*

The buoyant densities and genome sizes of mtDNA in the various *petite*-negative yeasts analysed together with *Sacch. cerevisiae* for comparison are shown in Table 1. Consideration of these data reveals that the *petite*-negative yeasts differ amongst themselves, as well as differing from *Sacch. cerevisiae,* both in the density and size of mtDNA.

Each organism should be discussed separately, as few generalizations are possible.

Kluyveromyces lactis is probably the most thoroughly studied species, although the density difference between nuclear and mtDNA in this species is not as great as in other yeasts (see Table 1), which does present some preparative problems. Smith *et al.* (1968) used caesium sulphate gradients containing Hg^{2+} to magnify the density difference between mtDNA and nuclear DNA. From the density of *K. lactis* mtDNA (1.692 g/ml) and its T_m value of 82.7°C (Smith *et al.*, 1968) a

TABLE 1. Characteristics of mtDNA from *petite*-negative yeasts

Species	Density of nuclear DNA in CsCl (g/ml)	Density of mtDNA in CsCl (g/ml)	Genome[b] size of mtDNA (daltons ×10⁻⁶)	References
Saccharomyces cerevisiae	1.698	1.683	45–55	See text
Kluyveromyces lactis	1.700	1.692	24	Smith *et al.* (1968) O'Connor *et al.* (1975) Sanders *et al.* (1974)
Kluyveromyces fragilis	1.699	1.683	N	Luha *et al.* (1971)
Candida parapsilosis	1.700	1.698	23	O'Connor *et al.* (1975)
Hansenula wingei	1.701	1.686	17	O'Connor *et al.* (1975)
Schizosaccharomyces pombe	1.696	1.695[a]	13	O'Connor *et al.* (1975)
Schizosaccharomyces pombe	1.695	1.689[a]	17	Bostock (1969) Tabak and Weijers (1976)

[a] For a discussion of the discrepancy in the density of mtDNA, see text.
[b] N indicates no data yet reported.

base composition close to 33% G + C is calculated. However, Sanders *et al.* (1974) suggest from their T_m data that the mtDNA is of 24% G + C. Direct base-composition analysis, or perhaps more detailed processing of the temperature–absorbance curve (cf. Bernardi *et al.*, 1970), may clarify this question. Closed circular DNA of density 1.692 g/ml has been isolated from mitochondria of *K. lactis* using CsCl–EthBr gradient centrifugation, and by electron microscopy the molecular length has been estimated as 11.4 μm (Sanders *et al.*, 1974; O'Connor *et al.*, 1975), corresponding to about 23 × 10⁶ daltons in mass. Renaturation kinetic analysis of *K. lactis* mtDNA (Sanders *et al.*, 1974) shows that this mtDNA shows a similar type of temperature dependence of renaturation rate to that shown by *Sacch. cerevisiae* (Christiansen *et al.*, 1974), and it was inferred that *K. lactis* mtDNA fragments adopt a secondary structure at temperatures below 10°C below the T_m value which slows down the renaturation rate. The kinetic complexity at a temperature of $T_m - 10°C$ was consistent with a genome size of about 20×10^6 daltons for *K. lactis* mtDNA (Sanders *et al.*, 1974), which agrees with the physical size measurements.

O'Connor *et al.* (1975) isolated closed circular DNA from subcellular fractions enriched in mitochondria of four yeast species. Results for *K. lactis* have already been discussed. In *Candida parapsilosis*, the circular molecules (presumably mtDNA) were of lengths corresponding to 23×10^6 daltons. The mtDNA in this species is not resolved from nuclear DNA in CsCl gradients as the density is so close. O'Connor *et al.* (1975) report the size of circular mtDNA molecules of *Hansenula wingei* to correspond to 17×10^6 daltons, and that of *Schizosaccharomyces pombe* at 13×10^6 daltons. However an uncertainty exists as to the density of *Schizosacch. pombe* mtDNA. The density of the *Schizosacch. pombe* circular DNA is reported by O'Connor *et al.* (1975) to be 1.695 g/ml, very close to that of nuclear DNA. However, Bostock (1969) describes a light satellite DNA in cells of *Schizosacch. pombe* of density 1.689 g/ml, which was found in the mitochondrial fraction of the cells. In another study, Bandlow and Kaudewitz (1974) examined the CsCl-gradient profile of *Schizosacch. pombe* whole-cell DNA labelled with radioactive adenine; a light satellite DNA was also noted. This DNA (1.689 g/ml) presumably represents the mtDNA of the particular strains of *Schizosacch. pombe* used by Bostock and by Kaudewitz. The estimate of 13×10^6 daltons made by O'Connor *et al.* (1975) for the *Schizosacch. pombe* mtDNA size should be considered as only tentative until the discrepancy in the density of *Schizosacch. pombe* mtDNA is clarified. Tabak and Weijers (1976) suggest, on the basis of restriction-enzyme digestion, that mtDNA (1.689 g/ml) of *Schizosacch. pombe* is 17×10^6 daltons in size. An interesting case of a *petite*-positive yeast (*Torulopsis glabrata*) containing mtDNA, 13×10^6 daltons in size, has been reported by O'Connor *et al.* (1976).

2. *Sequence Relationships with* Saccharomyces cerevisiae

Consideration of Table 1 (p. 171) shows the *petite*-negative yeasts in general to have smaller mitochondrial genome sizes than in *Sacch.*

cerevisiae. It may be asked if this decrease in size is also accompanied by divergence in the sequences present. This question was approached by Groot *et al.* (1975) who carried out DNA–DNA hybridization experiments between mtDNA of *Sacch. cerevisiae*, *Sacch. carlsbergensis*, *K. lactis* and *C. utilis*. The results showed that *Sacch. cerevisiae* and *Sacch. carlsbergensis* were very similar by the hybridization tests, while the hybridizations involving the *petite*-negative yeasts showed both *K. lactis* and *C. utilis* to have only about 10–20% sequence homology with *Sacch. cerevisiae*. Hybridization between *K. lactis* and *C. utilis* mtDNAs showed a similarly large sequence divergence between those two *petite*-negative yeasts, only 10–20% homology being measured. The melting temperature of the heteroduplexes (i.e. those hybrid molecules formed between mtDNAs of different origin) was considerably lower than that of homoduplexes, which indicates considerable mismatching of bases even in the regions able to form hybrids under the annealing conditions.

Bicknell and Douglas (1970) showed that, although nuclear DNAs of *Sacch. cerevisiae* and *petite*-negative yeasts had very little sequence homology overall, the nuclear rRNA cistrons had considerably less sequence divergence. For example, under very stringent annealing conditions, *K. lactis* nuclear DNA hybridized to 25S cytosolic rRNA of *Sacch. cerevisiae* to about 90% of the extent of the hybridization using *Sacch. cerevisiae* nuclear DNA. The same situation applies to mtDNA of these two types of yeasts. Groot *et al.* (1975) showed that the 15S and 21S mitochondrial rRNA of *Sacch. carlsbergensis* hybridized extensively to *K. lactis* mtDNA. The T_m value of the heterohybrid was 6°C less than that of the isohybrid, suggesting an extent of base mismatching of about 9% (Groot *et al.*, 1975). A lower level of hybridization of *Sacch. carlsbergensis* mitochondrial rRNA was found with *C. utilis* mtDNA compared with *Sacch. carlsbergensis* mtDNA. An extensive sequence divergence of rRNA cistrons is emphasized by the finding that the heterohybrids constructed by Groot *et al.* (1975) were highly unstable to heating above 25°C.

III. Synthesis of mtDNA

The results of several approaches with a range of organisms showed beyond doubt a high degree of independence of mtDNA synthesis from the processes of DNA replication within the nucleus. However, it is now clear that most of the factors responsible for mtDNA replication and its regulation are not synthesized by the mitochondrion, and are probably coded by nuclear DNA. This has led to a re-appraisal of earlier ideas concerning the degree of autonomy of mtDNA.

A. mtDNA SYNTHESIS *in vivo*

1. *Replication*

Attempts to show semiconservative replication of yeast mtDNA by standard density-lebelling techniques have not been successful. The difficulty is that the rate of density shift in the Meselson and Stahl (1958) type of experiment is far slower than expected (Corneo *et al.*, 1966). Recently Williamson and Fennell (1974) provided an interpretation for the unusual behaviour of the density shifts in mtDNA, proposing a process of dispersive DNA replication. In this scheme, each newly synthesized length of mtDNA is made in the authentic semiconservative mode, but extensive molecular recombination occurs between newly replicated duplexes of mtDNA and the homologous regions of other mtDNA molecules, which may or may not have already been replicated. This model explains why, after density transfer of cells from [^{15}N] into [^{14}N] medium, the modal density of mtDNA as analysed in a CsCl gradient decreases slowly and linearly with respect to time. The conventional pattern is for a gradual transfer from a discrete, fully heavy duplex band to a discrete hybrid (light-heavy) band after one cell generation time; the nuclear DNA exhibited this behaviour. The data of Williamson and Fennell (1974) appeared to rule out a significant level of incorporation of [^{15}N] into mtDNA after the transfer, and also suggested turnover of mtDNA to be an unlikely possibility in these experiments.

2. *Repair*

In addition to replicative mtDNA synthesis, it is known that repair systems exist in yeast mitochondria, but the molecular processes involved are poorly understood. The existence of repair activity in *Sacch. cerevisiae* mitochondria is inferred from genetic studies on *petite* induction, the ability of potential *petites* to recover from ultraviolet irradiation and from treatments with mutagens such as EthBr.

The phenomena of dark repair of ultraviolet-induced damage to mtDNA (*petite* induction) are complex. Whereas exponentially growing cells will show some recovery of respiratory competence if held in the dark after ultraviolet irradiation (Heude and Moustacchi, 1973), some strains grown to the stationary phase show an increased *petite* frequency under these conditions, termed "negative liquid holding" (Moustacchi and Enteric, 1970). Analyses of *petite* induction and repair in mutants affected in the sensitivity of the nuclear genetic system to ultraviolet-induced damage (Moustacchi, 1969, 1972a; Moustacchi and Enteric, 1970) suggest the separate nature of many steps in the mitochondrial and nuclear repair systems. Some mutants of *Sacch. cerevisiae* have been reported to be specifically sensitive to ultraviolet irradiation as regards *petite* induction, but not lethality (i.e. nuclear genetic damage; Chanet *et al.*, 1973). Whilst photoreactivation of ultraviolet-induced damage to mtDNA occurs (Pittman *et al.*, 1959; Maroudas and Wilkie, 1968), thymine dimers are not removed from mtDNA by dark-excision processes (Waters and Moustacchi, 1974; Prakash, 1975). The phenomena of EthBr-induced mutagenesis of mtDNA and its repair will be discussed separately in Section IV.C (p. 205).

The molecular events in ultraviolet-induced mutagenesis and repair of yeast DNA would be much more amenable to investigation if it were possible to achieve labelling of DNA to high specific activity with a precursor specific for DNA. The DNA labels which are readily incorporated, namely [^{3}H]- or [^{14}C]-adenine or uracil, or [^{32}P]-orthophosphate, enter RNA equally readily. This renders precise estimates of incorporation into, or loss of label from, DNA very difficult. Procedures partially to overcome these difficulties with the above precursors are described by Hatzfeld (1973), and approaches to find specific mutants which take up exogenous thymidine or its derivatives from the medium are considered by Fath *et al.* (1974) and Wickner (1974).

B. mtDNA SYNTHESIS *in vitro*

1. *DNA Synthesis in Isolated Mitochondria*

Mitochondria from all sources so far tested *in vitro* have been able to incorporate exogenous deoxyribonucleoside triphosphates into DNA. This type of experiment, in which incoporation of [^{3}H]-TTP or [^{3}H]-

dATP was measured, has been carried out in several laboratories using mitochondria from *Sacch. cerevisiae* (Wintersberger, 1966, 1968; Iwashima and Rabinowitz, 1969; Zeman and Lusena, 1974b; Mattick and Hall, 1977). The reactions are characterized by requirements for the presence of all four deoxynucleoside triphosphates, magnesium ions, a source of ATP (usually through a continuously regenerating enzyme system, such as pyruvate kinase–phosphoenol pyruvate or creatine kinase–phosphocreatine). The isolated mitochondria have been shown to incorporate TTP or TDP but not TMP into mtDNA in an *in vitro* DNA synthetic system (Zeman and Lusena, 1974b). In general, there is no response of the incorporation either to externally added DNAse or DNA, indicating high template specificity within the organelle. The product is shown to be exclusively mtDNA as judged by banding characteristics of *in vitro* labelled DNA in CsCl density gradients. Moreover, mitochondria from *petites* lacking mtDNA (rho^0) were unable to carry out significant incorporation (Mattick and Hall, 1977), but *petites* retaining mtDNA were as active as the wild type, provided the *petite* cells were grown in continuous culture to minimize the effects of glucose repression on mitochondrial development.

The rates and extents of mtDNA synthesis in the studies of Wintersberger (1966, 1968) and of Iwashima and Rabinowitz (1969) were rather low. Both of these groups carried out the incubations at 37°C and over a period of 15 minutes and observed up to 100 pmoles TTP incorporated per mg mitochondrial protein; that is, a synthesis of up to 2% of the pre-existing mtDNA occurred.

The kinetics of DNA synthesis were found by Wintersberger (1968) to consist of a fairly linear increase in incorporation up to 10 minutes. After this period, a decay in the accumulated acid-insoluble radioactivity was observed, suggesting the presence of an active nuclease within mitochondria which degrades some of the product DNA to acid-soluble components. This factor makes the system of Wintersberger not really suitable for a deeper analysis of the mtDNA synthetic process in mitochondria. A preferential digestion of A + T-rich sequences has been suggested (Zeman and Lusena, 1975).

A more recently described system (Nagley *et al.*, 1975a; Mattick and Hall, 1977) does, however, show good evidence that replicative synthesis of mtDNA takes place within isolated mitochondria. The kinetics of synthesis are such that the reaction proceeds steadily for up to 60 min at 30°C, albeit at a somewhat decreasing rate. Nevertheless, routine preparations show at least 200 pmoles TTP incorporated/mg protein during the first 15 minutes, and a total incorporation of about 600 pmoles TTP per mg protein overall. This extent of new synthesis (at least 10% of pre-existing mtDNA) is so great that it is not likely to reflect a repair type of polymerization. Mattick and Hall (1977) obtained direct evidence for a process of replicative synthesis within the mitochondria by replacing the TTP with 5-bromodeoxyuridine triphosphate and using [^{3}H]-dATP as labelled precursor. The product DNA was analysed for its size in sucrose velocity gradients, and for density in CsCl gradients. The product molecules contained a population of molecules of a density expected for duplexes consisting of a fully substituted strand (BU), with an unsubstituted strand (T), at a size of greater than 6×10^6 daltons (about 10% genome length). The appearance of such molecules is clear evidence for a replicative process, rather than a repair process. This mtDNA-replicating system has the potential for analysis of the topology of mtDNA replication, as replicating intermediates can be collected by collecting partially density-labelled molecules. For example, it may be asked if the *in vitro* system shows any properties of dispersive replication. This *in vitro* mtDNA-synthesizing system has already been used to study the role of the mitochondrial membrane in mtDNA synthesis (Hall *et al.*, 1975) and the effect of EthBr and berenil in inhibiting yeast mtDNA replication (Nagley *et al.*, 1975a).

2. *Mitochondrial DNA Polymerases*

Enzymes capable of catalysing DNA synthesis from deoxyribonucleoside triphosphates using a suitable DNA template (usually salmon sperm DNA) have been isolated from mitochondria of many organisms. In the case of *Sacch. cerevisiae*, two preparations have been reported. Wintersberger and Wintersberger (1970b) extracted and partially purified a DNA polymerase from mitochondria which were subjected to vigorous blending in a buffer containing 0.5 M ammonium chloride. Further purification was carried out on a DEAE-cellulose column. The molecular weight of the enzyme is about 150,000 (sedimentation coefficient, 7.5S). This mitochondrial enzyme was shown to differ by several criteria from two other yeast enzymes, presumably of nuclear origin (Wintersberger and Wintersberger, 1970a). The criteria used included the conditions for elution of the enzymes off DEAE-cellulose columns, the optimum Mg^{2+} concentration, the degree of stimulation by salts such as sodium or ammonium chloride, and the preference of the enzymes for various types of DNA template.

Iwashima and Rabinowitz (1969) extracted a DNA polymerase from yeast mitochondria by stirring in 1 M KCl. The enzyme was purified by ammonium sulphate fractionation, DEAE-cellulose chromatography, and finally a hydroxyapatite column was used. By most criteria, the properties of this purified mitochondrial enzyme were identical to those of a DNA polymerase isolated from the cell supernatant fraction (presumably of nuclear origin). The only real difference observed was in the sedimentation properties of the two enzymes in glycerol density gradients. The supernatant enzyme contained a major component with a slower minor component. The mitochondrial enzyme cosedimented with the minor supernatant component. It is conceivable that the technique of 1 M KCl extraction of mitochondria with little or no physical stress as carried out by Iwashima and Rabinowitz (1969) failed to release the mitochondrial enzyme, and what was purified was mainly a contaminating polymerase of non-mitochondrial origin.

It is not known whether the Wintersberger enzyme is indeed the replicative DNA polymerase in yeast mitochondria, or whether it carries out repair synthesis *in vivo*, or has some other role. The difficulties involved in experimentally approaching this question are considerable; the reader should consult Gefter (1975) for a recent review on the bacterial literature.

C. mtDNA REPLICATION AND THE MITOCHONDRIAL MEMBRANE

Bacterial studies have suggested that some events in the replication of the bacterial chromosome, and probably also of episomal DNA molecules, take place at sites on the bacterial membrane, although the evidence is by no means conclusive (Gefter, 1975). Structures positively identifiable as membrane fibres within mitochondria have been observed in yeast cells in certain growth states (Yotsuyanagi, 1966; Swift *et al.*, 1968). In electron microscopic studies of isolated yeast mitochondria lysed by osmotic shock, or other means, directly onto a protein monolayer, the ends of many DNA fibres can be seen intimately associated with membrane fragments (Avers *et al.*, 1968).

However, numerous intact DNA circular molecules liberated from mitochondria under these conditions are completely membrane-free (see e.g. Hollenberg et al., 1970). It is thus not certain whether the membrane fragments are fortuitously on top of otherwise free DNA molecules, or whether a certain proportion of mtDNA is bound to membranes *in vivo*.

There is both genetic and biochemical evidence in *Sacch. cerevisiae* for membrane involvement in replication of mtDNA. Alterations in the lipid composition of mitochondrial membranes have been shown to have significant effects on mtDNA. The well-known *petite*-inducing effect of growing yeast at temperatures above 36°C can be prevented by addition of exogenous ergosterol and unsaturated fatty acids to the culture (Sherman, 1959; Parks and Starr, 1963). The role of unsaturated fatty acid and ergosterol in replication of yeast mtDNA is further emphasized by the significant decrease (50%) in the cellular level of mtDNA in anaerobically-grown cells depleted of these lipid constituents (Nagley and Linnane, 1972b). The proteins of mitochondrial membranes have also been implicated in maintenance of mtDNA in yeast by the report of Juliani et al. (1973) that guanidine hydrochloride, a protein denaturant, quantitatively induces *petite* mutants in *Sacch. cerevisiae*.

The role of membranes in mtDNA replication in *Sacch. cerevisiae* has been investigated in a series of experiments with strain KD115 (a fatty acid desaturase mutant) which was grown in continuous culture in the presence of different concentrations of unsaturated fatty acid (Marzuki et al., 1975; see also Linnane et al., 1973b). The level of unsaturated fatty acid in the cells and mitochondria can be reproducibly controlled in this manner. It was found by Marzuki et al. (1974) that, when the unsaturated fatty-acid concentration fell below 35% of the total fatty acids, significant quantities of *petite* mutants accumulated in the cultures. The extreme case was that, at 12% unsaturated fatty acid, the steady-state level of *petite* mutants was 80% of the total cells in culture. The majority of these induced *petites* lacked mtDNA, implying that mtDNA replication was affected by the unusual membrane composition.

A positive demonstration of the role of membrane lipid in mtDNA replication was made by Hall et al. (1975) who examined the kinetics of replication of mtDNA by mitochondria isolated from cells of varying unsaturated fatty-acid content, grown in continuous culture as described by Marzuki et al. (1975). For each preparation of mitochondria, in whose membranes a determined level of unsaturated fatty-acyl residue was present, data were obtained for the initial rate of mtDNA synthesis at different temperatures of incubation. Arrhenius plots of the data revealed discrete discontinuities at a temperature for each preparation which was dependent upon the unsaturated fatty-acyl content of the mitochondrial membranes. The interpretation of these results is that a rate-limiting reaction in mtDNA replication takes place in a membrane environment such that the activation energy of this step is dependent upon the physical state of the lipid in the membrane.

Another approach which may be useful in further examining this functional relationship between mtDNA replication and the mitochondrial membrane is the use of specific membrane-DNA complexes (so called M-bands). Such complexes are prepared with sodium lauryl sarcosinate and Mg^{2+} and have been isolated from rat-liver mitochondria and shown to be able to carry out mtDNA synthesis *in vitro* (see Shearman and Kalf, 1975).

D. CELLULAR ORIGIN OF COMPONENTS INVOLVED IN mtDNA SYNTHESIS

Since the mitochondrion contains DNA and the ability to transcribe it and to translate the RNA (see Linnane et al., 1972b), it might be expected that some of the components involved in the replication of mtDNA would be specified and synthesized by the mitochondrion. Contrary to this expectation there is considerable evidence for the components involved in mtDNA synthesis being made on the cytosolic ribosomes and also being coded by nuclear DNA.

Saccharomyces cerevisiae treated with chloramphenicol (an inhibitor of mitochondrial protein synthesis) shows no inhibition of labelling of mtDNA (Grossman et al., 1969). There is no diminution in the cellular level of mtDNA of cells grown aerobically in erythromycin (Williamson, 1970) or anaerobically in the presence of lipid supplements with chloramphenicol for up to six generations (Linnane et al., 1972a). High concentrations of chloramphenicol were, however, reported to inhibit mtDNA synthesis on aeration of lipid-depleted anaerobically-grown cells (Linnane et al., 1972a) but it remains to be seen whether chloramphenicol is exerting a direct effect on the activity of the mitochondrial ribosomes, as suggested, or indirectly on some other membrane-associated function.

Petite mutants of yeast do not have a functional mitochondrial protein-synthesizing system (Kuzela et al., 1969; Schatz and Saltzgaber, 1969; Kellerman et al., 1971) so that the presence of high levels of mtDNA in petite mutants provides direct evidence for the mitochondrial protein-synthesizing system not contributing to replication of DNA within mitochondria. Wintersberger and Wintersberger (1970b) showed that their petite mutant contained exactly the same mitochondrial DNA polymerase as the grande parent, although the petite lacked mitochondrial ribosomes (Wintersberger, 1967; Wintersberger and Veihhauser, 1968). The nuclear coding of the mitochondrial DNA polymerase is suggested by the ability of petite cells to replicate mtDNA with a G + C content of only about 4% (Mehrotra and Mahler, 1968; Bernardi et al., 1968; Hollenberg et al., 1972a). Such a DNA is unlikely to code for any protein with biological activity, although such coding ability has never been formally excluded. Notwithstanding the apparent lack of contribution of mitochondrial protein synthesis to synthesis of mtDNA, some recent results have suggested that the mitochondrion does have some role in the grande cells in maintenance of the rho^+ state. Cells of some, but not all, strains of Sacch. cerevisiae, when grown through many generations in the presence of chloramphenicol or erythromycin, have been shown to be converted to petite mutants (Williamson et al., 1971a; Carnevali et al., 1971). On the other hand, some temperature-sensitive strains of Sacch. cerevisiae are exquisitely sensitive to these antibacterial drugs, and are rapidly and quantitatively converted to petites by chloramphenicol even at the permissive temperature (Weislogel and Butow, 1970, 1971). In one particular grande strain (not temperature sensitive) studied by Williamson et al. (1971a), no effect of erythromycin was apparent until after eight generations, at which time an abrupt onset of petite induction occurred. A simultaneous loss of mtDNA from the cells took place, which implied some active destruction process of mtDNA. Williamson et al. (1971a) propose that a particular product of mitochondrial protein synthesis (perhaps a membrane component) is required for maintenance of rho^+ mtDNA and, during growth in the presence of erythromycin, this product is not made, and the molecules are diluted out. This explanation is strictly only applicable to the particular strain studied, as the general pattern of chloramphenicol and erythromycin action is a more gradual accumulation of *petite* mutants in growing cultures.

As well as a possible mitochondrial role for the maintenance of the rho^+ state, there is a considerable nuclear contribution to the maintenance as demonstrated by genetic studies. Some of the nuclear mutants of the p series (Sherman, 1963; Sherman and Slonimski, 1964), now termed *pet*, have altered functions involved in maintenance of mtDNA since some obligatorily lose a functional *rho* factor (i.e. become extranuclear *petites* as well). Moreover, the mutant locus in the low temperature-sensitive strain of Weislogel and Butow (1970; *petites* induced at 180°C) is in the nuclear DNA.

Grossman et al. (1969) studied *Sacch. cerevisiae* treated with cycloheximide, or in which cytosolic protein synthesis was arrested by amino-acid starvation or incubation of a temperatue-sensitive (ribosome) mutant at the restrictive temperature. Whereas nuclear DNA replication was immediately strongly inhibited, mtDNA replication continued for another two to three hours leading to an approximate doubling in the cellular content of mtDNA. This result demonstrated the independence of mtDNA synthesis from nuclear DNA synthesis in yeast and, at the same time, showed that products of cytosolic ribosomal synthesis enabling mtDNA replication to proceed need not continuously be made for mtDNA replication. This observation has been extended by Nagley and Linnane (1972b) to show that lipid-depleted anaerobically-grown cells, aerated in the presence of cycloheximide, undergo a substantial burst of mtDNA synthesis comparable to that occurring in the absence of this drug.

Subsequent to these analyses of the effects of cycloheximide on mtDNA synthesis in yeast, it has been shown that cycloheximide blocks only the initiation of a new round of nuclear DNA replication, but cells already in S phase at the time of addition of cycloheximide complete replication of the nuclear DNA (Hereford and Hartwell, 1973; Williamson, 1973). A re-interpretation of the mtDNA studies is that initiation of mtDNA replication is a separate process from that of nuclear DNA, and that the components involved in initiation of mtDNA replication are probably accumulated within mitochondria to some extent. In Section III.E (p. 186), evidence is described for the cellular level of mtDNA (i.e. the rate of initiation of mtDNA synthesis) being controlled by nuclear genes, with no apparent contribution from mtDNA itself.

One approach which has been used by several authors (but so far with little success), in attempts to find components involved in synthesis of mtDNA, has been to investigate mtDNA replication in mutants of yeast known to be temperature-sensitive for certain processes of nuclear DNA replication (Hartwell, 1971). The two mutant types that have been most studied are strain 314D5, carrying the *cdc4-1* allele (temperature-sensitive for initiation of nuclear DNA synthesis) and strain 198D1 carrying the *cdc8-1* allele (temperature-sensitive for propogation of nuclear DNA replication).

Cryer *et al.* (1973) and Cottrell *et al.* (1973) studied synthesis of mtDNA and nuclear DNA in strain 314D5 (*cdc4-1*) after transfer of cells from 23°C to 36°C. Whereas nuclear DNA synthesis was markedly, but not completely, inhibited after the temperature shift (as expected for an initiation mutant), mtDNA synthesis continued at a normal rate. A similar high level of labelling of mtDNA relative to nuclear DNA occurred in strain 198D1 (*cdc8-1*) after a shift from 23°C to 36°C (Cryer *et al.*, 1973). In our own laboratory, mitochondria of a strain carrying the *cdc8-1* allele, and one with the *cdc21-1* allele (another DNA chain-propagation mutant), have been examined for temperature sensitivity of mtDNA replication *in vitro*. Neither batch of mitochondria showed diminished synthesis of mtDNA at 36°C compared with lower temperatures of incubation (R. M. Hall, unpublished observations). Taken together with the *in vivo* studies, these results suggest that the components altered by mutations at the *cdc4*, *cdc8* and *cdc21* loci are not involved directly in replication of mtDNA.

The results obtained from other studies appear to contradict the above conclusion. Wintersberger *et al.* (1974) interpreted their study of mtDNA synthesis in strain 198D1 (*cdc8-1*) to suggest that mtDNA synthesis is slowed to about one-third the normal rate after shifting cells from 25°C to 36°C. These authors did not observe the strong preferential labelling of mtDNA reported by Cryer *et al.* (1973). However, solubilized mitochondrial DNA polymerase of strain 198D1 did not show the temperature sensitivity (Wintersberger *et al.*, 1974). Further evidence concerning the possible influence of *cdc8* mutations on mtDNA synthesis comes from a study of DNA synthesis in Brij58-treated cells of *Sacch. cerevisiae* (Banks, 1973). Banks showed that *Sacch. cerevisiae* strain A364A (the parent of the strains carrying the mutated *cdc4* and *cdc8* loci), after treatment with Brij58 to render them permeable to exogenous deoxynucleoside triphosphate (Hereford and

Hartwell, 1971), could incorporate [^{3}H]-TTP only into mtDNA, but not into nuclear DNA, as judged by CsCl density-gradient analysis of the labelled product DNA. Hereford and Hartwell (1971) had previously reported that uptake of labelled TTP in Brij58-treated cells of strain 198D1 (*cdc8-1*) was temperature sensitive. Although Hereford and Hartwell (1971) interpreted this finding to mean that nuclear DNA synthesis is blocked in the Brij58-treated cells at 36°C, the data of Banks (1973) requires a re-evaluation of this conclusion. Therefore, the data of Banks (1973) and of Hereford and Hartwell (1971) taken together suggest that mtDNA synthesis is temperature sensitive in the cells carrying the *cdc8-1* mutation.

More definitive experiments by Newlon and Fangman (1975) showed that cells carrying mutant alleles of the *cdc8* and *cdc21* loci cease to synthesize mtDNA *in vivo* at the non-permissive temperature. The contradictory results for the *cdc21* mutants have been clarified by the demonstration of Game (1976) that the *cdc21* gene determines the ability of the cells to make thymidylate. The function of the *cdc8* gene is not yet known.

The mitochondrial repair systems in *Sacch. cerevisiae* do appear to involve components specified by mtDNA and made on mitochondrial ribosomes, as well as those specified by nuclear DNA and synthesized on cytosolic ribosomes. Chanet *et al.* (1973) studied the properties of two mutants which are more sensitive than the parent to ultraviolet induction of *petite* cells. One mutant (*uvs rho*5) is described as showing Mendelian segregation of the *uvs* phenotype (i.e. a nuclear gene), whilst the second mutant (*uvs rho*72) shows behaviour corresponding to a mitochondrial mutation (see Chanet *et al.*, 1973). Studies on the repair of ultraviolet-induced *petite* mutations in a wild-type strain, as influenced by inhibition of mitochondrial or cytosolic protein synthesis systems, showed that both systems contribute components to the dark recovery in exponentially growing cells (Heude and Chanet, 1975; Heude *et al.*, 1975).

E. LEVEL OF mtDNA AND NUMBER OF mtDNA MOLECULES IN CELLS OF DIFFERENT STRAINS

It is possible to calculate the number of mtDNA molecules in one yeast cell if the following parameters are known: the total cellular content of DNA, the percentage of total DNA which is made up by mtDNA, and the size of each individual mtDNA molecule. Assuming the total DNA content in a haploid to be 2.2×10^{-14} g (Ogur et al., 1952; Ciferri et al., 1969; Schweizer and Halvorson, 1969), and mtDNA representing 18% of the total cell DNA, a value commonly obtained (Williamson et al., 1971b; Nagley and Linnane, 1972a), it can be calculated that each haploid cell would contain approximately 50 molecules of mtDNA each of about 50×10^6 daltons. It is evident that, while different strains generally contain different levels of mtDNA, variations in mtDNA levels can be attributed in some cases to the physiological state of the cells. Systematic analysis of these differences may provide information concerning regulation of the cellular level of mtDNA.

The first question which may be asked is whether diploid *Sacch. cerevisiae* contain the same number of mtDNA molecules as haploids, or whether there are twice as many molecules of mtDNA in diploids. The cellular volume, dry weight, protein content, RNA content and total DNA content are all strictly proportional to the ploidy of the cell (Ogur et al., 1952; Ciferri et al., 1969; Schweizer and Halvorson, 1969). Inspection of the data of Williamson (1970) shows that diploids contain a similar percentage of mtDNA relative to total DNA as do haploids, namely 13–28% in different strains under the different growth conditions tested. This suggests that diploid cells contain twice as many mtDNA molecules as haploid cells. This question has been analysed in more detail by Grimes et al. (1974) who examined a pair of isogenic haploid and diploid strains of *Sacch. cerevisiae*. Cells were studied during the exponential phase of growth in a lactate-containing medium (i.e. in the absence of glucose repression). The cellular level of mtDNA (expressed as percentage of total DNA) was 13.5% for the haploid and 12.6% for the diploid strain. The double number of mtDNA molecules in the diploids compared with the haploids correlated with a doubling of the number of mitochondria (see p. 187). Mitochondria accounted for about 13% of cellular mass and cellular volume both in haploid and diploid cells (Grimes et al., 1974).

Hall et al. (1976a) carried out a study of the cellular levels of mtDNA in a series of strains of different ploidy all grown under standard conditions, namely to the end of the fermentative phase in a medium containing glucose (2%) as carbon source. A series of haploids, diploids, and one tetraploid all showed mtDNA to fall in the range 16–24% of total cellular DNA, thus confirming the results previously discussed, namely that the number of mtDNA molecules per cell is a ploidy-dependent character. Hall et al. (1976a) considered the differences in cellular levels of mtDNA in different strains in more detail. Two haploids, L411 and L2200, were shown to contain reproducibly mtDNA levels of 17% and 24% of total DNA, respectively. A genetic analysis of the determination of these particular cellular levels was carried out by constructing new strains, both diploid and haploid, in which the different mitochondrial genomes were placed in isogenic nuclear backgrounds, or by the converse, where a given mitochondrial genome was placed in different nuclear backgrounds. The results showed that the difference between the mtDNA levels of strains L411 and L2200 is controlled by at least two nuclear genes, and apparently not by any determinants in mtDNA. This conclusion was confirmed by the demonstration that a series of stable *petite* clones (shown to be free from rho° cells by appropriate tests) derived from strain L2200, contained a mtDNA level of 20–25% of total DNA, close to that of the parent L2200, namely 24% (Hall et al., 1976a). It had been previously shown by Nagley and Linnane (1972a) that a corresponding series of stable *petites* derived from L411 (17% mtDNA) contained a mtDNA level averaging 16% of the total DNA. Both of these groups of *petites* could be expected to retain a range of different segments of the *grande* mtDNA and also to lack mitochondrial protein synthesis. The nuclear genome and the products of cytosolic protein synthesis thus appear to control the cellular level of mtDNA.

Hall et al. (1976a) proposed a model for regulation of the cellular level of mtDNA in which the *number* of mtDNA molecules is controlled through the frequency of initiation of synthesis of individual mtDNA molecules. This initiation frequency is determined by the availability of some ploidy-dependent component, and this was suggested by Hall et al. (1976a) to be a mitochondrial membrane component or series of components synthesized by cytosolic ribosomes, and coded by nuclear DNA. The level of mtDNA in *petite* cells would be controlled by a similar process to that in *grandes*.

The final point to be made in this section concerns the number of mtDNA molecules present in each mitochondrion. Related to this question is the problem of what is meant by a mitochondrion. Hoffman and Avers (1973) took serial thin sections of individual cells of *Sacch. cerevisiae* strain Iso-N and concluded that the mitochondria of each cell consisted of a single continuum of membrane forming a giant branched structure occupying the cytoplasm. Grimes *et al.* (1974) used a similar technique, and found that the total number of discrete mitochondrial particles per cell was about 10 per haploid and twice as many per diploid. The number of mtDNA molecules per cell was estimated at 47 per haploid and 83 per diploid (Grimes *et al.*, 1974). There are thus about four mtDNA molecules per mitochondrion in these haploid and diploid cells. Bleeg *et al.* (1972) counted mitochondrial particles stained with Janus Green B in *Sacch. carlsbergensis* containing various cellular levels of mtDNA. There was a good correlation between the amount of mtDNA and the number of particles stained, and an estimate of 3–4 genomes per mitochondrion was made by Bleeg *et al.* (1972). Interestingly, a similar number of about five μm circles per mouse L-cell mitochondrion was determined by Nass (1969).

F. SYNTHESIS OF mtDNA DURING THE CELL CYCLE

In the lower eukaryotes, species of *Tetrahymena* and *Physarum*, mtDNA is synthesized continuously through the cell cycle, whilst in some mammalian cells the period of mtDNA synthesis occupies a discrete phase, often separate from the S phase of nuclear DNA synthesis (see Borst, 1972 for references). Somewhat surprisingly, various authors report both continuous and discontinuous synthesis of mtDNA during the cell cycle of different strains of *Sacch. cerevisiae*.

The cell cycle of *Sacch. cerevisiae* can be summarized as follows (see Hartwell, 1974; Hartwell *et al.*, 1974). For technical reasons the zero/unity point is most conveniently set at the time of cell separation; this point is half-way through the G1 phase which commences 0.1 fractional units before cell separation. The G1 phase continues for a further 0.1 units after cell separation, to be followed by the initiation of nuclear DNA synthesis at 0.1 units, and this S phase continues for about 0.3 units. The subsequent G2 phase runs from 0.4 to 0.75 units when mitosis commences. Mitosis lasts from 0.75 to 0.9 units. A new G1 phase then follows, and cell separation ensues one cycle later than the previous separation.

The first study on timing of mtDNA replication in yeast was carried out by Smith *et al.* (1968) using *Kluyveromyces lactis*. Unbudded cells were separated on sorbitol gradients and used as the inoculum for synchronous cultures. The cellular level of mtDNA was examined at different times during synchronous growth, and it was concluded that mtDNA synthesis occupied a discrete portion of the cell cycle separate from the S phase.

Using *Sacch. cerevisiae*, Cottrell and Avers (1970) reached a similar conclusion for the timing of mtDNA synthesis, but using a different technique. Cells were synchronized by a cyclic growth and starvation procedure, and DNA synthesis was followed by chemical analysis of total cell DNA. After a major rise in cellular DNA (corresponding to the S phase), a minor rise in the level of DNA followed, amounting to about 20% of the S phase increment. Cottrell and Avers (1970) interpreted this smaller increase as being due to mtDNA replication.

Williamson and Moustacchi (1971) used a similar synchronization technique, and analysed the DNA of cells in CsCl gradients. The amount of mtDNA in the population increased steadily with time, whereas nuclear DNA showed stepwise increases at the time of S phase. Pulse-labelling studies showed mtDNA to be labelled continuously over two cell division cycles, in contrast to the labelling of nuclear DNA, and this supported the contention that mtDNA is replicated continuously in the cell cycle.

Wells (1974) avoided any possible artefacts of synchronization techniques by taking an exponential culture of cells and separating cells according to age by a brief sedimentation in a sucrose gradient in a zonal rotor. Older cells, with large buds, sediment faster, and can be separated from younger cells with small buds, and from the youngest cells with no buds. Timing of nuclear and mitochondrial DNA synthesis was monitored by comparing long-term radioactive prelabelling patterns with pulse labelling occupying about 0.15 units of cell cycle, and DNA labelled in this way was analysed on preparative CsCl gradients. The results were interpreted by Wells (1974) to suggest that mtDNA replication takes place at the same time as nuclear DNA synthesis, and is perhaps completed before the end of S phase. However, our own evaluation of the data presented by Wells (1974) does not enable us to agree with the conclusions drawn. Substantial labelling of nuclear DNA occurred in the older three out of the four fractions of the population taken, suggesting that appreciable mixing of cells of different ages has occurred, because nuclear DNA synthesis which occupies less than one-third of the cell cycle would be expected to be pulse-labelled substantially in one fraction only and to a lesser extent in the two adjacent fractions. Moreover, Wells (1974) states that the cellular level of mtDNA is found to be about 5.8% of the total DNA in the youngest fraction, rises slowly to 6.6% in the third fraction, and then jumps to 8.5% in the fourth (oldest) fraction. This would suggest that substantial mtDNA synthesis occurs after nuclear DNA synthesis has finished (i.e. probably in the G2 phase).

Dawes and Carter (1974) obtained indirect evidence for synthesis of mtDNA in *Sacch. cerevisiae* in a discrete part of the cell cycle subsequent to the S phase. These authors used nitrosoguanidine as a mutagen which is known to act at the replicating forks on DNA molecules in bacterial systems. Mutagen-treated cells were separated according to size (hence age) by zonal rotor centrifugation. By examining the fractionated population for the presence of induced nuclear or mitochondrial mutations, the timing of nuclear and mtDNA synthesis can be deduced. As expected, nuclear mutations accumulated during the second quarter of the cell cycle, corresponding to the S phase of nuclear DNA replication. Mitochondrial mutants resistant to erythromycin or oligomycin were found in the oldest cells, implying that mtDNA replication is a late event in the cell cycle. Sena *et al.* (1975) used a similar cell-separation procedure, and using a density label to monitor the extent of DNA synthesis during the cell cycle concluded that mtDNA synthesis takes place continuously throughout the entire mitotic cycle.

G. EFFECTS OF CHANGES IN CELL PHYSIOLOGY ON SYNTHESIS OF mtDNA

1. *Mating*

The biochemical processes involved in mating between haploid *a* and α cells are beginning to be elucidated. Hartwell (1973) showed that synchronization of the cell cycles of the two cells was required before fusion occurs. One factor known to be involved in this process is a small polypeptide secreted by α cells which arrests nuclear DNA synthesis in *a* cells (Bucking-Throm *et al.*, 1973). This substance, termed α factor, has been tested for its effect of mtDNA synthesis (Petes and Fangman, 1973), which continued at near-normal rates even though initiation of nuclear DNA replication was prevented. A similar finding was reported by Cryer *et al.* (1973). The existence of a comparable factor secreted by *a* cells is currently under investigation.

The fate of mtDNA after zygote formation is not well understood at a molecular level, but is the subject of a substantial volume of study in mitochondrial genetics.

2. Meiosis

Diploid cells, heterozygous for the mating-type alleles a and α, can be induced to undergo meiosis (sporulation) when transferred from a nutritionally rich medium to a poor medium. As with meiosis in other organisms, a round of premeiotic nuclear DNA replication occurs. The two reduction divisions of meiosis then follow finally yielding four haploid spores. Timing of mtDNA replication relative to the premeiotic nuclear DNA replication has been investigated by Pinon et al. (1974) and Kuenzi and Roth (1974). Nuclear DNA synthesis begins about four hours after transfer to sporulation medium, and lasts for a further four hours or so (Pinon et al., 1974). Using mutants blocked in the ability to form spores, but which permit nuclear DNA replication to commence, synthesis of mtDNA can be examined without the complication of asci being present in the culture, since ascal DNA is not as easily extracted as DNA from vegetative cells. The results of both laboratories showed that mtDNA synthesis commences immediately after transfer to sporulation medium, and continues after nuclear DNA replication has been completed. Kuenzi and Roth (1974) found that, although incorporation of precursors into the mtDNA of their mutant continued between the period 24 and 48 hours in sporulation medium, no net accumulation of mtDNA was observed. A similar situation was noted by Grossman et al. (1969) for *Sacch. cerevisiae* incubated in the presence of cycloheximide for prolonged periods. The labelling may represent repair synthesis or turnover.

The overall effect of DNA synthesis during sporulation is to double both the nuclear DNA and mtDNA contents of the diploid cells before miotic divisions occur. Spores contain a similar level of mtDNA to the original diploid cells (Tingle et al., 1974), namely 11% of total DNA in the strains used. Synthesis of mtDNA during spore gemination has been studied by Tingle et al. (1974), who showed mtDNA replication to commence early in germination, before nuclear DNA synthesis begins. However the rate of mtDNA synthesis must subsequently be coordinated with nuclear DNA synthesis so as to restore the cellular level for vegetatively dividing cells.

3. *Glucose Repression*

Control of mtDNA replication in *Sacch. cerevisiae* and *Sacch. carlsbergensis* has been the subject of considerable investigation in view of the great variations in development of mitochondrial function manifested by these organisms under different growth conditions (for reviews see Linnane and Haslam, 1970; Linnane *et al.*, 1972a, b). Respiratory-competent cells, grown on non-fermentable substrates such as ethanol, glycerol or lactate, contain high levels of mitochondrial cytochomes $a + a_3$, b, c_1 and c, whose activity is necessary for oxidation of these substrates. The same cells grown in the presence of relatively high concentrations of glucose are catabolite repressed; the mitochondrial cytochromes $a + a_3$, b, c_1 occur only in small amounts while synthesis of cytochrome c is somewhat less affected. Energy for cellular processes is obtained largely by fermentation. The extent of catabolite repression is related in part to the rate of utilization of the carbon source for fermentation.

There is general agreement that *Sacch. cerevisiae* grown to the stage of completion of the fermentative phase of growth in glucose-containing media do not show significantly different cellular contents of mtDNA compared with cells grown on non-fermentable substrates (Fukuhara, 1969; Williamson, 1970; Nagley and Linnane, 1972b). A different situation is encountered when cells are grown exponentially in media containing a high concentration of glucose. Although Fukuhara (1969) reported only a slight decrease in the mtDNA content of the *Sacch. cerevisiae* strain used by him, Williamson (1970) found a decrease in the cellular content of mtDNA by amounts up to 50% in the strains of *Sacch. cerevisiae* he examined. It is quite likely that some of the differences relate to the strains used and also to the methods used for cell rupture, extraction and analysis of the DNA. However, that a further factor may be operative is suggested by some early observations of Moustacchi and Williamson (1966). *Saccharomyces carlsbergensis* was inoculated into a medium containing 5.4% glucose. During the first few generations of growth (about six hours) a drastic decrease in the content of mtDNA to less than 3% of total DNA was observed. The mtDNA level recovered to 20% of total DNA after 24 h (stationary phase). This behaviour of cells grown in the presence of high concentrations of glucose has recently been confirmed by Goldthwaite *et al.* (1974) for *Sacch. cerevisiae*. Exponentially growing cells (in 10%

glucose) were found to contain 2.2% mtDNA, while the mtDNA level rose to about 12% in cells harvested in the stationary phase of growth. Growth in the presence of 1.5% glucose showed similar effects: exponentially growing cells contained 6% mtDNA whilst, in the early stationary phase (just when growth stopped), 16.2% mtDNA was detected. If cells were left for a further day, the mtDNA level rose to 25.5% of the total cellular DNA; the high level in cells from the extended stationary phase is reminiscent of cells treated with cycloheximide (Grossman et al., 1969). In the absence of glucose repression during the logarithmic phase of growth (using 3% glycerol as carbon source), the mtDNA level was 16.7% of total DNA, rising to about 34% in the stationary phase (Goldthwaite et al., 1974).

Continuous-culture experiments re-inforce the conclusion that the mtDNA level is repressed in cells actively dividing in the presence of glucose concentrations sufficient to induce glucose repression both for *Sacch. cerevisiae* (Mian et al., 1973) and *Sacch. carlsbergensis* (Bleeg et al., 1972). However, it was pointed out by Mian et al. (1973) that the continuously cultured cells maintained in higher glucose concentrations divide more rapidly than cells grown in the presence of lower glucose concentrations, and thus would contain an increased quantity of nuclear DNA (cf. Grimes et al., 1974). This could make some contribution to the decreased proportion of mtDNA compared to nuclear DNA.

4. *Anaerobiosis*

Saccharomyces cerevisiae and *Sacch. carlsbergensis* can be grown anaerobically and, under these conditions, none of the cytochromes of the terminal respiratory system is synthesized (Linnane et al., 1972a, b; Linnane and Crowfoot, 1975). An important feature to be noted for cells growing in media free of oxygen is that exogenous sources of unsaturated fatty acid and ergosterol are critical in determining the status of the anaerobically grown cells; neither of these components can be synthesized in the absence of oxygen. Lipid-supplemented cells have a membrane structure more closely approximating that of aerobically grown cells, whilst lipid-depleted cells contain cellular membranes of grossly altered chemical composition. A final type of growth condition which has proved very useful in studies on the biogenesis of yeast mitochondria has been a study of the effect of aeration on anaerobically-grown cells, in order to stimulate a rapid regeneration of mitochondrial function to yield fully respiring organelles. It is often convenient to perform these aeration experiments under conditions where little cell division occurs, and this can be achieved by using high cell densities (about 10 mg dry weight of cells/ml).

Anaerobic growth to stationary phase in the presence of excess lipid supplements using glucose (Fukuhara, 1969) or galactose (Nagley and Linnane, 1972b) as carbon source does not decrease the cellular content of mtDNA as compared with fully derepressed aerobically grown cells. However, in cells partially depleted of unsaturated fatty acid and ergosterol by anaerobic growth on galactose, the cellular level of mtDNA is decreased to about half of the level maintained in cells grown in medium fully supplemented with lipid (Nagley and Linnane, 1972b). This change in the mtDNA level might conceivably arise from a derangement in control due to the altered lipid composition of the membranes in such cells, particularly in view of the recently demonstrated effect of various levels of membrane unsaturated fatty acid on mtDNA replication (Marzuki *et al.*, 1975).

Aeration of anaerobically grown yeast cells leads to a burst of mtDNA synthesis, although the extent of mtDNA synthesis is greatly influenced by the nature of the anaerobic cells. Lipid-supplemented cells, following aeration, show an increment in mtDNA of 10–30% over the initial level (Mounolou *et al.*, 1968; Rabinowitz *et al.*, 1969). Rabinowitz *et al.* (1969) reported that high concentrations of glucose in the aeration medium suppressed the extent of mtDNA synthesis on aeration. Nagley and Linnane (1972b) studied synthesis of nuclear DNA and mtDNA in lipid-depleted cells of *Sacch. cerevisiae*, anaerobically grown on galactose, the cells being aerated at high cell density in a galactose-containing medium. A three-to-four-fold increase in the quantity of mtDNA occurred during four hours of aeration, while the nuclear DNA increased in amount by only 30–50% during this period.

IV. Petite Mutants of *Saccharomyces cerevisiae*—Molecular Aspects

A. STRUCTURE AND ORGANIZATION OF mtDNA IN *petite* CELLS

1. *General Nature of* Petites *as Deletion Mutants*

It is clear from consideration of all of the evidence concerning extranuclearly inherited respiratory-deficient *petite* cells that the *petite* mutation is not unitary. There exists a multiplicity of mutants, showing a very wide range of properties of mtDNA and cytoplasmic genetic properties. In this section, discussion is centred on the physical nature of the *petite* mitochondrial genome and the arrangement of the mtDNA sequences. The processes involved in *petite* induction are discussed in the following paragraphs.

Whilst the genetic properties of *petites* are treated in some detail in later parts of this review, it should be noted that one of the most important approaches leading to the present understanding of the nature of *petite* mtDNA were the genetic observations concerning retention or loss in *petites* of mitochondrial antibiotic-resistance genes located on *grande* mtDNA (see e.g. Gingold *et al.*, 1969; Saunders *et al.*, 1971; Bolotin *et al.*, 1971; Deutsch *et al.*, 1974). These studies directed biochemical analysis, after 1970, towards a rationalized interpretation of a set of confusing observations on the properties of mtDNA in *petites*. This understanding of *petite* mutants as deletion mutants of mtDNA led to an intensified investigation of *petite* mtDNA which shows no signs of diminishing (cf. Linnane *et al.*, 1972b; Faye *et al.*, 1973; Gillham, 1974).

One further aspect of *petites* which still renders study of *petite* mtDNA subject to difficulties is the inherent instability of *petite* cells which retain mtDNA. *Petites* can be broadly divided into two classes namely: (i) the rho° class, the extreme state of deletion, in which cells lack all detectable mtDNA (Nagley and Linnane, 1970; Goldring *et al.*, 1970); and (ii) the rho^+ class in which cells retain mtDNA, albeit in a form different from *grande* (rho^+) cells. Unless steps are taken to subclone carefully *petite* cultures before genetic or biochemical analysis, it is found that rho^- clones not only may be mixed regarding the nature of the mtDNA in the cells comprising the culture, but a proportion of rho° cells may be present (Nagley and Linnane, 1972a). Although repeated subcloning steps generally lead to a relatively stable population of cells inferred to contain a homogeneous distribution of mitochondrial genomes, this is not always the case. *Petites* sometimes undergo profound changes, even on simple storage at 4°C or below. A frequent monitoring of stocks and cultures is required. Many a useful *petite* strain has been lost due to this instability problem.

2. Early Studies on Petite *mtDNA*

In spite of the difficulties just mentioned, early studies described the physical properties of mtDNA in cultures, which were generally unselected *petite* clones. The first striking observation was that *petite* mtDNA may have a very large difference in buoyant density in CsCl from that of *grande* mtDNA (Mounolou *et al.*, 1966). These differences, if present, are usually towards a lower density, down to 1.672 g/ml in extreme cases (Carnevali *et al.*, 1966; Mehrotra and Mahler, 1968). Recently, *petites* retaining mtDNA of increased density were reported (Michaelis *et al.*, 1973) when *petites* retaining genetically selected regions of mtDNA were analysed (*c.f.* Nagley *et al.*, 1975b).

Buoyant density changes in *petite* mtDNA, which have been described by authors too numerous to cite individually, have been correlated with changes in base composition in *petite* mtDNA. In extreme cases, mtDNAs have been shown to have G + C contents of as low as 4% (Bernardi *et al.*, 1968) and, more recently, 3% G + C for strain RD1A mtDNA (Mol *et al.*, 1974). *Petites* with little gross base-composition changes from *grandes* have also been reported (Bernardi *et al.*, 1970). Probably all possibilities between the 17% G + C of *grande* mtDNA and 3% G + C of RD1A mtDNA occur. In addition, *petites* with mtDNA of increased density have also been reported, implying higher G + C contents than 17%.

A third type of observation made by early workers was the rapid renaturability of *petite* mtDNAs (Mehrota and Mahler, 1968; Bernardi *et al.*, 1970). These findings can be interpreted in terms of the greatly decreased complexity of the mtDNAs. In more recent work, discussed later, the rate of renaturation has been used as a measure of genome size in many *petite* mtDNAs.

One particular *petite* clone, RD1A, containing mtDNA of very low density (1.672 g/ml), has been studied in detail by Borst and his colleagues for several years (see Borst, 1971, 1974). This *petite* was isolated following EthBr treatment of the parent *grande* and subcloning the derived *petites* (Hollenberg *et al.*, 1972a). The base composition was originally measured as 6% G + C, but later estimates revised this to 3% G + C (Mol *et al.*, 1974). Strain RD1A was the first *petite* to have its mtDNA subjected to quantitative renaturation kinetic analysis. The sequence length was calculated from the renaturation rate as less than 300 nucleotide pairs (Borst, 1971; Hollenberg *et al.*, 1972b). From the observed stimulation of the renaturation rate of labelled RD1A by addition of excess unlabelled *grande* mtDNA, it was calculated that the sequences in RD1A mtDNA were homologous to about 0.5% of *grande* mtDNA (about 400 base pairs) (Hollenberg *et al.*, 1972b). Hybridization experiments between RD1A mtDNA and the parent *grande* mtDNA confirmed this sequence homology between the DNAs

(Sanders *et al.*, 1973). However heteroduplexes between RD1A and *grande* mtDNAs melted at a temperature 6°C below the melting temperature of native or renatured RD1A mtDNA, indicating a slight degree of mismatching of base pairs. This suggested a small degree of sequence divergence between the RD1A sequence and the A + T-rich sequence of *grande* mtDNA from which it presumably was derived (Sanders *et al.*, 1973; Borst, 1974). Another estimate of the sequence length of RD1A mtDNA (70 nucleotide pairs) comes from pyrimidine-tract analysis and base-composition studies of the two complementary strands of RD1A mtDNA (Van Kreijl *et al.*, 1972; Mol *et al.*, 1974). Interestingly, RD1A showed little sequence homology with two other A + T-rich *petites* of a similar base composition (Bernardi *et al.*, 1968; Mehrota and Mahler, 1968), suggesting that the three *petites* represent different A + T-rich segments of *grande* mtDNA (Sanders *et al.*, 1973).

3. *Mitochondrial Gene Purification in* Petite *Mutants*

The studies just described dealt with *petites* enriched for very small largely non-informational regions of mtDNA. Early genetic studies initiated by Linnane and his colleagues first indicated that *petites* could contain meaningful mitochondrial genetic information (Gingold *et al.*, 1969; Saunders *et al.*, 1971). These studies opened the way for a greatly improved understanding of the nature of *petite* mtDNA. Nagley and Linnane (1972a) investigated the action of EthBr on *grande* cells in terms of the nature of the *petite* cells generated. Ethidium bromide had previously been shown to cause fragmentation of *grande* mtDNA (Goldring *et al.*, 1970; Perlman and Mahler, 1971a). From their studies of the nature of the *petites* produced by EthBr action, Nagley and Linnane (1972a) concluded that the fragments of sequences of mtDNA were distributed into various cells of *petite* progeny, and that cells containing mixtures of fragments could be purified for individual segments of mtDNA by successive subcloning steps. Study of relatively stable *petite* clones retaining mtDNA showed that the cellular mtDNA levels were similar to those of the parent *grande* cells (Nagley and Linnane, 1972a). It was concluded that, following partial deletion of mtDNA, the retained mtDNA sequences must be re-iterated in order to restore the cellular level of mtDNA to its constant level (Nagley and Linnane, 1972a; Nagley *et al.*, 1974a). The proposed scheme predicted that, in *petites* retaining a region of mtDNA defined by the presence of an antibiotic-resistance gene, it should be possible to demonstrate increased numbers of these genes in each *petite* cell as compared with *grande* cells. The efficiency of elimination by EthBr of the *ery1-r* locus (erythromycin resistance; termed ER at that time) present in a *grande* strain L411 and two derived stable *petite* clones, namely K55 and K5272, was examined by Nagley *et al.* (1973). It was found that, whilst in general mtDNA was eliminated from all three strains to give rise to many rho^0 and new rho^- cells, there was considerable difference between the extents of elimination of *ery1* loci from the three original strains (Nagley *et al.*, 1973, 1974a). Elimination of *ery1* loci was more difficult in the *petites* than in the *grande*. These results were interpreted to indicate that cells of the parent L411 *grande* strain contain one copy of the *ery1* locus per 50×10^6 dalton unit of mtDNA, whilst cells of *petite* K5272 and K55 contain more copies of *ery1* loci per 50×10^6 dalton units of mtDNA. In the following sections, we discuss the molecular nature of sequence organization in *petite* genomes and the physical form in which the gene amplification is manifested.

4. *Nature of Sequences in* Petite *mtDNA Studied by Techniques of Molecular Analysis*

Arising from the model of gene purification in *petite* described in the preceding section, several predictions can be made concerning the properties of *petite* mtDNA. Firstly, one expects to see a decrease in the complexity of *petite* mtDNA compared with *grande* mtDNA. Secondly, molecular hybridization experiments between *petite* and *grande* mtDNA should show the *petites* to be homologous to only a portion of *grande* mtDNA sequences. Thirdly, specific sequences in mtDNA, for example those specifying rRNA or tRNAs, should be found in some *petites* and lost from others. Moreover, in those *petites* where these sequences are found, the model predicts that they may account for a larger proportion of total mtDNA than in the *grande* mtDNA, that is to say, the *petite* mtDNA may be *enriched* for such sequences.

Recently obtained evidence discussed in this section in general supports all of the above predictions, and demonstrates at a molecular level the heterogeneity of some *petite* clones. The molecular analyses yield two further pieces of information: (i) some sequences retained in *petite* genomes occur in different relative proportions than they appear in *grande* mtDNA; we term this phenomenon *differential amplification* of sequences; (ii) some *petites* contain a proportion of their sequences which are unable to hybridize to parental *grande* mtDNA. These are termed *new* sequences.

The kinetic complexities of mtDNA in the genetically uncharacterized *petite* mutant D243-2B-R1-6, and a series of subclones (termed R1-6/1, R1-6/2, etc.), have been studied by Rabinowitz and his colleagues (Fauman and Rabinowitz, 1972; Casey et al., 1974a). In these related *petites*, the overall kinetic complexity of the primary *petite* clone (R1-6) was about half that of the *grande* (Fauman and Rabinowitz, 1972). Amongst the R1-6 subclones, the complexities ranged from 77% that of the *grande* to 27% (Casey et al., 1974a), indicating heterogeneity in the primary *petite*. All of the subclones described showed biphasic second-order kinetics of renaturation, indicating the presence differentially amplified sequences, as defined above. Physicochemical analyses of the mtDNAs of a series of genetically characterized *petites* derived from a *grande* (IL8-8C) in Slonimski's laboratory have been made. Measurements of kinetic complexities (Michel et al., 1974; Locker et al., 1974b) and thermal denaturation profiles (Michel et al., 1974) support the concept of *petite* mtDNA representing discrete segments of the parent *grande* mtDNA.

It is evident now that the changed buoyant densities of some *petite* mtDNAs reflect retention in these *petites* of relatively short segments of *grande* mtDNA whose average base distinction is different from that of *grande* mtDNA as a whole. Indeed, the very short segments of the *grande* genome available in some *petite* mutants enable very detailed analyses of the base distribution within the retained sequences, as for example, the studies on RD1A mtDNA, 70 base pairs in length (Mol et al., 1974). A new technique has been described by Michel (1974) which investigates clustering of A + T-rich and G + C-rich sequences in *petite* mtDNAs by monitoring, using ultraviolet absorption, the denaturation and speed of renaturation of partially denatured molecules, over the course of short-range heating and cooling cycles. This technique is akin to denaturation mapping by electron microscopy, and Michel (1974) was able to identify, in the *petite* C42 (containing 3000 base pairs), two relatively G + C-rich clusters separated by long homogenous A + T-rich sequences.

Other techniques illustrating the simplified nature of *petite* sequences with respect to the *grande* mitochondrial genome are restriction-enzyme analysis and DNA–DNA hybridization. Restriction enzyme technique is only beginning to be applied to *petite* and *grande* mtDNA, but will clearly become a major attack on the problem. Bernardi *et al.* (1975) have already shown the simplified nature of *petite* fragmentation products compared with those of *grandes*. Analyses of *petites* showed that, in general, their mtDNA digests showed fewer bands, with some *grande* bands missing, and others re-inforced, as well as the appearance of some novel bands not present in *grande* digests (Morimoto *et al.*, 1975; Di Franco *et al.*, 1976).

Hybridization (DNA–DNA) experiments using *petite* mtDNAs confirmed the idea of gene purification in *petites,* and demonstrated the possible appearance of new sequences in *petites*. Different *petites* analysed retained variable proportions of the *grande* mitochondrial genome, ranging from more than 80 units (Gordon and Rabinowitz, 1973) to four units (Fauman and Rabinowitz, 1974). Grande mtDNA is designated as 100 units, and the units refer to the fraction of *grande* mtDNA which is homologous to *petite* mtDNA, with the data obtained under conditions where labelled *grande* mtDNA is exhaustively hybridized to filter bound *petite* mtDNA. When compared with kinetic complexities as measured by renaturation rates (Michel *et al.*, 1974; Locker *et al.*, 1974b), the number of units of mtDNA in the IL8-8C series of *petites* (Lazowska *et al.*, 1974) show good correlation. However, the length of the segment of mtDNA retained in a *petite* appears larger from the hybridization tests as compared with measurement of the renaturation kinetics of the *petite* mtDNA. This result can be explained because renaturation kinetic analysis is only followed through to about 80% renaturation, so that the complexity of the slowest renaturating components of a particular *petite* is not measured. Thus, the interpretation of renaturation kinetics is biased towards the more rapidly renaturing components. On the other hand, in the DNA–DNA hybridization assay, the mass of filter-bound *petite* mtDNA is kept in large excess over the mass of labelled *grande* mtDNA in solution so as to ensure participation in the hybridization reaction of all sequences represented in the *petite* mtDNA, including the most poorly represented sequences. Thus, a more complete estimation of homology between *petite* and *grande* mtDNA is obtained (Sriprakash *et al.*, 1976b).

A complicating factor in DNA–DNA hybridization studies is the possible formation of concatenates of labelled DNA (Flavell *et al.,* 1974), which may overestimate sequence homology between *petite* and *grande* mtDNA (Lazowska *et al.,* 1974). This effect occurs if labelled *grande* DNA molecules become bound to the filter by base pairing, with the tails of other labelled molecules only partially hybridized to *petite* mtDNA on the filter. Concatenation can be considerably minimized by decreasing the length of labelled mtDNA by sonication.

Complementary hybridization experiments to those just described, namely where labelled *petite* mtDNA is incubated in solution with filters containing unlabelled *grande* mtDNA, yielded information concerning the presence of new sequences in *petites* not present in *grande* mtDNA. The technical problems here are great. Firstly, the rapid self-renaturability of *petite* mtDNA tends to prevent quantitative exhaustion of label from solution even with the same *petite* mtDNA on the filter. Secondly, it may be difficult for *grande* mtDNA to exhaust all of the radio-active counts in often highly re-iterated *petite* DNA sequences. For example, it may be envisaged that a *petite* may retain, say, 22 units overall of *grande* mtDNA, of which two units are differentially amplified 10-fold over the remaining 20 units. Labelled *petite* mtDNA thus would contain 50% of its radio-activity in the shorter over-represented, sequences. Although *grande* mtDNA filters may easily exhaust the longer under-represented sequence (50% of the *petite* label), it may be difficult to get sufficient *grande* mtDNA onto a filter to exhaust the remaining 50% of the label. This type of problem may explain why some, but not necessarily all, labelled *petite* mtDNAs fail to be exhaustively hybridized to *grande* mtDNA on filters (Lazowska et al., 1974). Nevertheless, a series of careful studies by Rabinowitz and his colleagues (Gordon and Rabinowitz, 1973; Gordon et al., 1974) showed that one *petite* (R1-6/1, 70 units in length) contained about 15–20% of its mass unable to hybridize the *grande* mtDNA, even under the most exhaustive conditions.

Hybridization experiments using RNA transcribed from *petite* mtDNA *in vivo* (Fukuhara et al., 1969, 1974; Fauman et al., 1973; Faures-Renot et al., 1974) or *in vitro* (Michaelis et al., 1972) support the conclusions from DNA–DNA hybridization studies that *petite* mtDNAs contain less information than *grande* mtDNA. The particular RNA sequences in these types of transcriptional products are not well defined, and probably represent only a restricted region of the *petite* genome (Lamb and Rojanapo, 1973). More valuable information is to be obtained using defined RNA species such as tRNAs or rRNAs. Rabinowitz and his colleagues carried out extensive studies on retention or loss of individual tRNA genes from mtDNA of different *petite* mutants (Rabinowitz *et al.*, 1974). Cohen *et al.* (1972) established that the leucyl-tRNA gene was retained in some *petite* mtDNAs and lost from others, and that some *petite* mtDNAs showed a four-fold higher level of hybridization than with *grande* mtDNA. The significance of this finding as regards the gene purification concept in *petites* already discussed was later investigated in more detail by Casey *et al.* (1974a). In particular, these authors enquired whether enrichment of various *petite* mtDNAs (of the R1-6 series) for the leucyl-tRNA cistron resulted from correspondingly large deletions of other segments of mtDNA, or whether the DNA sequences containing the leucyl-tRNA cistron were differentially amplified with respect to other sequences in the *petite* genome. The second possibility was found to explain the observed enrichment. As already discussed, the *petite* mtDNA of the R1-6 subclones showed genomes about 50–70 units in size with two components of different renaturation rate (see Rabinowitz *et al.*, 1974). In one strain (R1-6/1), leucyl-tRNA genes were enriched in the *petite* mtDNA 2.5-fold over the *grande*, and these cistrons were shown to be contained in the fast-renaturing component of this mtDNA (Casey *et al.*, 1974a). A wider study of tRNA hybridization to *petite* mtDNA was made by Casey *et al.* (1974b) who analysed 11 *petites* of the IL8-8C series from Slonimski's laboratory for hybridization of mtDNA to 11 different amino-acyl-tRNA species. The results showed retention or loss of these genes in the various *petites*. Importantly, in some cases, even after correction of hybridization levels of a particular tRNA for complexity of the *petite* mtDNA, elevated hybridization levels were found compared with the *grande*. Differential sequence amplification in *petites*, derived from this result, is strengthened by observations exemplified in *petite* F11. The mtDNA of this *petite* clone hybridized to lysyl-, glutamyl-, histidyl- and leucyl- tRNAs to levels of 0.3, 4.3, 7.3 and 17, respectively, as compared with *grande* mtDNA. It may be concluded that both sequence deletion and differential amplification lead to enrichment of *petite* mtDNAs for particular tRNA genes.

For rRNA genes, retention of 21S rRNA sequences has been detected in some *petites*, and a loss seen in others, by Nagley *et al.* (1974b) and Faye *et al.* (1974). Enrichments of some *petite* mtDNAs for 21S rRNA sequences were observed, with the highest enrichment (9-fold) seen in *petite* (EP1) of mtDNA density 1.687 g/ml (Nagley *et al.*, 1974b, 1975b). In some *petites*, an incomplete fraction of the 21S rRNA gene was retained (Faye *et al.*, 1974), with estimates ranging from 1 to

0.1 fractional lengths in different *petites*. Loss or retention of the 15S rRNA gene in different *petites* has also been described (Faye *et al.*, 1975; Spriprakash *et al.*, 1976a).

5. *Physical Arrangement of Sequences in* Petite *mtDNA*

It may be asked whether the shorter nucleotide sequences of *petites* are present in DNA molecules of correspondingly smaller lengths than the 50×10^6-dalton *grande* mtDNA molecules, or whether the size of the DNA is maintained at this full length. It will be recalled that the total mass of mtDNA in *petite* cells is, in general, very similar to that in the parent *grande* strain, irrespective of the sequences retained in the *petite*. A model for cellular regulation of the mtDNA level was put forward by Hall *et al.* (1976a) in which it is envisaged that it is the number of mtDNA molecules (in *petites* or *grandes*) which is controlled. Although early experiments on the physical size of *petite* mtDNA molecules showed a population of molecules very much smaller than 50×10^6 daltons (Billheimer and Avers, 1969; Goldring *et al.*, 1971), more recent studies support the proposition that *petite* and *grande* mtDNAs are of similar size. Careful release of mtDNA from mitochondria (Blamire *et al.*, 1972a) or sphaeroplasts (Michels *et al.*, 1974) revealed the presence of mtDNA molecules sedimenting at a rate consistent with a size of 40–50×10^6 daltons from four *petite* strains, each carrying a different segment of mtDNA. In another study, Weth and Michaelis (1974) analysed the electrophoretic mobility in agarose gels of single strains of mtDNA from IL8-8C *petite* E41 (mtDNA of density 1.688 g/ml, retaining about 2–4 units of mtDNA). The mobility was consistent with a double-stranded length of 54×10^6 daltons for E41 mtDNA, and 58×10^6 daltons for the parent *grande* IL8-8C mtDNA, implying that intact molecules were being examined.

The studies of Locker *et al.* (1974b) have shown that a proportion (usually 5–20% by mass) of *petite* mtDNA exists in the form of small circular molecules and, in some *petites*, these molecules form a distinct oligomeric series, although in others a number of different size groups are found. What is significant is that, where a unique series is observed, the length of the smallest (monomer) member of the series is directly proportional to the kinetic complexity of the mtDNA (measured at 50% renaturation on a $C_0 t$ plot). This linear relation between monomer circle length and complexity, when extrapolated to the complexity of *grande* mtDNA, predicts a circle length of 25 μm for *grande* mtDNA (Locker *et al.*, 1974b; Rabinowitz *et al.*, 1974). These observations suggest that the monomer circles contain one copy of the mtDNA segment comprising the *petite* genome. By constructing partial denaturation maps of *petite* mtDNAs (in which the sequence length is known), regular repeating patterns in the denaturation map of long molecules can be observed; the periodicity of the repeating pattern corresponds in length to about the size of the small circular molecules in the strain (Locker *et al.*, 1974a, b). The authors interpreted the data to indicate head-to-tail repeats of the small-circle equivalents. The situation with another *petite* was more complex where denaturation maps suggested the presence of inverted (head-to-head) repeats with short spacer sequences between the repeats (Locker *et al.*, 1974a).

The metabolic relationship between small circles and long molecules is not understood at present. It remains to be determined whether the small circles represent precursors of the longer molecules (e.g. rolling-circle replication processes could be envisaged) or whether the small circles are excision products from long molecules, arising by some process of internal recombination. This may perhaps occur during replication of longer molecules containing the repeated sequences in linear array, and which thus have many regions of internal homology.

B. GENERAL ASPECTS OF *PETITE* INDUCTION

1. *Influence of Nuclear and Mitochondrial Genes on* Petite *Induction*

Spontaneous *petite* frequency varies for different strains of *Sacch. cerevisiae* from 0.01% to greater than 50%. Petite frequency in a particular strain is determined, in part, by the nuclear genetic complement of the strain. Some examples of gross effects are presence of the recessive nuclear genes, *pet 3*, *lys 6* or *lys 8*, which render the *rho* factor unstable and the cultures *petite* (Sherman, 1963). Moreover, Negrotti and Wilkie (1968) described a nuclear mutation, denoted g_i, whose presence leads to the cells becoming *petite* when they are grown on glucose. Similarly, yet another nuclear mutant, in which the *rho* factor is unstable at 18°C, has been described by Weislogel and Butow (1970, 1971).

Handwerker *et al.* (1973) described an extranuclearly inherited temperature-sensitive mutant, cells of which are normal at 23°C but at 36°C become quantitatively converted to the *petite* state.

2. Types of Petite-Inducing Agents

In addition to the influence of nuclear genes, many agents both of a specific and apparently non-specific nature have been reported to induce the *petite* mutation (Nagai et al., 1961; Linnane et al., 1972b). The specific mutagens include acridines (Ephrussi, 1953), EthBr (Slonimski et al., 1968), berenil (Mahler and Perlman, 1973) and ICR 170 (a frame-shift mutagen; Werkheiser and Pittman, 1972); among the non-specific agents may be mentioned Mn^{2+} (Putrament et al., 1973) and heat (Sherman, 1959). The molecular mechanisms for many of these *petite*-inducing agents are not understood, although generally these are known to affect DNA metabolism or membrane synthesis.

3. Target Analysis and General Models

A survey of literature on target analysis, using different *petite*-inducing agents, shows that the number of targets is small, lying in the general range 1–10. Variation in the apparent number of targets is, to some extent, influenced by the physiological state of the cells, e.g. whether they are catabolite repressed or otherwise (Wilkie, 1963; Maroudas and Wilkie, 1968; Allen and MacQuillan, 1969, Dujon et al., 1975; Sherman, 1959; Sugimura et al., 1966; Slonimski et al., 1968; Mahler et al., 1971; Deutsch et al., 1974; Uchida and Suda, 1973).

Various suggestions have been made to account for the apparently small number of targets compared with the number of mtDNA molecules in each cell (about 50). Thus, it was proposed some years ago, that not all of the molecules of mtDNA are genetically competent (cf. Roodyn and Wilkie, 1968). On the other hand, Williamson (1970) outlined a system of regulatory circuits responsible for maintenance of integrity of mtDNA in which inactivation of a small number of the regulatory molecules will suffice to upset the regulatory circuit. Rank and Person (1969) proposed that the defective mtDNA molecules, which arise soon after *petite* induction, replicate faster than normal mtDNA molecules. Recently, Clark-Walker and Miklos (1974b) proposed a general hypothesis which involves excision and insertion events between circular DNA molecules to explain *petite* induction and two other yeast mitochondrial genetic phenomena, viz. suppressiveness and polarity of recombination. These authors consider that the *petite*-inducing agents promote frequently-occurring spontaneous processes which involve excision (Lusena and James, 1976), insertion and structural re-arrangements in mtDNA. In view of the relationship between the mitochondrial membrane and mtDNA, the membrane should be considered in any model which involves processes maintaining the integrity of mtDNA.

C. MECHANISM OF *PETITE* INDUCTION BY ETHIDIUM BROMIDE

1. *Factors Modulating* Petite *Induction by Ethidium Bromide*

Ethidium bromide is one of the most efficient *petite*-inducing agents known. At concentrations of 10 µg/ml, within about 1–2 hours of treatment, the drug can virtually quantitatively convert the cells to the *petite* state. Before discussing in detail the molecular events accompanying *petite* induction, it is worthwhile briefly to consider the types of situations in which the rate of EthBr *petite* induction can be considerably affected. The presence of nalidixic acid (Hollenberg and Borst, 1971; Vidova and Kovac, 1972; Whittaker *et al.*, 1972), and very low levels of cycloheximide (Whittaker and Wright, 1972) have been reported substantially to decrease the *petite*-inducing effect of EthBr. Treating cells at 45°C (Perlman and Mahler, 1971b; Mahler and Perlman, 1972b) after exposure to EthBr apparently rescues many of the cells from becoming *petite*. The carbon source in the media in which the cells were grown and, consequently, cell physiology exert varied effects on the *petite* induction by EthBr (Hollenberg and Borst, 1971; Mahler and Perlman, 1972b; Wallis *et al.*, 1972). Lipid composition of the mitochondrial membranes also influences the susceptibility of cells to *petite* induction by EthBr (Mahler and Perlman, 1972b). Arguments for the involvement of a membrane site in the action of EthBr in inducing *petite* mutants are presented by Mahler and Perlman (1972b), Mahler (1973a, b, c) and by Bech-Hansen and Rank (1972).

Genetic factors also influence the susceptibility of cells to the action of EthBr. Bech-Hansen and Rank (1972) described stable nuclear mutants leading to enhanced resistance of cells to the effects of EthBr. Another class of mutants showing extranuclear inheritance of EthBr resistance have also been reported (Bech-Hansen and Rank, 1972; Gouhier and Mounolou, 1973). These mutants in both laboratories were rather unstable, showing high rates of reversion to sensitivity. Involvement of a gene product of mtDNA in resistance to EthBr was emphasized by Gouhier and Mounolou (1973) who showed that inhibition of mitochondrial protein synthesis in the extranuclearly inherited mutants resulted in loss of the EthBr-resistant phenotype. The nature of the altered components in the EthBr-resistant mutants is not yet identified, but it may be significant that recombination between mitochondrial loci is enhanced in crosses involving mitochondria carrying an EthBr-resistance allele (Gouhier-Monnerot, 1974).

The study of other classes of mutants known to be altered in different aspects of metabolism of DNA in yeast has given some information on gene products involved in EthBr-induced *petite* induction. Two strains isolated by Moustacchi as being sensitive to *petite* induction by ultraviolet radiation, presumably altered in repair of ultraviolet damage, have been studied by Mahler and Perlman (1972b). In *uvs rho 5* (a nuclear mutant), EthBr was found to be less effective than in the parent (wild-type) strain. By contrast, in *uvs rho 72* cells (considered to be extranuclearly determined), an enhanced rate of *petite* induction by EthBr was observed (see also Mahler, 1973a). In a strain affected in (nuclear) mitotic intragenic recombination (carrying the *rec5* gene), susceptibility to *petite* induction by EthBr was considerably decreased (Moustacchi, 1973). The inference can be drawn that normal products of the *uvs rho 5* and *rec5* genes are required in certain steps of pathway of *petite* formation caused by EthBr. On the other hand, the normal product of the *uvs rho 72* gene is antagonistic to the *petite*-inducing effects of EthBr, presumably carrying out some repair of EthBr-generated damage. These inferences are supported by study of the molecular events accompanying EthBr action to be described in the next section.

2. Action of Ethidium Bromide on mtDNA

There are two effects of EthBr on mtDNA in *Sacch. cerevisiae*—one is inhibition of mtDNA replication (Goldring *et al.*, 1970; Perlman and Mahler, 1971a; Nagley and Linnane, 1972a) and the second is fragmentation of pre-existing mtDNA (Goldring *et al.*, 1970; Perlman and Mahler, 1971a). Nagley *et al.* (1975a) proposed that the two effects take place at different sites on the mitochondrial membrane. We will concentrate on the fragmentation effects here, but will consider briefly inhibition of mtDNA replication where it is relevant.

Goldring *et al.* (1970) studied the effect of EthBr in cells in which mtDNA has been specifically prelabelled with radio-active adenine in the presence of cycloheximide. The sedimentation rate of the mtDNA was measured as a function of time after exposure to the prelabelled cells to EthBr (10 μg/ml), and it was found that a gradual decrease in the size of mtDNA occurred during EthBr treatment. This was paralleled by a broadening of the band formed by mtDNA in equilibrium CsCl gradients. The same phenomena were observed by Perlman and Mahler (1971a) who also showed that mtDNA degradation continued for about one hour after removal of EthBr from the cells. Thereafter, synthesis of mtDNA resumed and newly synthesized DNA was heterogeneous in density, tending towards a modal density less than that of normal *grande* mtDNA. These observations reflect the appearance of *petites* replicating different segments of *grande* mtDNA, amongst which A + T-rich pieces tend to be better represented than G + C-rich segments. Studies have been made of the nature of *petite* clones generated by EthBr, and the results show the population of *petite* cells to consist of both the rho^- and rho^O classes (Nagley and Linnane, 1970, 1972a; Goldring et al., 1970, 1971; Michaelis et al., 1971; Hollenberg et al., 1972a). Some of the newly formed rho^- clones are highly unstable, and can throw off many rho^O subclones, and also be heterogeneous for the particular segments of mtDNA retained (Nagley and Linnane, 1972a).

Mahler and his colleagues conducted a detailed study of the events which culminate in fragmentation of mtDNA to very small pieces. Mahler (1973a) refers to the first observable change in mtDNA which occurs after exposure of cells to EthBr as a "registration" event, in which the size of the mtDNA as extracted (25×10^6 daltons) is approximately halved to 12×10^6 daltons pieces, presumably by one double-strand scission event. The inability to isolate 50×10^6 pieces (intact molecules) in control cells renders this interpretation of an initial single-cut event somewhat uncertain, but for simplicity we will accept Mahler's interpretation pending evidence to the contrary. Studies using [^{3}H]-labelled EthBr (Bastos and Mahler, 1974a) showed that this initial scission event was accompanied, or preceded, by a covalent binding of EthBr to the mtDNA (Mahler and Bastos, 1974a). These studies on whole cells opened the way to an investigation of the action of EthBr on mtDNA in isolated mitochondria (Mahler and Bastos, 1974b; Bastos and Mahler, 1974b). In these studies, it was shown that, in isolated mitochondria, EthBr can bind covalently to mtDNA and that this is accompanied by a single scission event which halves the size of the mtDNA (Mahler and Bastos, 1974b) just as occurs *in vivo*. The characteristics of binding of EthBr to mtDNA have been analysed to show that there is, on average, about one molecule of EthBr bound per 100 nucleotides of mtDNA (Bastos and Mahler, 1974b). It is not known whether the EthBr molecules are evenly spread along the mtDNA or clustered. The initial binding of EthBr to mtDNA requires mitochondrial membranes to be energized, but does not require a direct source of ATP. This is evidenced by inhibition of the initial binding of EthBr by uncouplers such as DNP or CCCP, and respiratory inhibitors such as cyanide, but not by agents which block the supply of ATP, for example, oligomycin or hexokinase together with glucose. Divalent cations are also involved in the initial binding.

A second series of reactions with more stringent energy requirements follows covalent binding of EthBr to mtDNA. The 12.5×10^6 daltons EthBr–mtDNA complex acts as a substrate for an ATP-dependent endonuclease which degrades mtDNA to acid-soluble fragments. This reaction requires a source of ATP and, in addition, the nuclease reaction requires transfer of energy through a coupled energy-transducing system, since both oligomycin, Dio9 (an inhibitor of the F_1-ATPase) and CCCP (an uncoupler) block the nucleolytic reaction. Because ATP hydrolysis is obligatorily coupled to nuclease activity, Mahler and Bastos (1974b) describe this series of reactions as an (EthBr–mtDNA)-dependent ATPase.

The best correlations arising from analyses of these reactions come from measurements of the ability of mitochondria from different genetically-defined strains to bind EthBr to mtDNA, as well as catalysis of ATP-dependent degradation of the EthBr–mtDNA complex. Mitochondria from a strain carrying the *uvs rho 72* gene (which is sensitized to *petite* induction by EthBr *in vivo*) produce an enhanced level of the EthBr–mtDNA complex (three times normal); moreover, degradation of the complex occurs at a faster rate than normal. The strains in which *petite* induction proceeds more slowly than normal show a corresponding decrease in the reactions carried out by their isolated mitochondria. Thus, *uvs rho 5* mitochondria show a normal rate and extent of formation of EthBr–mtDNA complex, but no degradation. This may reflect the absence of an excision enzyme, which leads to increased susceptibility to ultraviolet induction. The *rec5* strain (defective in mitotic intragenic recombination) shows a very low level of covalent binding of EthBr to mtDNA, but degradation-ATPase reactions appear normal. The nature of the affected component is not known.

Respiratory-deficient cells carry out EthBr incorporation and subsequent degradation but at decreased rates (see Bastos and Mahler, 1974b). This is consistent with the ability of EthBr to cause mutagenic damage to mtDNA of *petites* in a similar manner to that occurring in *grande* cells (Nagley *et al.*, 1973), and might explain why higher concentrations of EthBr are required for elimination of mitochondrial genetic information (e.g. suppressiveness) from *petites* than to induce *petites* in *grande* cells (Nagley *et al.*, 1973; Bech-Hansen and Rank, 1973a). However, genetic analyses are complicated by the presence in *petites* of re-iterated copies of the mitochondrial genetic information retained. Inhibition of mtDNA replication by EthBr in *petites* is less effective than in *grandes* (Nagley *et al.*, 1973, 1975a) and different *petites* differ amongst themselves in this respect (Nagley and Mattick, 1977).

The final biological correlation from the *in vitro* studies of Bastos and Mahler (1974b) on isolated mitochondria concerns *petite*-negative yeasts whose mitochondria were shown to be completley incapable of carrying out covalent binding of EthBr to mtDNA. This aspect is discussed more on p. 214.

An EthBr-activated endonuclease has been described in yeast mitochondria by Paoletti *et al.* (1972), but its relationship to the activities studied by Mahler and his colleagues is not clear. The physiological changes in EthBr-treated cells which probably result from inhibition of mitochondrial transcription (Fukuhara and Kujawa, 1970) and translation have been discussed by Mahler and Perlman (1972a).

Two other approaches are being used to probe the mechanism of action of *petite*-inducing drugs. Mahler (1973b) studied the *petite*-inducing ability of various drugs structurally related to EthBr, or to euflavine, but with modified substituent groups. In this way, functional groups on the EthBr molecule can be identified (as well as on the euflavine molecule), and it is hoped that future studies will clarify the importance of structural features in the various steps of mutagenesis. In a novel approach to the problem, Hixon *et al.* (1975) used a photosensitive azide derivative of EthBr which, when exposed to light, binds covalently to DNA. Cells of *Sacch. cerevisiae* treated with the derivative, when exposed to light, suffered *petite* induction as well as selective binding of the drug to mtDNA as compared to nuclear DNA.

This type of approach should enable a deeper analysis to be made of the target site within mitochondria at which *petite* induction is initiated, as well as clarifying the details of the interaction between EthBr (and other drugs) and mtDNA, and the various steps in mutagenesis. One target site at which diazo-ethidium bromide becomes covalently bound is a 7800 dalton proteolipid which seems to be subunit 9 of the mitochondrial ATPase (Bastos, 1975).

3. *Events in Cells Exposed to Very High Concentrations of Ethidium Bromide*

A novel aspect of EthBr mutagenesis of mtDNA of *Sacch. cerevisiae* was recently discovered by Criddle and his colleagues, and details of the process have been extended in our own laboratory in a collaborative effort. Whilst most studies previously had used EthBr at 20 μg/ml or less, little attention had been paid to the events occurring when the EthBr concentration was in the range 80–100 μg/ml. Wheelis *et al.* (1975) showed that, with experimental cultures in 100 μg EthBr/ml, a very rapid rate of *petite* induction occurred, so that within 15 minutes of exposure the *petite* frequency rose to about 90%. However, after 30 minutes, the *petite* frequency decreased to reach about 40% by 120 minutes. This recovery phase was followed by a slower rise in *petite* frequency, back towards 100% *petites* so that, if the culture was examined after an overnight treatment, quantitative *petite* induction was observed. There was no loss of viability during the *petite* induction or recovery phases. Wheelis *et al.* (1975) proposed the existence of a transient intermediate state of the mitochondrial genome, termed rho^*, which is susceptible to repair *in the presence of EthBr* but, if cells are plated immediately, leads to the committed *petite* state. In this scheme, the rho^* state does not accumulate to an appreciable extent in the presence of EthBr at the concentrations conventionally used (10 μg/ml), and is not generally seen. It should be recalled however that Mahler's studies described above have shown that cells treated with EthBr in starvation buffer are susceptible to rescue from the *petite* state by various treatments such as heat or antimycin A or euflavine. These conditions may reflect a repairable state similar to rho^*.

The molecular and genetic events that occur in the formation of rho^* and subsequent recovery to rho^+ state have been investigated by Criddle *et al.* (1976). Measurement of the size of mtDNA at various time intervals after treatment with high concentrations of EthBr showed that mtDNA was partially degraded in parallel with the initial rapid phase of *petite* induction, and subsequently an increase in sedimentation rate was observed during the recovery phase. Further exposure of the cells to high concentrations of EthBr caused an irreversible degradation of mtDNA. During the recovery phase, [^{3}H]-adenine was incorporated into mtDNA, the extent of which is consistent with repair synthesis occurring.

Genetic analyses of the mitochondrial genome in the rho^* state yielded interesting results (Criddle *et al.*, 1976). Transmission of genetic markers from cells in the rho^* state to a tester *grande* strain decreased towards zero, but increased as the cells recovered to the rho^+ state. This indicates that, although mtDNA is physically present in the cells at the first peak of *petite* induction, the fragments are genetically incompetent as regards transfer of markers in the cross. Measurement of the suppressiveness of the *petites* at the peak of induction gives further insight into the unusual nature of the mitochondrial genome in this rho^* state. If the cells are mated for a suppressiveness measurement directly after reaching the peak of induction, then a high suppressiveness value is recorded, indicative of a population of mtDNA fragments (rho^-) genetically active by this parameter. If, however, cells are taken from the culture at the peak of induction, and plated to obtain single colonies, nearly all of the *petites* are found to be rho°, lacking mtDNA and measurable suppressiveness. Overall description of the rho^* state is one of a fragmented set of mitochondrial genomes which, in the short term of presence of EthBr, are repaired to the rho^+ state. However, if EthBr is removed, the mtDNA becomes completely degraded giving rise to rho° cells. Long-term exposure of cells to EthBr results in quantitative conversion to the *petite* state (some rho^- but mainly rho°).

Criddle *et al.* (1976) also examined rho^+ cells which had arisen from repair of rho^* cells for gene transmission. In most cases, the cells were indistinguishable from the original parent but, in a few instances, cells showed an altered degree of transmission of a marker conferring resistance to paromomycin. These findings suggest that the repair of rho^* does not always restore the genome of the rho^+ strain exactly to its original state. What precisely the altered marker-transmission frequencies reflect is not certain. However, it is possible to envisage from the model proposed by Prunell and Bernardi (1974) for organization of mtDNA, in which genes (G + C-rich regions) are separated by A + T-rich sequences (spacers), that a gene could be translocated within different spacers during repair of mtDNA fragments. This concept allows for reconstruction of the complete genetic information albeit in a new order. It would then follow that the gene in the new position could be transmitted differently in a genetic cross.

Hall et al. (1976b) studied genetic and physiological influences on the *petite* induction pattern in the presence of high concentrations of EthBr. In general, the recovery phase (Wheelis et al., 1975) is favoured by higher temperatures of incubation and by derepression of mitochondrial function. Furthermore, the concentration of EthBr needed to cause rapid *petite* induction and subsequent recovery can be decreased from 100 µg/ml to 10 µg/ml in the presence of the detergent sodium dodecyl sulphate. This effect is attributed to the increased permeability of cells to EthBr. It is also suggested that at least one of the steps involved in the action of EthBr at high concentrations is associated with mitochondrial membranes. By using inhibitors of energy metabolism, it was shown that all of the steps in the pathway of $rho^+ \rightarrow rho^* \rightarrow rho^+$ have energy requirements. However, inhibition of cytosolic or mitochondrial protein synthesis was without effect (Hall et al., 1976b).

The enzymes and other factors involved in the rapid induction and recovery phases are not known. One possible candidate is an endonuclease present in mitochondria of cells demonstrating the rapid induction and recovery phenomena (strain D234-4A {Hall et al., 1976b}). This endonuclease activity was stimulated by low concentrations of EthBr, but was inhibited at higher EthBr concentrations. Mitochondria from another strain (S288C), which did not show the rapid induction and recovery phenomena, did not contain an endonuclease activity as described for D243-4A. Diploids were constructed by crossing S288C and D243-4A which were sporulated to give tetrads. Tetrad analysis showed segregation of the rapid induction plus recovery phenotype to be 2:2, indicating single nuclear gene inheritance. Mitochondria of the only two spores showing this induction-recovery phenotype also contained the endonuclease. There thus appears to be a connection between this endonuclease and the ability of cells rapidly to be induced to the rho^* state, and subsequently be repaired. These studies on recovery from the rho^* state bring a new dimension which must be considered in the schemes of molecular events accompanying EthBr action.

4. *Comparative Studies of the Action of Ethidium Bromide and of Other* Petite-*Inducing Agents*

The modes of action of euflavin and of berenil, an unrelated drug although not as extensively studied as EthBr, have been the subject of a number of investigations. The action of euflavine differs from that of EthBr in two respects; firstly this agent mutates only the daughter cell and not the mother cell (Ephrussi, 1953) and, secondly, in contrast to EthBr it does not initiate in the "registration" event leading to cleavage of mtDNA into two halves (Mahler, 1973a, b). Euflavine shares with EthBr the ability to inhibit *in vivo* mtDNA replication. The apparent inability of euflavine to fragment mtDNA is accompanied by production of rho^0 *petites* in far greater proportion compared to rho^- *petites*, than is the case with EthBr or berenil (Mattick and Nagley, 1977).

The trypanocidal drug, berenil, a non-intercalating DNA-binding compound, which is structurally unrelated to EthBr, has also been shown to be a *petite* inducer (Mahler and Perlman, 1973). Both berenil and EthBr show similar responses to the mutations *uvs rho 72* and *uvs rho 5*. Thus, the presence of the *uvs rho 72* allele renders cells more susceptible to *petite* induction by these drugs, whereas the presence of the *uvs rho 5* allele retards the rate of *petite* induction (Perlman and Mahler, 1973). Energy is evidently required for *petite* induction by berenil as the *petite* mutation is not induced in cells held in salt solutions (Perlman and Mahler, 1973) in contrast to the effect of EthBr. Furthermore, berenil does not produce the transient rho^* state as do high doses of EthBr (Nagley *et al.*, 1975a). These results indicate that the initial events taking place in mitochondria are different for EthBr and berenil, although at some later stage a common pathway seems to be followed (Nagley *et al.*, 1975a); berenil has been reported to induce the "registration" cleavage process in mtDNA (Mahler, 1973a, b).

Genetic studies on retention of specific regions of mtDNA in *petites* induced by ultraviolet radiation and EthBr show that considerable differences exist between these mutagens (Molloy *et al.*, 1976). However, it is of interest that *petite* induction by EthBr in ultraviolet-sensitive strains as already described shows that the *uvs rho 72* mutation stimulates EthBr-induced *petite* mutagenesis. On the other hand, the *uvs rho 5* (nuclear) mutation leads to a lowered susceptibility to

EthBr. It is suggested that ultraviolet and EthBr mutagenesis proceed by different routes although probably sharing some common features, since a synergistic effect of these two *petite*-inducing agents has been observed (Hixon and Yielding, 1975).

D. *Petite* NEGATIVITY OF YEASTS OTHER THAN *Saccharomyces cerevisiae* AND CLOSELY RELATED SPECIES

Saccharomyces cerevisiae and related species are unique amongst living organisms in giving rise to *petite* mutants which have bizarre alterations in the mtDNA that have been described. Apart from the dyskinetoplastic strains amongst the trypanosomes (see Simpson, 1972) there are few, if any, well documented instances where stable cell lines exist in which mtDNA is substantially altered, and the cells lack the mitochondrial respiratory system.

In the *petite*-negative yeasts in particular, it has been asked whether absence of viable *petite* mutants arises because mutation to respiratory deficiency would prevent cellular growth, and perhaps may be lethal. Earlier work showed that there is a good correlation between the *petite*-negativity of yeasts and their inability to grow anaerobically (Bulder, 1964). Thus, it was thought that, in the absence of energy supplied by a functional mitochondrial respiratory system, growth would be prevented. However, recent studies have shown this not to be true, and, indeed, respiratory-deficient (but not extranuclearly inherited *petite*) mutants have been found in various species (Heslot *et al.*, 1970; Wolf *et al.*, 1971; Crandall, 1973b; Goffeau *et al.*, 1972, 1973; Bandlow *et al.*, 1974; Subik *et al.*, 1974). Some of the *petite*-negative species (for example *Candida parapsilosis*) continue to grow, albeit at a decreased rate, in the presence of antibiotics which block the activity or development of the mitochondrial respiratory system (Kellerman *et al.*, 1969; Subik *et al.*, 1974) by use of oxidative pathways insensitive to cyanide or antimycin. It has not been possible to induce *petite* mutants in these strains with EthBr. Although growth of *petite*-negative yeasts is considerably inhibited by EthBr, the cells continue to divide slowly. Development of the respiratory system is blocked probably through inhibition of mitochondrial transcription and translation (Mahler and Perlman, 1972a). On removal of EthBr (or acriflavin), development of a completely normal level of respiratory activity resumed. This behaviour has been described for *C. parapsilosis* Kellerman *et al.*, 1969), *Schizosacch. pombe* (Heslot *et al.*, 1970;

Schwab et al., 1971). *Kluyv. lactis* (Luha et al., 1971), and *H. wingei* (Crandall, 1973a). Moreover, mtDNA replication was shown to be inhibited by EthBr in *Kluyv. lactis* (Luha et al., 1971), an effect observed with every eukaryote so far studied (see Borst, 1972, for a bibliography).

More recent work has shown the action of EthBr on some *petite*-negative yeasts to be more complex than at first thought. After prolonged exposure to EthBr of *Schizasacch. pombe* (Bandlow and Kaudewitz, 1974), or *Kluyv. fragilis* (Luha et al., 1974), the cultures break out of the inhibition caused by EthBr and begin to resynthesize mtDNA. However, when such "resistant" cells are grown in the absence of EthBr, and re-inoculated into EthBr-containing medium, they are found to be as sensitive as they were before adaptation to EthBr (Bandlow and Kaudewitz, 1974). A killing effect of EthBr is also observed in some species. Although a few cell divisions of *Kluyv. lactis* (and *Zygosaccharomyces fermentati*) take place in the presence of EthBr, the cells, at the end of this growth period, are found to be almost completely non-viable (Subik et al., 1974; R. M. Hall and A. W. Linnane, unpublished observations). The lethality of EthBr in these species may not be related to its activity against the mitochondrial genome, since acriflavin had no lethal effects. It is possible that EthBr affects some other function essential to viability. However, in no study so far has EthBr or acriflavin been shown to give rise to stable respiratory-deficient mutants, save in one species *Bretanomyces anomalus* (Subik et al., 1974). However *Br. anomalus* can be grown anaerobically, and there is no indication that the respiratory-deficient mutants are inherited in an extranuclear manner.

The absence of a *petite*-inducing effect of EthBr in most *petite*-negative yeasts has been reported by Bastos and Mahler (1974b) to be correlated with the inability of isolated mitochondria to carry out the initial step involved in *petite* induction in *Sacch. cerevisiae*, namely covalent attachment of EthBr to mtDNA. It must be remembered that *petites* arise spontaneously in *Sacch. cerevisiae* as well as being able to be induced by a wide variety of agents other than EthBr. Thus, there could be some unique activity within the mitochondrial membranes of *Sacch. cerevisiae* which permits degradation of mtDNA, as well as a system for maintaining the fragments of mtDNA which so arise.

It is possible that mitochondrially determined respiratory-deficient mutants of *petite*-negative yeasts can exist, but the best mutagenic procedures may not have been used. Indeed, use of EthBr may even be counterproductive. In *Tetrahymena*, EthBr has been shown to "freeze" replicating mtDNA molecules in forms able to be analysed by electron microscopy (Upholt and Borst, 1974). This inhibition of processes involved in mtDNA replication may well protect mtDNA against the introduction of errors which could generate the desired mutants. Different kinds of mutagens should be used; the better selection procedures would help, because it is likely that the respiratory-deficient mutants would arise at very low frequencies. Such mutants (Wolf *et al.*, 1976b) resemble the *mit*$^-$ types in *Sacch. cerevisiae* in which mtDNA is only slightly changed and in which one function in the respiratory metabolic system coded by mtDNA is affected, rather than having the pleiotropic changes and extensive alterations in mtDNA as occur in *rho*$^-$ and *rho*^{0} strains of *Sacch. cerevisiae*.

V. Mitochondrial Genetics in Yeast

The complexities of mitochondrial genetics are such that an exhaustive review of all aspects in detail would be inappropriate in this review. Nevertheless, we attempt to survey the field in such a way as clearly to explain certain important aspects, such as the types of mutants currently known and their phenotypic properties. We consider typical mitochondrial genetic crosses, and procedures for mapping mitochondrial genetic loci using both genetic and physical approaches. Finally, we deal with the genetics of *petites* in considering the suppressiveness phenomenon. Details of practical aspects of yeast mitochondrial genetics, including selection of mutants and performance of crosses, can be found in Coen *et al.* (1970) and Linnane *et al.* (1975).

TABLE 2. Criteria of mitochondrial inheritance in *Saccharomyces cerevisiae*

Mitochondrial genes	Nuclear genes
1. Mitotic (vegetative) segregation of alleles after zygote formation, i.e. heteroplasmic state	1. No mitotic segregation of alleles
2. No segregation of alleles on meiosis of a homoplasmic diploid (4:0 or 0:4 tetrads)	2. Segregation of alleles on meiosis (2:2 tetrads)
3. a. Loss of gene associated with *rho*^{0} state b. Retention or loss of gene in different *rho*$^-$ isolates	3. No loss of nuclear genes associated with extranuclear *petite* mutation

A. CRITERIA OF MITOCHONDRIAL INHERITANCE

Three main criteria of mitochondrial inheritance of a particular trait have been propounded (Linnane et al., 1968a, 1972b; Bolotin et al., 1971). The behaviour of mitochondrial genes, compared with nuclear genes, is summarized in Table 2. The first two criteria, namely mitotic segregation and non-segregation at meiosis, are characteristic of any extranuclear gene. It is thus important for the unequivocal characterization of a determinant as being on mtDNA that the third criterion be applied. This relies on the loss of all mitochondrial genes concomitant with the rho^0 state, and the loss or retention of mitochondrial genes in different rho^- isolates, in which partial sequence deletion from mtDNA occurs. A case has recently been reported of a drug-resistance gene obeying the first two criteria above, but not properly satisfying the *petite*-deletion requirements (Griffiths et al., 1975a, discussed on p. 226).

B. GENETIC DETERMINANTS ON mtDNA

Although the cytoplasmic *petite* mutation was discovered by Ephrussi and his colleagues in the 1940s, a systematic analysis of the fundamental properties of the cytoplasmic system was not possible due to the lack of other mutants. A detailed genetic analysis became possible following our recognition that many of the antibacterial antibiotics (chloramphenicol, erythromycin, spiramycin, lincomycin, tetracycline, oleanodomycin) were specific *in vivo* and *in vitro* inhibitors of mitochondrial protein synthesis, and that inhibited cells resembled a phenocopy of the cytoplasmic *petite* mutation (Huang et al., 1966; Clark-Walker and Linnane, 1967; Wilkie et al., 1967; Lamb et al., 1968; Linnane et al., 1968b; Linnane and Haslam, 1970). These studies led us to the idea that genetic analysis of derived antibiotic-sensitive and resistant strains could provide a tool for study of extranuclear (mitochondrial) genetics. An erythromycin-resistant strain showing patterns of inheritance consistent with extranuclear localization was the first such mitochondrial gene described (Linnane, 1968; Linnane et al., 1968a, b). The locus was termed ER, but is now designated *ery1* following the three-letter convention. Following this discovery, many new loci conferring resistance to antibiotics, affecting both ribosomes and membrane-associated functions such as ATPase, have been identified. These mutants are discussed in the following section. We then consider other types of determinants in mtDNA, which lead to some limited defects in respiratory function (e.g. mit^- loci).

1. *Allelism Tests*

In many cases, more than one isolate has been obtained which shows a particular phenotype, for example, resistance to a given drug. The allelism test provides information as to whether the independent isolates carry mutations at the same locus, or whether a number of loci confer a particular phenotype. In the case of drug-resistant phenotypes, such tests require crosses to be performed between two haploid strains resistant to the same antibiotic. The diploid progeny are examined for the appearance of antibiotic-sensitive recombinant clones. Sensitive diploids are found only if the haploids carried resistance mutations conferred by separate loci in each case. The mitotic segregation characteristic of mitochondrial genetic behaviour permits distribution of the recombination products into individual diploid clones.

Unfortunately, direct comparison of the phenotypes of related mutants does not necessarily indicate whether the same or different loci are involved. Thus, for example, the *ery1* and *ery2* mutations were selected as independently isolated mutants from the same parent strain, having developed resistance to erythromycin. Recombination could not be detected between the two mutations, and thus they are considered to be different alleles at the same locus (Trembath *et al.*, 1973). Nevertheless, phenotypes of the two mutations differ in that *ery1-r* cells are cross-resistant to spiramycin, oleandomycin, carbomycin and lincomycin, whilst *ery2-r* cells are cross-resistant to the first two of these drugs but sensitive to carbomycin and lincomycin (Trembath *et al.*, 1973).

2. *Drug-Resistance Loci*

About ten different loci conferring resistance to a range of antibiotics have been identified in the *Sacch. cerevisiae* mitochondrial genome. The loci can be broadly classified according to the drug involved as well as the mitochondrial component affected (summarized in Table 3). Some details of the properties of the various types of drug-resistant mutants now follow.

a. *Erythromycin and Spiramycin*. Mutants resistant to the macrolide antibiotics erythromycin and spiramycin have been obtained at a minimum of two loci by three research groups. The erythromycin-resistance locus *ery1* detected in our own laboratory is mentioned on p. 217. Thomas and Wilkie (1968) described mutations at two loci, designated E and S, conferring resistance to erythromycin and spiramycin, respectively. Slonimski and his colleagues (Bolotin *et al.*, 1971; Netter *et al.*, 1974) have isolated a number of such mutants classified as belonging to one of the two loci *RIB2* and *RIB3* (formerly called R_{II} and R_{III}, respectively). Representative markers at these two loci are S^R_{551} and E^R_{553} at *RIB2*, and E^R_{514} and S^R_{352} at *RIB3*. Spiramycin-resistant mutants isolated by Trembath *et al.* (1973) were found to occur at two loci, although mutants at one locus (*spi3* and *spi4*) were not allelic but were very tightly linked. The correspondence between the many loci described by the three laboratories is not clear. Nevertheless, it seems likely that the following loci are allelic: *ery1*, *ery2*, *spi2* (Linnane) = *RIB3* (Slonimski) = E (Wilkie).

Strains carrying mitochondrial erythromycin-resistance genes contain a mitochondrial protein-synthesizing system which is resistant to erythromycin (Linnane *et al.*, 1968b). The resistance is apparently expressed at the level of the peptidyl transferase in the large 50S subunit of mitochondrial ribosomes (Grivell *et al.*, 1971). Mutants resistant to macrolides show complex patterns of cross resistance to other macrolides *in vivo*, which serve to distinguish phenotypes not only at different loci but often of different alleles at the same locus. Interpretation of these results is rendered even more difficult when considering the cross-resistance patterns of mitochondrial protein synthesis *in vitro* (Linnane and Haslam, 1970; Linnane, 1971; Linnane *et al.*, 1972b, c; Trembath *et al.*, 1973). It would appear that the different allelic forms of the drug-resistance genes modify the ribosomes in very subtle ways. Some of the effects also may involve the mitochondrial membrane, exemplified by comparison of the resistance phenotypes of various mutants (*ery1-r, ery2-r, spi2-r*) which are alleles of the same locus (Trembath *et al.*, 1973). Mitochondrial protein-synthesis studies were interpreted to mean that the various mutations at this locus can represent changes in either ribosome or membrane, thus emphasizing the close relationship between the ribosomes and membrane. It is to be emphasized that, in no case, has it been definitively demonstrated whether any of these drug-resistance

TABLE 3. Drug-resistance determinants located in mtDNA of *Saccharomyces cerevisiae*[a]

Drug[b]	Reference	Altered component	Reference
Erythromycin	Linnane et al. (1968a) Thomas and Wilkie (1968) Coen et al. (1970) Rank and Martin (1972) Kleese et al. (1972b) Uchida and Suda (1973) Goldthwaite et al. (1974)	Ribosomes (large subunit)	Linnane et al. (1968b) Grivell et al. (1971, 1973)
Spiramycin	Thomas and Wilkie (1968) Bolotin et al. (1971) Trembath et al. (1973) Callen (1974a)	Ribosomes (and membranes?)	Grivell et al. (1973) Trembath et al. (1973)
Chloramphenicol	Coen et al. (1970) Kleese et al. (1972a) Rank and Bech-Hansen (1972a) Molloy et al. (1973) Wilkie and Thomas (1973) Uchida and Suda (1973) Goldthwaite et al. (1974)	Ribosomes (large subunit?)	Molloy et al. (1973) Grivell et al. (1973)
Paromomycin	Thomas and Wilkie (1968) Kleese et al. (1972b) Kutzleb et al. (1973)	Ribosomes (small subunit?)	Wilkie (1970) Kutzleb et al. (1973)

Mikamycin/Antimycin A	Bunn et al. (1970) Linnane et al. (1973a) Groot Obbink et al. (1976) see also text	Membrane component (complex III)	Bunn et al. (1970) Linnane et al. (1973a) Groot Obbink et al. (1976) see also text
Oligomycin	Avner and Griffiths (1970, 1973a, b) Stuart (1970) Wakabayashi and Gunge (1970) Rank and Martin (1972) Mitchell et al. (1972) Shannon et al. (1973) Wilkie and Thomas (1973) Callen (1974a) Goldthwaite et al. (1974) Colson et al. (1974)	ATPase (membrane component)	Griffiths and Houghton (1974); Shannon et al. (1973)
Venturicidin	Griffiths et al. (1974)	ATPase	Griffiths et al. (1974)
Ethidium bromide	Bech-Hansen and Rank (1972) Gouhier and Mounolou (1973)	Not known (*petite* induction)	
D-Glucosamine	Elliot and Ball (1975)	Not known (glucose repression)	

[a] Determinants (genes) listed in this table have been shown to satisfy some or all of the criteria of mitochondrial inheritance (Table 2; p. 216) and, in all cases except possibly EthBr and D-glucosamine, it is certain that these genes lie on mtDNA.
[b] The drug indicated is that most frequently used in identifying the presence of the locus, although cross resistance to other drugs may also be manifested.

mutations leads to modification of the ribosomal-RNA, or of a ribosomal protein, in spite of much intensive investigation. The 21S rRNA cistron is found to be very close to the *eryl* locus (see p. 252).

b. *Chloramphenicol.* Resistance to chloramphenicol is determined at two (or possibly more) loci. Two laboratories report non-allelic chloramphenicol-resistance mutants (Kleese et al., 1972a; Molloy et al., 1973). Other laboratories have reported isolation of chloramphenicol-resistant strains (see Table 3). In our laboratory, we reported two tightly linked loci *cap1* and *cap2* (Molloy et al., 1973) and, in Slonimski's group, about seven chloramphenicol-resistant mutants mapped at a single locus, termed *RIB1* (formerly R_I) (Bolotin et al., 1971; Netter et al., 1974). One representative allele is C_{321}^R. The correspondence between loci of different laboratories is not clear, but it seems that *cap1* or *cap2* corresponds to the *RIB1* locus of Slonimski. The *cap1* locus is allelic with, or very tightly linked to, the *spi3* and *spi4* loci (Linnane et al., 1974, 1976). It is conceivable that the *RIB2* locus (Netter et al., 1974) may correspond to *spi3* and *spi4*.

Chloramphenicol-resistant mutants contain altered mitochondrial ribosomes since mitochondrial protein synthesis *in vitro* is resistant to chloramphenicol for strains carrying both the *cap1* and *cap2* loci (Molloy et al., 1973). The *RIB1* mutants also contain modified mitochondrial ribosomes (Grivell et al., 1973), in which a resistant peptidyl transferase (a large subunit function) is found.

c. *Paromomycin.* Paromomycin-resistance genes have been reported by three groups. In two cases, mitochondrial protein synthesis *in vitro* has been shown to be resistant to paromomycin (Table 3, p. 220). Paromomycin resistance is accompanied by resistance to neomycin. The probable site of action of this drug is the small ribosome subunit (Davey et al., 1970), so that the altered component in resistant ribosomes probably lies within the small (37S) subunit. Mapping studies (described on p. 252) place the *par1* locus close to the 15S rRNA cistron but, as with the *eryl* and *cap1* types, it is not known if the RNA or protein components of the ribosomes are directly modified by the mutation.

d. *Mikamycin/Antimycin A.* Study of mikamycin resistance in yeast, first described by Bunn et al. (1970), has presented many difficulties, illustrative of some of the problems that confront investigators characterizing mutants. The problems that were encountered include the existence of nuclear genes which act in concert with mitochondrial genes to determine the level of resistance to a particular drug, and the possible impurity of commercial drug preparations. Thus, it has finally been established that the mitochondrial mikamycin resistance-determining locus (originally called *mik1*) controls resistance to antimycin A (Groot Obbink *et al.*, 1977) and the locus has been renamed *ana1*.

Preliminary genetic characterization of mikamycin-resistant cells suggested mitochondrial inheritance of the *mik1* locus (Bunn *et al.*, 1970), but later studies showed a very complex inheritance pattern having both nuclear and cytoplasmic characteristics (Linnane *et al.*, 1973a). The problem was investigated in detail by Howell *et al.* (1974b) who demonstrated the existence of nuclear genes conferring moderate levels of mikamycin resistance (25 μg/ml) as well as a mitochondrial gene also conferring resistance to small concentrations of the drug (3–8 μg/ml). There is a synergistic action between the mutant loci such that, when both nuclear- and mitochondrial-resistance determinants are present in the same cell, cells are resistant to concentrations of mikamycin up to the limit of solubility of the drug (>800 μg/ml). Mitochondrial genetic crosses involving the *mik1* locus should therefore be carried out in the presence of nuclei lacking the chromosomal resistance genes. This situation is an extreme case of a general problem with mitochondrial drug-resistance genes. In other instances, as with paromomycin or chloramphenicol, scoring of resistant cells in attempts to monitor the presence of *par1* or *cap1* loci is complicated by the presence of nuclear genes in different strains which influence the concentration of drug needed to distinguish between resistance and sensitivity.

Although mikamycin is an inhibitor of the activity of bacterial ribosomes, it was evident from the first biochemical characterizations of mikamycin-resistant cells (Bunn *et al.*, 1970) that the mitochondrial membrane, rather than the mitochondrial ribosomes themselves, was involved in expression of resistance to the drug. This was shown by demonstrations that the protein-synthesizing ability of either isolated mitochondria (Bunn *et al.*, 1970) or purified ribosomes (T. W. Spithill and A. W. Linnane, unpublished observations) from *mik1-r* strains is sensitive to mikamycin. It was originally suggested (Bunn *et al.*, 1970; Linnane and Haslam, 1970) that *in vivo* mikamycin is prevented from entering mitochondria by a permeability barrier resulting from an altered membrane component specified by the *mik1* gene. It was envisaged that this membrane permeability barrier in *mik1-r* cells could be lost during the physical manipulations involved in isolating mitochondria for the protein synthesis assay, and could also be lost *in vivo* by modification of the membrane structure such as by growing *mik1-r* cells anaerobically. However, attention was focused on a site of action for mikamycin other than the ribosome by the demonstration in our laboratory that *mik1-r* cells are always cross-resistant to another antibiotic, namely antimycin A, which blocks respiration (as does mikamycin at moderately high concentrations) at the level of cytochrome *b* oxidation.

A link between the effects of mikamycin and antimycin A was established with the demonstration that the particular component of commercial mikamycin preparation that inhibited yeast cells was neither of the major components, mikamycin A or B, but a minor component which showed a similar mobility in thin-layer chromatographic systems to antimycin A (Groot Obbink *et al.*, 1977). It was then demonstrated that the mikamycin-resistant cells (carrying *mik1-r* loci, with no nuclear-resistance loci) were cross-resistant to antimycin A at a concentration of 0.005 μg/ml. Rationalization of these data is that we have been studying mutations in complex III function, where the mutated gene affects the site at which antimycin A acts. The locus has thus been renamed *ana1* (for antimycin A). It is interesting that mapping studies (described on p. 228) place the *ana1* locus in the same three-unit segment of mtDNA as a cytochrome *b*-controlling locus, *cyb1*. It is not known if these two loci, which both affect complex III function, are on the same gene. More mutants are required in this region of the genome. It is interesting that an extranuclearly inherited antimycin A-resistant strain of *Schizosacch. pombe* has been described (Wolf *et al.*, 1976a).

e. *Oligomycin, Venturicidin and Triethyl-tin.* Studies on oligomycin-resistant mutants demonstrated three mitochondrial genetic loci as well as another group of cytoplasmic loci not on mtDNA, and have yielded a very useful approach to the study of the function and assembly of the mitochondrial ATPase complex. Many laboratories have reported mitochondrially inherited oligomycin-resistant mutants (Table 3, p. 220), and three loci have so far been identified. Trembath *et al.* (1976) compared allelic relationships between the following loci: O_I, O_{II} (Avner and Griffiths, 1973a, b), O_{III} (Lancashire and Griffiths, 1975b), *oli1* (Mitchell *et al.*, 1972), *oli17* (Trembath *et al.*, 1976), *tso*

(Trembath et al., 1975); *OLG1*, *OLG2* (Callen, 1974a). The results indicated the following allelic identities:

$O_I = oli17 = OLG2$ (collectively designated class A)
$O_{III} = oli1 = tso = OLG1$ (class B).

O_{II} (the sole representative of class C) is not allelic with any of the others studied. Mapping studies (discussed on p. 251) showed O_I and O_{III} to be linked whilst O_{II} is unlinked to the others and probably lies within a different gene (Lancashire and Griffiths, 1975b; Trembath et al., 1976).

The cross-resistance patterns of oligomycin-resistant mutants are not simple, as in the case of the ribosome mutants. Mutants allelic with O_{III} are cross-resistant to venturicidin (another oxidative phosphorylation inhibitor), but mutants allelic with O_I and O_{II} show no cross-resistance. To illustrate the complexity of phenotypic behaviour, it can be noted that some mutants, resistant to venturicidin but not cross-resistant to oligomycin, map at the O_I locus (Griffiths et al., 1974; Lancashire and Griffiths, 1975b). A class of mutants resistant to both venturicidin and triethyl-tin have loci unlinked to any of the other mitochondrial loci; these loci probably lie on a cytoplasmic DNA molecule other than mtDNA (Lancashire and Griffiths, 1975b; see also discussion on p. 226).

Biochemical studies on oligomycin-resistant mutants (and related mutants resistant to venturicidin and triethyl-tin) indicate that the ATPase is modified, although in different mutants isolated ATPase is found to have various degrees of resistance to inhibition by the drugs (Griffiths and Houghton, 1974; Somlo et al., 1974; Griffiths et al., 1975b; Lancashire and Griffiths, 1975a). Reconstitution experiments between the membrane components on the one hand, and soluble ATPase (F_1 and OSCP) on the other, isolated from resistant and sensitive strains respectively, showed membrane components to be responsible for oligomycin resistance (Shannon et al., 1973). The membrane components of the ATPase complex are those synthesized by mitochondrial ribosomes (for reviews, see Schatz and Mason, 1974; Linnane and Crowfoot, 1975), and genetic data strongly support the conclusion that oligomycin-resistance mutations are located in genes coding for membrane polypeptides of the ATPase complex. The particular polypeptides controlled by class A and class B loci have been tentatively assigned as subunit 6 (size 20,000 daltons) and subunit 9 (size approx. 9000 daltons), respectively, of the ATPase. Cells carrying the *tso-r* allele (class A) are not only oligomycin resistant but also unable to grow at 19°C on non-fermentable substrates, because of improper attachment of the soluble ATPase components (F_1 and OSCP) onto the mitochondrial membrane (Trembath et al., 1975). A specific defect in synthesis of subunit 6 takes place in *tso-r* cells grown at 19°C (Groot Obbink et al., 1976). On the other hand, some class-B mutants show the property that the 9000 dalton proteolipid (subunit 9) is unable to aggregate into a 45,000 dalton component (Groot Obbink et al., 1976; Tzagoloff et al., 1976a) unlike the proteolipid from wild-type cells. The class C locus (O_{II}), as well as the non-mitochondrial gene in which the triethyl-tin-resistant locus lies, may code for other membrane polypeptides affecting ATPase function (see Lancashire and Griffiths, 1975a).

A different kind of extranuclear mutation resistant to venturicidin has recently been discovered by Griffiths et al. (1975a). These mutants are also cross-resistant to triethyl-tin and other drugs such as rhodamine G. The locus determining this phenotype was found to be unlinked to the other oligomycin-resistance loci (Griffiths et al., 1974), but showed some criteria of mitochondrial inheritance. The aberrant behaviour was that the gene was sometimes present in rho^0 cells lacking mtDNA (Griffiths et al., 1975a). These authors inferred that the resistance locus lies on a cytoplasmic DNA species other than mtDNA, and suggested that this DNA possibly has some sort of episomal role with mtDNA.

Mutants with similar patterns of cross resistance, but also oligomycin resistant, have been described by Guerineau et al. (1974). These authors concluded that resistance was dependent upon both nuclear gene and a cytoplasmic factor, not mitochondrial, which could act as an episome and become integrated into the nuclear genome. A positive correlation between retention of resistance determinants and the presence of the 2 μm circular DNA (*omicron* DNA) was reported by Guerineau et al. (1974). No such correlation was observed by Griffiths et al. (1975a) with their multiple drug-resistance cytoplasmic determinant. It appears that these two types of mutations may be part of a new class of cytoplasmic mutants not located on mtDNA, but related to mitochondrial function. However, the particular DNA species in which the mutations reside remains to be more positively established, and further work is required to clarify the situation.

3. Loci Conferring Limited Defects in Respiratory Function

An important recent development in the field of mitochondrial genetics has been the isolation of mutants defective in respiratory function, but which arise through point mutations (or very small deletions) and which do not in general show the extensive pleiotropic effects manifested by "classical" *petite* mutants. This class of mutants is defined as comprising those clones which are unable to grow on a non-fermentable substrate such as glycerol, but which still retain mitochondrial protein synthesis. The genetic term mit^- has been coined for such mutants, which were first isolated by Tzagoloff who used a selection procedure involving the direct assay, using suitable multiple-sample techniques, of the mitochondrial protein-synthesizing ability of clones unable to grow on glycerol following mutagenic procedures. Such an approach yielded of the order 20 to 100 extra-nuclearly-inherited mutants out of 1000 respiration-deficient colonies tested, depending on the mutagenic agent used. The nature of the lesions found cover a range of phenotypes, including the specific loss of cytochrome oxidase (*OXI*) activity (Flury *et al.*, 1974; Tzagoloff *et al.*, 1975b), coenzyme QH-cytochrome *c* reductase (*COB*) activity (Tzagoloff *et al.*, 1975c) or ATPase (*PHO*; Tzagoloff *et al.*, 1976b). In *OXI* and *COB* mutants, the lesions are usually accompanied by loss of the spectrally detected cytochromes *a* and *b*, respectively. Some, but not all, of the cytochrome-oxidase mutants show aberrations in synthesis of particular products of mitochondrial protein synthesis (Tzagoloff *et al.*, 1975b, c). A further class of mutants (termed *BOX*) in which cytochromes *a* and *b* are simultaneously deleted, has been described (Kotylak and Slonimski, 1976), but it is possible that these are double mutants, or cytochrome *b* mutants in which synthesis of cytochrome *a* is very sensitive to glucose repression (Cobon *et al.*, 1976). Other mutants show more pleiotropic effects, including one strain which has normal mitochondrial protein synthesis yet lacks NADH-cytochrome *c* reductase and cytochrome oxidase activities (Tzagoloff *et al.*, 1975a). It is thus quite clear that a great range of possible phenotypes exist, and identification and mapping of the loci will greatly increase our knowledge of the mitochondrial genome. It is, however, desirable to adopt mutant-isolation procedures which are more convenient than direct biochemical assay of mitochondrial protein-synthesizing activity previously used.

We have recently directed our attention to this question in our laboratory. A new selection procedure (Rytka et al., 1976) involves a simple genetic test for the presence of mitochondrial protein-synthesizing ability in clones unable to grow on non-fermentable substrates (ethanol, as used by ourselves). The test relies on the requirement for a functional mitochondrial protein-synthesizing system of the ability of certain strains to metabolize galactose. In such strains, in which the ability to grow on galactose as carbon source is induced by the presence of galactose (i.e. the enzymes are not made constitutively), mitochondrial protein synthesis is required for a successful induction (Puglisi and Algeri, 1971). Our screening test therefore involves starting with a parent *grande* inducible for galactose-metabolizing enzymes. Following mutagenesis, usually with Mn^{2+}, cells unable to grow on ethanol are tested for ability to grow on galactose. The cells that grow on galactose are found to be mit^- in many of the clones tested. An important feature of the screening of the repiratory-incompetent clones is their crossing with a collection of physically defined *petites* in which the particular segments of mtDNA retained have been determined (Linnane et al., 1976; Nagley et al., 1976). A true mit^- clone will give rise to respiratory-competent recombinants when crossed with those *petites* which retain in their mtDNA the wild-type allele of the mit^- locus, but not where the locus is deleted from the *petite*. In this way, mit^- clones are identified by their ability to give *grande* diploids with some, but not all, *petites*. The power of this technique is that knowledge of the particular mtDNA segments retained in the various *petites* enables the physical map position of the mit^- locus to be immediately determined (Rytka et al., 1976; see also Section V.D, p. 249).

Using these procedures, three loci controlling the activity of cytochrome oxidase (*cya1, cya2, cya3*) and one locus controlling cytochrome *b* (*cyb1*) have been identified and physically mapped (Groot Obbink et al., 1976; Cobon et al., 1976; Rytka et al., 1976). From the approximate map position of the Tzagoloff loci as determined by genetic (but not physical) techniques (Slonimski and Tzagoloff, 1976; Tzagoloff et al., 1976b), it is possible that *cyb1* co-incides with *COB*. The loci, *cya2* and *OX13*, lie within the same segment of the genome (between O_{II} and *par1*) and *cya1, cya3, OX14* and *OX12* lie within another segment of the genome (between *par1* and *cap1*; see Fig. 1, p. 229).

Other mutants have been reported which can be described as

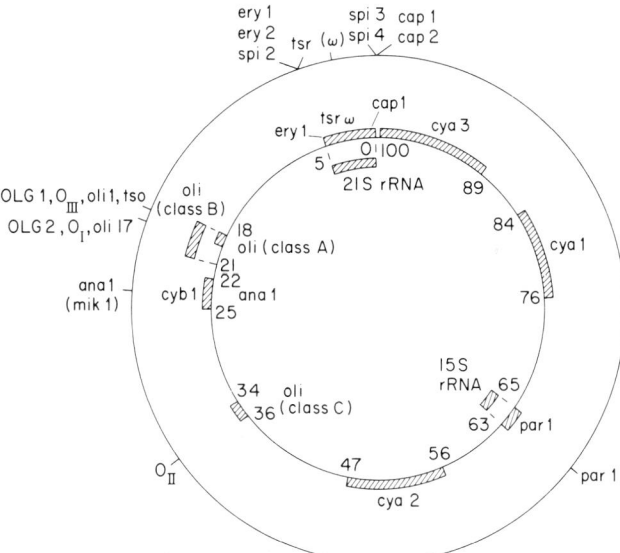

FIG. 1. Genetic and physical map of the *Saccharomyces cerevisiae* mitochondrial genome.

Outer circle—genetic map: The relative positions of the markers shown were determined in our laboratory by genetic procedures, namely coretention and deletion analysis using *petites* (Molloy *et al.*, 1975; Trembath *et al.*, 1976), or by transmission and recombination analysis for markers in the region of *omega* (ω) (Linnane *et al.*, 1974; Howell *et al.*, 1974; C. Oliver, K. English, H. B. Lukins and A. W. Linnane, unpublished data). The positions of some of the markers have been adjusted slightly from those previously published so as to align them with the physical map locations. Groups of markers which are very tightly linked as established by recombination studies (Trembath *et al.*, 1973, 1976) are shown as a single locus here.

Inner circle—physical map: The bars on this circle represent the regions within which the markers were physically mapped by using DNA–DNA hybridization analysis of genetically characterized stable *petite* clones, as described in the text (Sriprakash *et al.*, 1976a, b; Trembath *et al.*, 1976; Linnane *et al.*, 1976; Nagley *et al.*, 1976; Rytka *et al.*, 1976; Choo *et al.*, 1977). Numbers inside circle are map units.

conditional mit^-. Storm and Marmur (1975) described a temperature-sensitive mutant which specifically lost cytochrome oxidase activity after transfer to 36°C. A cold-sensitive mutant (*tso*) was reported by Trembath *et al.* (1975) which lost the activity of membrane-bound ATPase at 19°C. In both cases, at the restrictive temperature growth did not occur on non-fermentable substrates but mitochondrial protein synthesis was still normal.

We have recently characterized (K. J. English, T. W. Spithill, P. Nagley, H. B. Lukins and A. W. Linnane, unpublished data) a further cold-sensitive mutant (*tsr*) which grows very slowly on ethanol at 18°C (normal at 28°C), and is considerably denuded of the products of mitochondrial protein synthesis (cytochromes $a + a_3$, cytochrome oxidase, cytochrome b, NADH cytochrome c reductase, oligomycin-sensitive ATPase) at this temperature, yet it retains chloramphenicol-sensitive, amino acid-incorporating ability *in vivo* at 18°C. The *tsr* locus (formerly *CS*) has been mapped physically to within a segment of mtDNA very close to the *ery1* and 21S rRNA genes (see Fig. 1, p. 229) and is considered to affect the function of mitochondrial ribosomes (T. W. Spithill and A. W. Linnane, unpublished data), possibly disturbing the fidelity of translation.

C. TRANSMISSION AND RECOMBINATION OF MITOCHONDRIAL GENES IN GENETIC CROSSES

1. *General Aspects*

Isolation of the antibiotic-resistant mutants discussed in the preceding section permitted a study of recombination between mitochondrial genes. Following a cross between two haploid yeast strains carrying different markers, recombination is genetically indicated by the appearance of diploids containing the markers in combinations different from the haploids. This was first demonstrated by Thomas and Wilkie (1968). That this process does not represent a simple reassortment of independent genetic units, but rather results from molecular recombination involving mtDNA molecules, was suggested first by Gingold *et al.* (1969). These authors isolated *petites* from a *grande* strain carrying the *ery1-r* allele. Some of the *petites*, when crossed with an *erys grande*, gave rise to *grande* diploids resistant to erythromycin.

Direct molecular evidence for physical recombination was provided by analysis of diploids constructed from two haploids in which the mtDNAs were of different buoyant density in CsCl. Shannon *et al.* (1972) crossed a *petite* with a *grande* whilst Michaelis *et al.* (1973) crossed two *petites*. In both studies, diploids were produced in which the mtDNAs were of buoyant densities intermediate between those of the haploid parent strains. Molecular recombination was demonstrated in crosses between two *grande* strains by comparing restriction enzyme patterns of diploid clones with those of the two parent strains (Bernardi, 1975).

Two separate approaches have been used for a study of mitochondrial gene recombination in crosses between *grande* cells. Firstly, diploid progeny cells from a mass cross have been analysed after 24 to 48 hours growth on non-selective medium (random diploid analysis). Secondly, individual zygotes produced from a cross have been studied either by analysis of the genetic constitution of the entire zygotic clone or by pedigree analysis of the daughter buds (zygote cell lineages) separated from the zygote by micromanipulation.

2. Random Diploid Analysis

It is generally accepted that, upon fusion of haploid cells to form a zygote, a pool of mitochondrial genomes, donated by both haploid cells, is established. In the zygote (and possible also in early buds), several rounds of recombination occur and small numbers of genomes are sequestered to bud cells in a more or less random fashion. After six generations, most of the progeny cells are pure for a single mitochondrial genotype (Gingold *et al.*, 1969; Coen *et al.*, 1970; Kleese *et al.*, 1972b; Callen, 1974a, b), although in some cases diploids remain heteroplasmic for a much longer period (Rank and Bech-Hansen, 1972b; Forster and Kleese, 1975b). Analysis of the frequencies of occurrence of parental and recombinant genotypes (by testing the phenotypes of many individual diploid clones produced in this way) constitutes random diploid analysis.

Whilst the results of crosses between any two particular *grande* strains under controlled conditions are highly reproducible, it is clear that there are many influences on the relative proportions of parental and recombinant types issuing from crosses involving the same genetic markers, when they are now placed in different strains. In addition to the many nuclear influences recognized, specific mitochondrial factors have also been identified, the best known of which is *omega* (ω, symbolized as *ome*), first defined by Bolotin *et al.* (1971). The *omega* locus, in its most common allelic forms *ome*$^+$ and *ome*$^-$, controls transmission and recombination properties of a group of markers close to *omega* on the mitochondrial genome. It is important to stress the limited role of *omega*, because its influence extends to markers in its direct vicinity only, comprising a region about 5% of the genome length (the *cap-ery* region), and thus has little or no influence on the behaviour of markers in the other 95% of the genome. The emphasis that mitochondrial geneticists once placed on *omega* as a key determinant in studies of recombination and mapping of mitochondrial genes came about because the first discovered drug-resistance loci were *ery* and *cap*, both in the region of *omega*. However, in 1973, it was found that loci such as *oli* (Avner *et al.*, 1973) and *par* (Wolf *et al.*, 1973) were not influenced by *omega*, and could not be mapped by transmission or recombination analysis. As described in Section V.D (p. 245), since 1975 analysis of *petites* has permitted construction of a genetic (Molloy *et al.*, 1975) and physical map (Linnane *et al.*, 1976) of the genome, and it became possible to place *omega* in a better perspective.

In the following pages, we describe the properties of mitochondrial genetic crosses, considering the interpretation of the relative abundance in the diploid progeny of parental and recombinant types, firstly regarding the frequency of transmission of individual markers. Secondly, recombination between markers is considered from the points of view of the frequency of recombination, as well as the relative abundance of reciprocal recombinant types, that is, the *polarity* of recombination. To consider the role of *omega*, it is convenient to classify mitochondrial genetic crosses as homosexual ($ome^+ \times ome^+$, or $ome^- \times ome^-$) or heterosexual ($ome^+ \times ome^-$).

The data obtained in two typical crosses are presented in Tables 4 and 5, including data from a homosexual cross (i), and a heterosexual cross (ii). In these crosses, a strain (863-2C) carrying five resistance markers was crossed with two different antibiotic-sensitive strains (L2300 and D253-9D). Each line in Table 4 is designated by an italized letter, with the bracketed symbols (i) and (ii) indicating the cross in question. To simplify the presentation of data from these crosses which involve five separate drug-resistance markers, we first treat the two crosses as three-factor crosses and show the frequencies of all parental (a, h) and recombinant classes (b through g) in Table 4. The transmission frequencies (j, k, l) recombination frequencies (m, n, o) and polarities of transmission (p, q, r) are shown for only the three markers *cap1*, *ery1* and *ana1*. In Table 5, these calculated parameters are shown for all five markers (i.e. including *oli1* and *par1*). The complete data for all genotypes, if presented, would comprise 32 different classes.

TABLE 4. Characteristics of non-polar and highly polar mitochondrial genetic crosses

Line number	Parameter	Cross (i) (homosexual)		Cross (ii) (heterosexual)	
		Number	Percentage of total	Number	Percentage of total
	Individual genotype				
a	$C^R E^R A^R$	380	41.4	248	31.3
b	$C^R E^S A^R$	49	5.3	31	3.9
c	$C^S E^R A^R$	15	1.6	0	0
d	$C^S E^S A^R$	74	8.1	0	0
e	$C^R E^R A^S$	47	5.1	178	22.4
f	$C^R E^S A^S$	50	5.4	250	31.5
g	$C^S E^R A^S$	15	1.6	2	0.3
h	$C^S E^S A^S$	288	31.4	84	10.6
i	Total number of diploids	918	100.0	793	100.0
	Total for single markers (transmission frequencies)				
j	$C^R (a + b + e + f)$	526	57	707	89
k	$E^R (a + c + e + g)$	457	50	427	54
l	$A^R (a + b + c + d)$	518	56	279	35
	Recombination frequency				
m	C, E $(b + c + f + g)$	129	14	283	36
n	C, A $(c + d + e + f)$	186	20	428	54
o	E, A $(b + d + e + g)$	185	20	211	27
	Recombination polarity	Ratio		Ratio	
p	C, E $[(b + f)/(c + g)]$	99/30 = 3.3		281/2 = 141	
q	C, A $[(e + f)/(c + d)]$	97/89 = 1.1		428/0	
r	E, A $[(e + g)/(b + d)]$	62/123 = 0.5		180/31 = 5.8	

Strains: 863-2C (rho^+ ome^+ $cap1$-r $ery1$-r $oli1$-r $ana 1$-r $par1$-r)
L2300 (rho^+ ome^+ cap-s ery-s oli-s ana-s par-s)
D253-9D (rho^+ ome^- cap-s ery-s oli-s ana-s par-s)

Cross (i) is 863-2C × L2300, cross (ii) is 863-2C × D253-9D. Crosses were performed by Dr N. Howell in our laboratory, using a random diploid-analysis procedure. The diploids were scored for sensitivity or resistance to all five drugs but, in this table, data are analysed for three out of five mitochondrial markers. See Table 5 for a summary of the five-factor analysis. Single-letter marker symbols are C, $cap1$; E, $ery1$; A, $ana 1$; superscripts R, S denote resistance and sensitivity, respectively.

(i) *Homosexual Crosses.* First consider homosexual crosses, as these crosses represent the most commonly reported crosses in the literature (Thomas and Wilkie, 1968; Coen et al., 1970; Bolotin et al., 1971; Kleese et al., 1972b; Rank and Bech-Hansen, 1972b; Rank, 1973; Waxman et al., 1973; Howell et al., 1973; Callen, 1974b; Goldthwaite et

TABLE 5. Further characteristics of non-polar and highly polar mitochondrial genetic crosses

Markers	Cross (i) (homosexual)		Cross (ii) (heterosexual)	
	Transmission frequency (%)			
cap1-r	57		89	
ery1-r	50		54	
oli1-r	61		33	
ana1-r	56		35	
par1-r	56		33	
	Recombination parameters			
	Frequency (%)	Polarity	Frequency (%)	Polarity
cap1, ery1	14	3.3	36	141
cap1, oli1	21	0.7	56	148
cap1, ana1	20	1.1	56	428/0
cap1, par1	19	1.4	56	443
ery1, oli1	22	0.3	26	7.8
ery1, ana1	20	0.5	26	5.8
ery1, par1	21	0.6	29	5.8
oli1, ana1	15	1.7	9	0.7
oli1, par1	19	1.6	18	1.0
ana1, par1	16	1.1	19	1.2

Crosses are as described in Table 4, namely:

cross (i) 863-2C (ome^+ cap1-r ery1-r oli1-r ana1-r par1-r) ×
L2300 (ome^+ cap-s ery-s oli-s ana-s par-s)
cross (ii) 863-2C (ome^+ cap1-r ery1-r oli1-r ana1-r par1-r) ×
D253-9D (ome^- cap-s ery-s oli-s ana-s par-s)

The data have been analysed for transmission and recombination of all five mitochondrial drug-resistance genes present in the cross. The frequencies of each genotype scored amongst the random diploids are not shown for the sake of clarity (see Table 4 for a partial detailed analysis of these crosses, using the data for the cap1, ery1 and ana1 markers only).

al., 1974). In the particular cross shown in Table 4 (i), an ome^+ × ome^+ cross is shown, but ome^- × ome^- crosses behave similarly (Bolotin et al., 1971).

Looking at the frequency of transmission of each marker individually, it is seen that slightly unequal representation of the R and S alleles of each marker is found in the diploid progeny (Table 5 (i)). The small difference between individual markers is probably not significant; all fall in the range 50–61% transmission of the resistant allele.

Recombination frequencies between marker pairs are rather high, falling in the range 14–22% (Table 5 {i}). There is close to equal representation of most reciprocal recombinant types, the polarities of recombination approximating one, ranging from not more than 3.3 or not less than 0.5. This cross is therefore described as *non-polar*.

Unequal distribution of the alleles of any one marker in the progeny (i.e. transmission frequency not exactly equal to 50%) was noted in the first single-factor crosses carried out (Linnane et al., 1968a; Saunders et al., 1971), and varies from cross to cross. It has been pointed out by Avner et al. (1973) that this unequal transmission in homosexual crosses should not be termed "polarity" of transmission, as these authors prefer to reserve this term for unequal transmission resulting from interactions between ome^+ and ome^- genomes in heterosexual crosses (see p. 237). Avner et al. (1973) proposed the term "asymmetry" when determination of the unequal transmissions is shown to be under the influence of nuclear genes, and "bias" when under influences of unknown origin. Some genetic determinants known to influence the parameters of crosses will be considered on p. 238.

Dujon et al. (1974) suggested that, because there is not a high polarity of recombination in a homosexual cross, and that all the markers in the cross showed a co-ordinate output (that is they all show similarly biased transmissions, as exemplified in Table 5 {i}), then frequency of transmission represents the relative input of markers into the cross. One factor controlling the relative input of each parent into the pool of mtDNA molecules in the zygote is the cellular level of mtDNA in each haploid. Goldthwaite et al. (1974) and Birky (1975a) could correlate the bias in transmission of markers in homosexual crosses (i.e. the input) with the cellular level of mtDNA in the haploids. These parameters were modified by growing the cells under different extents of glucose repression, and held for different periods of time in the stationary phase of growth. By crossing haploids and diploids, a similar correlation between the cellular level of mtDNA and the bias in marker transmission was reported by Gunge (1975). Further support for the suggestion of Dujon et al. (1974) comes from the ultraviolet irradiation of one parent entering a homosexual cross (Dujon et al. 1974, 1975), where it was found that there was a co-ordinate decrease in transmission of markers from the irradiated parent, suggesting a diminished input of mtDNA from that parent.

It has been noticed that the upper limit for recombination frequency between unlinked markers in a non-polar cross is 25%, or less if the input of parental genomes is unequal (Wolf et al., 1973). The upper limit of 25% probably results from a situation where multiple rounds of recombination occur in the pool of molecules in the zygote, and homologous pairs (i.e. molecules of the same parent pairing), as well as heterologous pairs, can be established leading to recombination. It is not possible from the data in Tables 4 (*i*) and 5 (*i*) to differentiate single, double (and triple) recombinant classes, since the recombination frequencies between different markers are found to be so similar (about 20%). This feature is termed positive co-incidence (or, alternatively, negative interference between linked markers), and has been suggested to be a consequence of the occurrence of multiple rounds of recombination (Kleese et al., 1972b; Dujon et al., 1974; Linnane et al., 1974). Unfortunately, it precludes mapping of markers in homosexual crosses by classical three point analysis.

(ii) *Heterosexual Crosses.* In Tables 4 (*ii*) and 5 (*ii*) are presented the results of a heterosexual cross between an ome^+ strain with five drug-resistance genes (863-2C) and an ome^- strain sensitive to the antibiotics (D253-9D). Heterosexual crosses have not been described as frequently as homosexual crosses, and typical data come from at least three laboratories (see e.g. Linnane et al., 1974; Dujon et al., 1974; Wakabayashi, 1974a).

Considering Table 4 (*ii*), it can be seen that a greater extent of recombination occurs than in the corresponding homosexual cross (Table 4 {*i*}), since the parental genotypes in (*ii*), *a* and *h*, account for only about 42% of the genotypes recovered whilst, in (*i*), *a* and *h*, they account for 73% of the genotypes, However, amongst the recombinant classes, three configurations are not recovered at all or are found in very low frequency (*c, d, g*). This gives rise to extremely high polarities of recombination in some cases. The polarities of recombination listed in Tables 4 (*ii*) and 5 (*ii*) show that all pairs including *cap1* show polarities exceeding 100, and pairs including *ery1* exceed 6 (significantly different from unity). However, pairs not including *cap1* or *ery1* show polarities not significantly different from unity. In this sense, one can refer to markers in a "polar" region of the genome (i.e. *cap1*, *ery1*) and markers in a "non-polar" region (e.g. *oli1*, *ana1*, *par1*). The vast majority of the genome is non-polar. The great inequalities in the abundance of the various genotypes lead to a great polarity of

STRUCTURE, SYNTHESIS AND GENETICS OF YEAST MITOCHONDRIAL DNA

transmission of the *cap1-r* marker (89%). The *ery1-r* marker shows 54% transmission whilst the remaining three (non-polar) markers (*oli1-r*, *ana1-r* and *par1-r*) show 33–35% transmission (Table 5 {*ii*}). By convention, transmission frequencies of the markers originating from the *ome*⁺ strain are used.

Considering the frequencies of recombination in Tables 4 (*ii*) and 5 (*ii*), it is seen that the upper limit of 25% recombination observed in homosexual crosses does not obtain for pairs involving markers in the polar region in the heterosexual crosses. Thus, the *cap1, ery1* recombination frequency is 36% compared with 14% in the homosexual cross. Recombination between *cap1* and the markers in the non-polar region is 56%. Recombinants between *ery1* and the non-polar markers are seen in 26–29% frequency. In all, the recombinant groups which involve polar and non-polar markers together, polarities are high and the majority recombinant type carries the polar marker of the *ome*⁺ parent (consider, e.g. $q(ii) = (e + f)/(c + d)$). In the *cap1, ery1* pair involving two polar markers, the majority recombinant classes $(b + f)$, carry the *cap1-r* allele which originated from the *ome*⁺ parent and the *ery-s* allele originating from the *ome*⁺ parent. The minority configuration *cap-s, ery1-r* (*c* and *g*) hardly turns up at all (2 out of 793 diploids analysed).

These extreme polarities of recombination are responsible for the high polarity of transmission of *cap1-r*, and to a lesser extent *ery1-r*, compared with the non-polar markers originating from the *ome*⁺ parent. It should again be noted (Avner *et al.*, 1973; Wolf *et al.*, 1973) that the 35% transmission of non-polar markers is not to be regarded as a polarity of transmission but as a biased distribution probably resulting from unequal inputs into the pool of molecules in the zygote as contributed by each parent haploid cell (Dujon *et al.*, 1974). Co-ordinate output of the non-polar markers, and frequencies of recombination amongst themselves (Table 5 {*ii*}), are very similar to the behaviour of all markers in a non-polar homosexual cross (Table 5 {*i*}).

Looking at the three markers followed in Table 4 (*ii*), it can be seen that a marker order is suggested by both the relative transmission frequencies of the markers as well as the recombination frequencies (Avner *et al.*, 1973; Wolf *et al.*, 1973; Netter *et al.*, 1974; Linnane *et al.*, 1974; Wakabayashi, 1974a). The order derived is: *cap1–ery1–ana1* from transmission frequencies. By comparing recombination frequencies, the *cap1, ana1* frequency (54%) is close to the sum of the frequencies of the other two pairs (36 + 27 = 63%), in agreement with this arrangement. Nevertheless, recombination frequencies bear no relation to the map distance; physical mapping of markers is considered in more detail later (Section V.D.3, p. 249) but it can be pointed out here that the 36% recombination between *cap1* and *ery1* is out of proportion to the short distance (five units) between these loci. Moreover, the order of non-polar markers cannot be obtained even in heterosexual crosses, using transmission and recombination analysis, so it is almost impossible to judge the map distance from recombination frequency.

3. *Genetic Factors Influencing Transmission and Recombination*

Transmission and recombination of mitochondrial genes are susceptible to influence by genetic factors located on mtDNA and nuclear DNA. The genetic factors described to date are summarized in Table 6.

The role of *omega* in determining the behaviour of certain mitochondrial genes has already been discussed. The ome^+ allele is found to be present in randomly selected strains more frequently than the ome^- allele. A third form of *omega* has been described by Dujon et al. (1976), which leads to essentially non-polar behaviour in crosses with both ome^+ and ome^- strains. This allele, termed ome^n (for neutral), was found by a study of particular mutants at the *RIB1* locus (chloramphenicol resistance). Similar changes in *omega* behaviour have been noticed in mutants at the *cap2* locus (N. Howell, H. B. Lukins and A. W. Linnane, unpublished data). The ability to obtain ome^n from ome^- strains, but not vice versa (Dujon et al., 1976), suggests ome^n may be a deletion of some or all of ome^-. It is very difficult to obtain ome^n from ome^+ strains (Dujon et al., 1976).

In addition to the meiotic non-segregation behaviour of *omega* (Boloton et al., 1971), the *omega* locus has been shown to be deleted or retained in different *petite* isolates (Molloy et al., 1975), thus verifying its location on mtDNA. The loss of *omega* from a *petite* (ome^0) can change the general pattern of transmission and recombination of loci retained in the *petite*, from that of the parent (Molloy et al., 1975). Some *petites* derived from an ome^+ strain and retaining only the *RIB1* (*cap*) locus show modified recombination properties with ome^+ and ome^- *grande* strains, but are totally unable to recombine with ome^n *grande* strains (Dujon et al., 1976; Dujon and Michel, 1976). In other cases, it is apparent that secondary deletions and perhaps other re-arrangements in *petite* mtDNA may alter the transmission and recombination

TABLE 6. Genetic factors influencing the transmission and recombination of mitochondrial genes

Type of cross	Loci[a] tested	Parameter[b] affected	Nature of change	Inheritance[d] of factor	References
1. OMEGA					
Homosexual	C, E	Pol. trans.	Defines sexuality	Mitochondrial	Bolotin et al. (1971)
Heterosexual		Pol. of recomb.			
2. OTHER FACTORS					
Homosexual	E, O	Pol. of recomb.	[c]	Mitochondrial	Linnane et al. (1974)
Homosexual	C, E	Freq. of recomb.	Decreased	Mitochondrial	Gouhier-Monnerot (1974)
Homosexual	S, O	Freq. of recomb.	Increased	Nuclear	Callen (1974a, b)
Heterosexual	C, E, O	Freq. of trans.	[c]	Nuclear	Wakabayashi (1974b)
Homosexual	E, O	Freq. of recomb.	Decreased	Nuclear	Linnane et al. (1974)
Homosexual	C, E	Freq. of recomb.	Decreased	Nuclear	Woll et al. (1973)
Homosexual	C, E	Freq. of recomb.	Increased	Complex	Avner et al. (1973)
Homosexual	C, E, O	Transmission of one parental	Increased	Nuclear	Avner et al. (1973)
Homosexual	C, E. O	Transmission of one parental Freq. of recomb.	Greatly increased Decreased	Nuclear	Waxman et al. (1973)
Homosexual	C, O	Transmission of one parental	Greatly increased	Nuclear	Birky (1975b)

[a] C, S, E, O designate loci conferring resistance to chloramphenicol, spiramycin, erythromycin and oligomycin, respectively.
[b] Pol. of trans. indicates polarity of transmission; Pol. of recomb. indicates polarity of recombination; Freq. of trans. indicates frequency of transmission; Freq. of recomb. indicates frequency of recombination.
[c] In some instances, "nature of change" is not applicable, namely where a normal or standard pattern does not obtain. In such cases, different alleles of the factor control particular levels of a parameter.
[d] In this table, the inheritance is judged mainly on the basis of the pattern of segregation of the factor at meiosis.

properties of loci in crosses between *petites* and *grandes* (Perlman, 1976; Sriprakash et al., 1976b).

Studies in our own laboratory have shown that *omega* is not the only determinant of recombination polarity, although it is the only one discovered to date which causes the very highly polar behaviour. Howell et al. (1973) observed a wide range of recombination polarity values which were arbitrarily grouped into three classes, namely nonpolar (polarity not significantly different from unity), low polar (polarities between 0.2 and 3) and highly polar (polarities less than 0.1 or greater than 10). Crosses between closely related strains were nonpolar or low polar, whilst crosses of unrelated strains displayed a wide spectrum of polarity values. It was concluded that polarity must be under the control of multiple genes. Linnane et al. (1974) examined this question further, and were able to show that one particular extranuclearly inherited factor causes crosses with standard tester strains to show polarities of about 0.5 and frequencies of recombination between the *oli1* and *ery1* markers to be 11–15%. The other allele of this (unnamed) locus leads to polarities of about 1.4 and frequencies of recombination of 20–23% for the same pair of markers. It is also apparent from studies on the frequencies of recombination involving mit^- loci determining cytochrome *b* (Kotylak and Slonimski, 1976; Rytka et al., 1976) that there may be highly localized regions of high recombination frequency at places on the genome other than the region controlled by *omega*.

The majority of the genetic factors listed in Table 6 affect the frequency of recombination, or the frequency of transmission of markers, without influencing the polarity of recombination. The mitochondrial factor increasing the frequency of recombination between the *RIB1* and *RIB3* loci (*cap* and *ery*) also confers EthBr resistance (Gouhier-Monnerot, 1974); the latter phenotype depends on mitochondrial protein synthesis (Gouhier and Mounolou, 1973).

Callan (1974a, b) described nuclear genes which determine transmission of markers in homosexual crosses. In the crosses studied by Callen, strains with nuclear mating types α always showed higher transmission of mitochondrial genomes to diploids than *a* strains. This correlation had been proposed earlier (Saunders et al., 1971; Coen et al., 1970) but was later discarded (Bolotin et al., 1971; Howell et al., 1973). However, Callen (1974a, b) used a set of almost completely isogenic strains constructed by repeated backcrosses to a given parent strain. This would eliminate other extraneous influences apart from the nuclear a and α factors studied by Callen. However, the data of Forster and Kleese (1975b), who also used isogenic strains, did not support the view that a or α play an important role in the transmission frequency.

Other factors listed in Table 6 affect the frequency of recombination in homosexual crosses. Avner et al. (1973), Waxman et al. (1973) and Birky (1975b) described nuclear factors which lead to an increased transmission of one particular parental type. It was found by Waxman et al. (1973) that the presence of cycloheximide (but not antimitochondrial antibiotics) caused abnormal crosses to revert to a more usual type of non-polar cross. This also indicates the role of a nuclear product in transmission of mitochondrial genes.

4. Analysis of Individual Zygotic Clones

(i) *Whole Zygotic Clones.* Zygotic clones can be isolated either by prototrophic selection of single colonies immediately after mating, or by micromanipulation of zygote forms under the microscope. Early studies on single-factor crosses showed that zygotic clones were mixed for cells carrying markers of either parent and that, among different clones, transmission frequency of individual markers varies widely (Linnane *et al.,* 1968a; Saunders *et al.,* 1971; Coen *et al.,* 1970). The large distribution amongst single zygotic clones is one of the features of mitochondrial genetics which is evidently a system of population genetics at a molecular level. Analogies between mitochondrial genetic behaviour and bacteriophage genetic systems have been drawn by many authors (Kleese *et al.,* 1972b; Wilkie and Thomas, 1973; Wolf *et al.,* 1973; Dujon *et al.,* 1974; Linnane *et al.,* 1974).

In a non-polar cross involving the *ery1* and *oli1* markers, a wide variation in the frequencies of parental and recombinant genotypes was observed in individual zygotic clones (Lukins *et al.,* 1973). Thus the polarities of recombination in a similar cross were distributed broadly, but equally, about the population mean (Linnane *et al.,* 1974). However, in a highly polar cross, a broad distribution was again found but in this case heavily skewed. Thirty-one out of forty-two clones possessed only one of the two reciprocal recombinant types in the highly polar cross (Linnane *et al.,* 1974).

Heterogeneity of zygotic clones in non-polar crosses was emphasized by zygotic clonal analysis of a non-polar three factor cross (*cap1, ery1, oli1*). In contrast to heterosexual highly polar crosses, where a unique marker order is found amongst the progeny diploids (cf. Table 5 {*ii*}), zygotic clones in the non-polar cross showed various orders of transmission, with any of the possible orders found (Linnane *et al.*, 1974). The net effect is for a co-ordinate output of markers from the cross (Table 5 {*i*}).

(ii) *Zygote Cell Lineages*. Micromanipulation techniques permit separation of bud cells from the mother. The genetic constitution of clones derived from individual buds arising from zygotes, of buds arising from other diploid bud cells (secondary, tertiary buds, etc.) and of the residual zygote, can be determined in an analogous manner to that of whole-zygote clones. These analyses potentially provide further information about recombination and segregation of mitochondrial genomes.

Saunders *et al.* (1971) analysed transmission of the *ery1* locus in a single-factor cross, and showed that clones derived from buds could be mixed for cells of different genotypes. Moreover, the proportion of cells carrying one allele did not necessarily correlate with the proportion of that allele in the residual zygote. A study of two-factor (Lukins *et al.*, 1973; Callen, 1974c) and three- and four-factor crosses (Wilkie and Thomas, 1973) extended these observations to show segregation of parental and recombinant types to early buds. In these studies, some buds were pure for just one parental type. Where a recombinant type was found, the reciprocal recombinant type was usually absent from that bud clone. In a few late buds in the two-factor crosses, both parental and both recombinant genotypes were found in the one clone. Wilkie and Thomas (1973) point out that some genotypes apparently pass through a bud without replication or expression and then, in a secondary or tertiary bud, these replicate to form the major mitochondrial type. Segregation of mitochondrial genomes from the zygotes into primary buds, and then into later buds, is thus a complex phenomenon of selection of perhaps a relatively small number of molecules from pools in the zygote (Forster and Kleese, 1975b), and possibly also in the early buds. Schemes have been proposed in which destruction of molecules as well as selection occurs (Birky, 1975a, b).

Callen (1974c) observed that buds separating from the extremities of a zygote frequently contained exclusively cells of a parental genotype, whilst buds issuing from the central constricted region of the dumb-bell-shaped zygote contained cells of both parental and recombinant genotypes. This may imply that the pool of recombining mtDNA molecules is formed in this central region, and complete mixing of all parental mtDNA molecules does not necessarily occur. However, the data of Aufderheide (1975) and Forster and Kleese (1975b) suggest that there is no relationship between the site of bud formation and the genetic constitution of the bud.

Fusion of two haploid cells to form a zygote has been studied by ultrastructural techniques (Smith *et al.*, 1973). It was found that degeneration of mitochondrial membranes occurs before nuclear fusion takes place, and persists until the time of bud formation when normal mitochondrial structure is restored. These events co-incide with establishment of the recombining molecular pool, followed by segregation of product molecules into buds, respectively. Timing of the recombination events is difficult to establish (Lukins *et al.*, 1973). Callen (1974c) suggests recombination events are restricted to the zygote, whilst Wilkie and Thomas (1973) interpret their results to imply that further recombination events occur in early buds.

5. *Molecular Models of Recombination*

In the absence of any direct experimental evidence concerning interaction of molecules of mtDNA in zygotes, molecular models of mitochondrial genetic behaviour are necessarily limited by the genetic evidence available. Instead of reviewing in detail the various models that have been proposed, we choose here to comment briefly on certain features of the models.

Much attention has been given to the highly polar behaviour of *cap* and *ery* loci controlled by *omega* which, as already pointed out (p. 231), represents a rather limited region of the genome. Early models sought to explain the polar behaviour in terms of bacterial conjugation analogies, with one genome acting as a donor, with an ordered transmission of markers to the other genome in the cross (Coen *et al.*, 1970; Bolotin *et al.*, 1971; Wakabayashi, 1974a). These schemes were later shown to be inconsistent with the genetic evidence obtained when the behaviour of non-polar markers in heterosexual crosses was examined (see Gillham, 1974). Moreover, they did not provide an adequate explanation of non-polar behaviour.

More recently, Dujon *et al.* (1974) formulated a general scheme to explain both polar and non-polar behaviour in which it is estimated that about four rounds of recombination occur in the zygote, involving both homologous and heterologous pairing. A key feature of the model of Dujon *et al.* (1974) is that each individual recombination event is non-reciprocal. The absence of polarity of recombination in homosexual crosses at a population level implies that events generating either one or the other of the reciprocal recombinant types occur with equal probability. Each individual event is proposed to involve a gene conversion process where one recombinant and one parental genome arise from each molecular pairing. Another feature of the scheme would be that pairing and hence the gene-conversion process could be initiated at many points in the genome since all markers showing non-polar behaviour have a co-ordinate output overall, yet in individual zygotic clones, a variety of gradients of transmission of markers as well as different marker orders are observed (Linnane *et al.*, 1974; Birky, 1975b).

The special nature of the polar behaviour in heterosexual crosses has been considered to result from some obligatory recombination event in the polar region of the genome (Linnane *et al.*, 1974). Dujon *et al.* (1974) propose a special case of their scheme of recombination proceeding through a gene-conversion process to explain the influence of *omega* in this situation. This specific mechanism provides for any pairing of ome^+ with ome^- resulting in an obligatory gene-conversion process being initiated at or near the *omega* site, in which the ome^- genome is effectively destroyed and is replaced by a copy of the ome^+ genome. This gene-conversion process has a greatly diminishing probability of continuing the further away from *omega* it proceeds. This model explains the steep gradient of polarity of transmission of polar markers through to non-polar markers in heterosexual crosses, and also explains the frequencies of appearance of the various majority recombinant types and parental types. However, the scheme fails to account for the appearance of minority recombinant types, as well as the distribution of alleles of the *omega* locus amongst them (Howell *et al.*, 1974a). It is likely that in heterosexual crosses the recombination events in non-polar regions of the genome proceed as in homosexual crosses. Thus the output of a heterosexual cross represents the sum total of these non-polar interactions together with the polar effects influenced by *omega*, which are largely restricted to the polar region of the genome and probably predominate in that region. Although it is likely that the various alleles of *omega* represent deletions or repetitions of base sequences (Perlman and Birky, 1974; Dujon *et al.*, 1976), molecular studies have yet to define the particular changes involved (Dujon and Michel, 1976).

D. GENETIC AND PHYSICAL MAP OF YEAST mtDNA

Mapping of the yeast mitochondrial genome is an area in which considerable success has been obtained only very recently. This has of course depended on the availability of genetic markers but, as already discussed, mapping procedures involving recombination between markers in crosses between *grande* strains have yielded only limited information about a region of the genome close to *omega*. Techniques using deletion mapping procedures with *petite* mutants first provided a genetic map (Molloy *et al.*, 1975), which was extended by use of molecular hybridization procedures to a physical map (Sriprakash *et al.*, 1976b; Linnane *et al.*, 1976; Nagley *et al.*, 1976) in which physical distances between the genetic loci were measured. Application of these types of approaches is discussed in the following section.

1. *Mapping by Transmission and Recombination*

In heterosexual crosses ($ome^+ \times ome^-$), analyses of the transmission and recombination properties of markers in the polar region of the genome allow a map order to be established within this limited region, as discussed above (Table 4 and 5 in Section V.C.2, p. 231). From data presented here, the order: *cap1, ery1, (oli1, ana1, par1)* is apparent. Linnane *et al.* (1974) in similar experiments determined the order: *spi4, spi3, ery1, (mik1, oli1)* (N.B. *mik1* has since been renamed *ana1*). In Slonimski's laboratory (Avner *et al.*, 1973; Wolf *et al.*, 1973; Netter *et al.*, 1974), the order of their loci was determined to be: *omega, RIB1, RIB2, RIB3* (O_I, O_{II}, P) (N.B. *RIB1, RIB2, RIB3* correspond to our *cap*, *spi* and *ery* loci respectively, and P is a paromomycin-resistance locus). In all of these map orders, the brackets enclose the non-polar markers which cannot be ordered using crosses between *grande* strains, except for rare exceptions where the markers are fairly tightly linked. For example, Trembath *et al.* (1976) determined the order *oli* (classes A, B)-*ana1(mik1)-oli* (class C) by recombination in non-polar crosses. More typically, however, the high negative interference causes most markers to behave as though they are unlinked.

Location of *omega* was assumed by Slonimski and his colleagues to be at the end of the array of polar genes, as part of their model (Dujon et al., 1974) in which polarity is manifested originating at *omega* and diminishing gradually through the polar region until the non-polar genes are reached. However, it is important to note that *omega* had not been mapped, as such, by these workers. Studies in our laboratory suggested that *omega* may lie amongst the polar markers. On the basis of the frequencies with which ome^+ and ome^- alleles were present in minority recombinant types from a heterosexual cross, Howell et al. (1947a) suggested that *omega* lies between the *cap1* and *ery1* loci. *Petite*-deletion analysis data (Molloy et al., 1975) and restriction-enzyme analysis of *petites* in the *cap1–ery1* region of the genome (Heyting and Sanders, 1976) support this conclusion. It is conceivable that the influence of *omega* may be spread bidirectionally in the terms proposed by Dujon et al. (1974).

2. Petite *Deletion Analysis*

The only successful approach that has so far been used to map markers on the *Sacch. cerevisiae* mitochondrial genome relies on the use of *petite* mutants as deletion mutants. The deletions are random in that any region of the genome can apparently be retained. The rationale behind *petite* deletion analysis is that the closer two loci are on mtDNA the less likely is there to be a fragmentation event between them during *petite* formation. Thus, two markers positioned close to each other will be either deleted together, or retained together, when a population of *petites* is isolated from a *grande* strain. This genetic mapping procedure involves examination of markers retained or lost in a large population of *petites* derived from a *grande* strain, and measurement of the frequencies with which different markers are retained or lost together.

In practice, it is convenient to analyse *petites* from a *grande* carrying several antibiotic-resistance genes. For a generalized antibiotic-resistance locus *ant*, the *grande* strain in this case has the *ant-r* allele. The *petites* will either be *ant-r* or *ant-o*, representing the locus retained or the locus deleted, respectively (see e.g. Gingold et al., 1969; Saunders et al., 1971; Bolotin et al., 1971; Deutsch et al., 1974). The test used for retention or loss of the *ant* locus is indirect, because *petites* cannot grow on non-fermentable substrates on which drug-resistance tests are usually carried out. The *petites* must be crossed with another *grande* carrying the *ant-s* allele. Appearance of antibiotic-resistance *grande* diploids shows that the *ant-r* allele was carried in the *petite*. If the *grand* diploids issuing from the cross are all *ant-s*, the *petite* was *ant-o*.

Molloy *et al.* (1975) analysed spontaneous *petites* from several *grande* strains involving five drug-resistance markers. Spontaneous *petites* were initially chosen because these would be expected to have fewer double-deletion events which could lead to *petites* retaining non-continuous segments of mtDNA, and thus interfere with the ability of the

TABLE 7. Frequencies of different genotypes among spontaneous *petite* isolates from a strain carrying five drug-resistance markers (863-2C)[a]

		Petite genotypes[b]			Frequency of[c] genotype (percent)
par1	*ana1*	*oli1*	*ery1*	*cap1*	
0	0	0	0	0	60
r	r	r	r	r	10
r	0	0	0	0	7.2
0	0	r	r	r	4.2
r	0	0	r	r	4.2
r	0	r	r	r	3.7
r	r	r	0	0	3.0
0	r	r	r	r	2.8
0	0	0	r	r	1.6
r	r	0	0	0	1.2
0	r	r	0	0	1.2
0	0	r	0	0	0.4

[a] Data from Trembath *et al.* (1976).
[b] The symbols r and 0 denote retention or deletion, respectively, of the particular locus.
[c] The spontaneous *petite* frequency was 3% of total cells in the culture; 427 *petites* were analysed. The frequency of single-marker retention amongst the *petites* is: *cap1*, 26%; *ery1*, 26%; *oli1*, 25%; *ana1*, 18%; *par1*, 29%.

procedure to yield a map. The type of data obtained by Molloy *et al.* (1975) is illustrated here in Table 7, which shows the patterns of gene retention in 427 spontaneous *petites* from a strain 863-2C carrying all five antibiotic-resistance markers. Genotypes of the *petites* have been arranged in order of decreasing frequency, with the top two lines representing complete loss of all markers (60% of *petites* analysed), and retention of all five markers (10%), respectively. The remainder of Table 7 shows the frequencies of *petite* genotypes with the combination of markers that were detected. Many of the theoretically possible combinations were not found. Considering the data, it is seen that a definite pattern is present with particular combinations of markers favoured over others. The marker columns have been arranged in Table 7 in order consistent with the unique map order derived from these analyses. The ability to obtain a unique order suggests that these spontaneous *petites* in general contain continuous segments of the *grande* mtDNA. In the arrays shown in Table 7, *petites* were not found which contained segments which would be considered non-continuous (i.e. arising from multiple deletion events), according to the order deduced. The ability to map in this way also depends on the analysis of a large number of freshly arisen *petites*. This should be contrasted with the small set of EthBr-induced and extensively subcloned *petites* considered by Faye et al. (1975), where unique map order could not be established. Subcloning of the *petites* to stability led, in many cases, to selection of *petites* with internal deletions (which may have arisen as secondary deletions after the initial mutagenic event).

It is important to note that the order shown in Table 7 should be considered as a circular array, in view of the fifth and sixth lines, which show that *par1-r cap1-r ery1-r petites* and *par1-r ana1-r oli1-r petites* occur in similar frequency. This circular genetic map (Molloy et al., 1975), and the order of markers (*cap1–par1–ana1–oli1–ery1–cap1*), provide a basis for the physical mapping studies discussed on p. 249. A similar coretention and loss analysis by Trembath et al. (1976) placed the O_{II} locus (class C) between *ana1* and *par1*, unlike *oli1* (class A) which lies between *ery1* and *ana1* (see Fig. 1, p. 229).

A similarly circular genetic map was later obtained by Schweyen et al. (1976b) who used a nuclear mutation (*tsp25*) which leads to production of a high *petite* frequency at 35°C. By introducing this *tsp25* locus into a strain carrying three mitochondrial antibiotic-resistance genes *cap*(C^R_{321}), *oli*(O^R_I class B) and *par* (P^R_{454}), and growing the strain at 35°C, large numbers of *petites* were produced. The *petites* were not only screened for retention of drug-resistance loci, but also by crossing each *petite* to three different *mit*⁻ strains, the presence or loss of the corresponding wild-type *mit*⁺ locus in the *petite* clones could be scored by the appearance or absence, respectivley, of *grande* diploids. The data obtained for the *petites* yielded the order of C–M_I–M_{III}–P–M_{IV}–M_{II}–O–C, where M_I is the temperature-sensitive locus *TSM1* (also called T^S_8) and M_{II}, M_{III}, M_{IV} are *mit*⁻ loci whose effects have not been biochemically characterized.

Essentially the same order of markers has been established by these and other studies (Schweyen et al., 1976a; Fukuhara et al., 1976; Slonimski and Tzagoloff, 1976) on *petite* mutants. Distances between markers can only be estimated on the basis of calculations of the frequency with which two markers are separated (e.g. for the marker pairs *cap1, par1* the frequency of separation would be [*cap1-r par1-o*] + [*cap1-o par1-r*]). For most marker pairs for the data shown in Table 7, the frequency of separation exceeds 10%, but a striking exception is the *cap1 ery1* pair (<0.4%). This very low frequency of separation between *cap* and *ery* loci has been observed elsewhere (Uchida and Suda, 1973; Suda and Uchida, 1974; Wakabayashi and Kamei, 1973; Deutsch et al., 1974; Michels et al., 1974; Molloy et al., 1975; Trembath et al., 1976) and implies that these two loci are very close to the genome. Schweyen et al. (1976b) applied a more complex formula to derive apparent map distances between their markers, but it is to be emphasized that the distances so estimated are only approximations. The method of estimation assumes that *petite*-deletion events are entirely random and that each mtDNA fragment produced has an equal chance of survival. The difficulties in estimating distances from genetic analysis of *petites* are apparent in the data of Schweyen et al. (1976a) who studied spontaneously arising subclones of ethidium bromide-induced *petites*. It was not possible to estimate the *oli* (class B)-*par* distance compared to the other distances (e.g. *par-cap*) from this analysis. It is thus evident that the only way to obtain accurate physical map distances is to combine genetic analysis of *petites* with physical measurements on the mtDNA that they contain. These studies are described in the following section.

3. *Physical Mapping Using* Petites *and Molecular Hybridization*

The technique we have used to solve this problem and establish a physical map involves isolation of stable *petites* retaining genetically defined segments of the mitochondrial genome, and determination by DNA–DNA hybridization analysis of the physical length and location of the segment of *grande* mtDNA retained in such *petites* (Sriprakash et al., 1976b; Linnane et al., 1976; Nagley et al., 1976).

a. *Establishing the Library of* Petite *Mutants with Physically Defined Segments of mtDNA.* A series of spontaneously arising *petites* retaining known antibiotic-resistance genes was prepared by repeatedly subcloning primary *petite* isolates of the desired genotypes, and selecting for the desired genotype in the subclones. Spontaneous *petites* were chosen to minimize the occurrence of secondary deletions. For each stable *petite*, the fraction of the *grande* genome represented by the sequences in the *petite* mtDNA was determined by DNA–DNA hybridization. Excess denatured *petite* mtDNA was bound to filters, and the ability of the filters exhaustively to hybridize a small amount of very highly labelled *grande* mtDNA in the liquid phase was determined. The fraction of labelled *grande* mtDNA bound to the filters represents the fraction of *grande* sequences retained in each *petite*. The hybridization procedure was chosen in order to obtain the best estimate of sequence homology between each *petite* and *grande* mtDNA, as even the least amplified sequences in a *petite* are detected by this procedure (see p. 199).

The next step was to position the various *petite* DNA segments with respect to one another, by measuring the extent to which different *petite* mtDNA segments overlap one another. In practice, this has been done by choosing a reference *petite*, and determining the extent of sequence homology between this *petite* and other *petites*, using DNA–DNA hybridization. For each such overlap measurement, the particular end of the reference *petite* which has sequences in common with any other *petite* was determined by consideration of the genetic loci retained in common between the *petites*. For correct orientation of the DNA segments with respect to one another, knowledge of the order of markers in the genome (from coretention and loss analysis) is prerequisite. The first group of 13 *petites* that we characterized (Sriprakash et al., 1976b) were positioned by measuring sequence homology of their genomes to a reference *petite* (U4) retaining four loci ($cap1$, $ery1$, $oli1$ (class A), $ana1$) in a segment of mtDNA 36 units long (36% of the *grande* genome). Hybridizations were carried out between labelled DNA containing the sequences of the U4 genome and filters bound with excess mtDNA from each of the other *petites*. In more recent work (Choo et al., 1977), we confirmed the validity of this procedure by using a second reference *petite* (Y1.2, retaining the $ana1$ and $par1$ loci in a segment of mtDNA 43 units long) and an alternative hybridization procedure involving measurements of the ability of labelled *grande* mtDNA to hybridize to two *petite* mtDNAs bound individually to separate filters, or to a double-DNA filter loaded with both *petite* DNAs together. The hybridization value of the double-DNA filter is compared to the sum of the hybridization values of the two individual DNA filters. The difference represents the extent of overlap between the *petite* mtDNAs, because the unlabelled DNA on the filter is always in vast excess over the labelled DNA. By this procedure, the positions of the first set of *petite* genomes has been confirmed, and a further seven molecularly defined *petites* have been added to the collection.

b. *Use of* Petite *Library for Mapping Genetic Loci and RNA Genes.* This collection, or library (Linnane *et al.*, 1976), of genetically and molecularly defined *petites* can be used for physically mapping genetic loci as follows (Nagley *et al.*, 1976). Using an *inclusion principle,* a locus is taken to lie within the segment of mtDNA common to all *petites* retaining that locus. Secondly, an *exclusion principle* is also used to map physically, whereby a locus is taken to lie outside the mtDNA segments of all *petites* from which this locus has been deleted.

Using these principles, Sriprakash *et al.* (1976b) determined the map positions (Fig. 1, p. 229) of the five drug-resistance loci for which the original group of 13 *petites* had been selected (*cap1, ery1, oli1, ana1, par1*). The 0/100 point on the map is one end of the U4 genome (closest to *cap1*). However, some of these early estimates of the segments of mtDNA within which loci were mapped were relatively long—thus *cap1* and *ery1* were found to lie within a 15-unit segment, and *par1* was located in a 12-unit segment of mtDNA (Sriprakash *et al.*, 1976b; Linnane *et al.*, 1976). A more recent evaluation of the data (Nagley *et al.*, 1976) and consideration of further *petites* enable estimation of the *cap1–ery1* segment to have a maximal length of 5 units (3750 base pairs) and the *par1* locus to lie within a segment 2 units long (1500 base pairs) (Choo *et al.*, 1977), as shown in Fig. 1. The position of the O_{II} (class C) locus has also been determined to lie within a two-unit segment (1500 base pairs) using the new group of *petites* (Choo *et al.*, 1977).

The great advantage of the molecularly defined library of *petites* is that the positions of other genetic loci not directly selected for in establishing the library can readily be determined. All that is required is a mutant carrying a defined locus which can be mated to each of the *petites* in the library. The *petites* are thus tested for their retention or loss of the wild-type allele of the locus in question. For a mutant carrying a drug-resistance locus, the appearance of *grande* diploid clones sensitive to the drug indicates the presence of the wild-type (sensitivity) locus in the *petites*. For *mit*⁻ mutants, the appearance of *grande* diploid clones as such indicates retention in a *petite* of the wild-type allele (*mit*⁺) of that locus. In both cases, failure to detect the wild-type allele of the locus in the diploids is taken to indicate deletion of that locus from a *petite*. The inclusion and exclusion principles of physical mapping can now be applied to map the locus, and it is clear that, using the molecular library of *petites*, the whole mapping process can take place in less than a week for any new locus as soon as it is discovered (Linnane *et al.*, 1976; Rytka *et al.*, 1976). This is a general solution to the physical mapping problem.

Using these methods, the map positions of the oligomycin-resistance loci class B (e.g. *OLG2*) and class C (O_{II}) were determined (Trembath *et al.*, 1976). The temperature sensitivity locus *tsr* has been mapped within the same five-unit segment of mtDNA as the *cap1*, *ery1* and 21S rRNA gene (see next paragraph; K. J. English, T. W. Spithill, P. Nagley, H. B. Lukins and A. W. Linnane, unpublished data), and the *mit*⁻ loci *cya1*, *cya2*, *cya3* and *cyb1* have also been physically mapped (Rytka *et al.*, 1976; Choo *et al.*, 1977; see Fig. 1, p. 229).

The molecular library of *petites* can also be used to map the positions of defined base sequences in mtDNA, such as ribosomal RNA genes. This is done by testing the ability of each of the *petite* mtDNAs to hybridize to labelled 15S and 21S rRNA molecules from wild-type mitochondria (Sriprakash *et al.*, 1976a). The map positions of the two rRNA genes are shown in Fig. 1, and it is apparent that the two rRNA genes are separated by at least 35 units, which agrees very well with the separation determined by hybridizing the two rRNAs to restriction-enzyme fragments of mtDNA (Sanders *et al.*, 1975b) whose relative positions around the genome had been determined (Sanders *et al.*, 1975a, 1976). This agreement gives strong support to the validity and accuracy of using the *petite* library.

Previous studies on the ability of genetically defined *petite* mtDNAs to hybridize to rRNA had shown that the 21S rRNA was close to the *cap–ery* region of the genome (Nagley *et al.*, 1974b, 1975b; Faye *et al.*, 1974), whilst the 15S rRNA was near the *par* region (Faye *et al.*, 1975). However, these studies cannot be considered as physical mapping because, at the time they were carried out, information on the physical distances between the genetic loci was not available. More importantly, the segments of *petite* mtDNA used had not been characterized for their lengths and relative positions with respect to one another, as required for proper physical mapping. Similar considerations apply to the tRNA genes that have been studied by measuring the ability of specific aminoacyl-tRNAs to hybridize to mtDNA from genetically defined *petites* (Casey *et al.*, 1974b; Martin *et al.*, 1976; Faye *et al.*, 1976; Bolotin-Fukuhara *et al.*, 1976). These studies represent elegant work in which the relative order of many tRNA genes, rRNA genes and drug-resistance and *mit*⁻ loci have been determined (see Bolotin-Fukuhara *et al.*, 1976). However, the map distances between the genes remain arbitrary until the overall segments of mtDNA in the various *petites* used are better characterized. What can currently be stated is that 12 tRNA genes lie between *cap* and *par* (a 35-unit segment on our physical map), whilst one tRNA gene (tryptophan) lies between *par* and O_{II}, and another (one of the two glutamyl-tRNAs) lies between O_I (class B) and O_{II} (class C).

c. *Other Methods of Physical Mapping.* Other approaches have yielded the physical map positions of a relatively small number of loci on mtDNA. These involve use of restriction endonucleases to characterize the mtDNA of *grandes* and *petites*. Borst and his colleagues have determined the fragment map of *grande* mtDNA for several restriction enzymes including EcoRI and HindII + III (Sanders *et al.*, 1975a, b, 1976). The positioning of rRNA genes (Sanders *et al.*, 1975b) has already been mentioned (p. 252). Localization of genetic loci has been made by Sanders *et al.* (1976) who obtained *petites* with very small segments of mtDNA that retained either the *OLI1* (O_I, class B), or *PAR1* (P_{454}^R) loci. *In vitro* transcription of the *petite* mtDNAs was carried out, and the radioactively-labelled complementary RNA was tested for its ability to hybridize to each of the restriction-enzyme fragments. The results showed the positions of *OLI1* (class B), and *PAR1* to be almost identical to those shown here in Fig. 1. The same conclusion was reached by restriction-enzyme analysis of the *petite* mtDNAs themselves (Di Franco *et al.*, 1976). This work revealed that some of the *petites* used showed evidence of re-arrangements which interfered with attempts to compare the *petite* bands with those of the *grande* parent (Morimoto *et al.*, 1975; Rabinowitz *et al.*, 1976). Sanders *et al.* (1976) suggest that the 15S rRNA gene is about 10 units away from *PAR1*, unlike in our own map where the 15S rRNA gene lies within the same two-unit segment of mtDNA as the *par1* locus (Fig. 1). This difference has yet to be resolved. As a general mapping procedure, the methods just described are not well suited to rapid mapping of new loci, because it is necessary to obtain a *petite* (or preferably a group of *petites*) retaining this locus and deleted for all others.

Use of a library of genetically and molecularly defined *petites* is thus preferred. It is possible to characterize a set of *petites* by restriction-enzyme analysis alone, but this approach is complicated by the existence of re-arrangements and amplification in *petite* genomes which do not affect our own hybridization measurements. Nevertheless, some limited success in one small region of the genome has been achieved. Heyting and Sanders (1976) studied 10 *petites* retaining one or both of the $cap(RIB1, C_{321}^R)$ or $ery(RIB3, E_{514}^R)$ loci, and were able to deduce their lengths and relative positions from the restriction enzyme-band patterns. From other genetic tests on the *petites* and inclusion and exclusion mapping principles, the order $ery-ome^+-cap-TSM1$ was deduced. However, the length of the segments of DNA within which ery, ome^+ and cap mapped were so large as to prevent any conclusions being made as to whether any of these loci lie inside or outside the 21S rRNA gene sequence. The upper limit for the degree of separation between cap and ery from the data of Heyting and Sanders (1976) is about 5800 base pairs, which corresponds to eight units, and is similar to our own estimate of five units as shown in Fig. 1 (p. 229).

4. *Possible Map Differences in Different Strains*

The question has already been raised in this review concerning possible sequence differences between the mtDNAs of different *grande* strains (Section II.A.6, p. 167). The evidence concerning this point comes from restriction-enzyme analysis and molecular hybridization studies. Further genetic evidence arises from analysis of the *petites* produced from different *grande* strains by various mutagens (Molloy *et al.*, 1975, 1976). The results showed that the frequency of retention of single markers in *petite* populations varies from strain to strain, and depends on the mutagenic treatment used, in contrast to the assertion of Deutsch *et al.* (1974) that loss of a marker is an intrinsic property of the locus itself. Molloy *et al.* (1975) studied *petites* derived from *grandes* carrying $ery1$ and $oli1$ loci, and analysed their data in terms of the ratio of frequencies of the genotypes: $(ery1\text{-}r\ oli1\text{-}o)/(ery1\text{-}o\ oli1\text{-}r)$ which we will denote here as the function r. The r values for *petites* from four strains isolated under three different conditions are shown in Table 8.

It is seen that the *r* values for spontaneous *petites* of different strains vary from 1.46 to 0.35. In general, the *r* value for EthBr-induced *petites* is similar to that for spontaneous *petites* although mutagenesis with EthBr is usually more severe than in spontaneous *petites*. However, the *r* values for ultraviolet-induced *petites* can be considerably different from the *r* values of spontaneous and EthBr *petites*. In one case (770-7B), the *r* value for ultraviolet is 11 times greater than in spontaneous *petites*, whilst in another case (829-5B), it is three times smaller. Thus, ultraviolet-induced *petites* differ in the probability of retention of specific fragments of the genome compared with spontaneous and EthBr *petites*. Molloy *et al.* (1976) extended this type of analysis to show that *r* values for spontaneous or EthBr-induced *petites* of different

TABLE 8. Relative retention of different regions of the mitochondrial genome in *petites* isolated from different grandes

Grande strain	$r = (ery1\text{-}r\ oli\text{-}o)/(ery1\text{-}o\ oli1\text{-}r)$		
	Spontaneous	Ethidium bromide	Ultraviolet
770-7B	1.46	0.71	16.5
432-31	0.50	0.46	2.1
761-7A	0.90	0.77	6.7
829-5B	0.35	0.31	0.12

Data taken from Molloy *et al.* (1975, 1976).

strains are a property of the mitochondrial genome of the particular *grande*, although some nuclear influence modifying the *r* value to a relatively small extent was observed in one case. These results provide genetic evidence for the view that sequence differences exist between different *grande* strains.

The question now arises that there may be differences in the physical separation of genetic loci in different *grandes*. Indeed, this has been demonstrated in the physical maps of three *grande* strains (Sanders *et al.*, 1976). A more extreme possibility is that the marker order may even be different, if the sequence differences arise as a consequence of deletions, amplifications and re-arrangements within the yeast mitochondrial genome (Bernardi *et al.*, 1975; Clark-Walker and Miklos, 1974b). The ability of mtDNA carrying gross translocations to specify correctly all of the components needed for production of a *grande* cell awaits future investigations.

E. SUPPRESSIVENESS IN *Petite* MUTANTS

1. General Aspects and Definitions

The original acriflavin-induced haploid *petite* clones described by Ephrussi (1953), when crossed with a *grande* haploid strain, yielded diploids virtually all of which had a respiratory-competent phenotype. Another class of *petite* cells was later described by Ephrussi *et al.* (1955), and these when crossed with a *grande* strain yielded a significant and often very high proportion of *petite* diploids. This partial dominance of the *petite* character was called suppressiveness; the earlier described *petites* were referred to as neutral, or of zero suppressiveness. The degree of suppressiveness is operationally defined as the percentage of respiratory-deficient zygotic clones (rather than diploids) out of the total zygotic clones formed when the *petite* is crossed with a haploid *grande* (tester) strain (Ephrussi *et al.*, 1955). Some corrections have to be made in order to allow for certain other factors which may lead to the appearance of *petite* zygotes, and this aspect is discussed in detail on p. 257. Unfortunately, some authors measure suppressiveness as the *petite* frequency amongst the diploid progeny issuing from the *petite* × *grande* cross, after allowing several cell divisions of the diploids to occur before plating out. It should be emphasized that, for consistency in the literature, suppressiveness should be measured by examining the frequency of *petite* zygotic clones, and this is done by plating out the mating mixture directly after zygote formation. Each culture of haploid *petite* cells has a characteristic value of suppressiveness in the range 0–99% (Ephrussi and Grandchamp, 1965), although this may vary depending on the *grande* tester strain used, and on environmental conditions as discussed on p. 260.

A substantial body of evidence is now available that, in general, the neutral *petites* of zero suppressiveness lack mtDNA (i.e. are $rho^°$; Nagley and Linnane, 1970, 1972a; Linnane *et al.*, 1972b; Michaelis *et al.*, 1971). The suppressive *petites* are rho^-, and contain the altered mtDNA characteristic of *petite* cells. It is also apparent that the *petite* mtDNA in some way controls the particular suppressiveness value manifested; justification for this assertion is discussed on p. 258.

Measurement of very low suppressiveness values is difficult. A rough estimate of suppressiveness (the observed frequency of *petite* zygotic clones) is an overestimate of the true suppressiveness. Firstly, the *grande* culture invariably contains a proportion of *petite* cells, commonly in the range 1–10%. Secondly, nuclear interactions can serve to destabilize the *rho* factor in zygotes and so yield *petite* colonies. One suggested correction has been subtraction from the crude suppressiveness figures of the frequency of *petite* colonies formed by the haploid *grande* tester strain at the time of crossing (Ephrussi *et al.*, 1955; Sherman and Ephrussi, 1962). It has been more recently proposed (Nagley and Linnane, 1972a; Nagley *et al.*, 1973) that a more useful correction procedure is to determine the percentage of *petite* zygotes formed when a reference rho^O clone, defined as having zero suppressiveness and which is chromosomally isogenic to the test *petite*, is crossed with the particular *grande* tester being used for suppressiveness measurements. This correction procedure implicitly carries the notion that suppressiveness is to be defined as the *petite* mitochondrial genome suppressing the *grande* genome in the zygotes. Nuclear contributions to the frequency of *petite* zygotes do not lie within the definition of suppressiveness.

This point is made in the light of a paper by Bech-Hansen and Rank (1973b) who describe a nuclear genetic mutation in a strain GR25a which leads to the *rho* factor being destabilized specifically during mating with another cell. Thus, while the parent rho^+ strain GR25 gives 3% petite zygotic clones on being crossed with a rho^O strain, strain GR25a (a spontaneous rho^+ isolate of GR25) gives about 30% *petite* zygotes on mating with the same rho^O strain. The spontaneous *petite* frequencies of haploids GR25 and GR25a are 2% and 5%, respectively. The increased frequency of *petite* zygotes observed when GR25a (rho^+) is used in a cross should not be termed suppressiveness. Firstly, the instability of *rho* factor is a property of GR25a and not the *petite* clone with which it is being mated. Secondly, if a given rho^+ strain is crossed with GR25a (rho^+), the frequency of *petite* zygotic clones is considerably elevated above that obtained when GR25 (rho^+) is used. The warning made by Bech-Hansen and Rank (1973b) is valid in that any *grande* strain used in suppressiveness measurements should be checked for the presence of an unusual instability of its rho^+ mtDNA in zygotes. It should be made clear, however, that, if suppressiveness is to be defined as a mitochondrial genetic phenomenon determined by mtDNA, then the phrase employed by these authors in reference to GR25a, "bivious suppressiveness of *petites* lacking mtDNA", is not consistent with the definition of suppressiveness adopted here.

It would be very convenient if all cells of apparently zero suppressiveness lacked mtDNA. It is conceivable, however, that some cells contain mtDNA conferring a suppressiveness so low as to be indistinguishable experimentally from zero suppressiveness. Indeed, the *petite* clone RDIA described by Hollenberg *et al.* (1972b) was found by Moustacchi (1972b) to be of apparently zero suppressiveness. We have examined the suppressiveness of RDIA in our laboratory in more detail (P. Nagley and A. W. Linnane, unpublished work). Five separate determinations of the suppressiveness of RDIA were made using crosses with a particular *grande* tester strain. Each time, a parallel cross was made between a rho^O derivative of RDIA (prepared by EthBr treatment of RDIA) and the *grande* tester. Over 2000 zygotic clones were scored in each case. Applying Student's t-test to compare the data obtained with each *petite*, it was concluded that RDIA has a suppressiveness of 2% as compared with its rho^O derivative ($p < 0.05$). Nevertheless, it is clear that, without this detailed analysis, RDIA would probably be regarded as having no detectable suppressiveness. The claim of Wintersberger and Hirsch (1973) that several methotrexate-induced *petites* contain mtDNA but lack suppressiveness is to be regarded with caution because only about 100 colonies were scored in each suppressiveness cross.

Experience has shown that clones such as RDIA are rare and that, under normal circumstances, zero suppressiveness can be taken to indicate a loss of mtDNA. Nevertheless, where interpretation of a critical experiment depends on a knowledge of whether a particular *petite* clone contains mtDNA or not, it is considered that demonstration of zero suppressiveness alone is insufficient unequivocally to establish the rho^O state. In such a case, direct analysis of the DNA of the cells or mitochondria is required.

2. *Suppressiveness and mtDNA*

In this section, we will discuss the evidence that suppressiveness is a function of the mtDNA in *petite* cells. Existence of other factors which influence the suppressiveness value is considered in the subsequent section.

The extranuclear nature of the postulated suppressive factor (Ephrussi *et al.*, 1955) was emphasized by the experiments of Wright and Lederberg (1957). The latter authors used cells of *Sacch. cerevisiae* var. *ellipsoideus* which form transient heterokaryons thus having the characteristics that, in crosses, cytoplasms of mating cells mix without any nuclear interactions. When a suppressive *petite* was crossed with a *grande* strain under these conditions, the suppressive factor of the *petite* could still act on the *rho* factor of the *grande*.

In a separate genetic study, the conclusion was drawn that the suppressive factor was only present in cytoplasmically inherited *petite* mutants, but not in nuclear respiratory-deficient mutants carrying a normal rho^+ factor (Sherman and Ephrussi, 1962). Thus, a damaged *rho* factor was required for suppressiveness, and in retrospect these findings suggest that suppressiveness is associated with the altered mtDNA of *petite* cells.

The conclusion that the presence of mtDNA is a necessary condition for a *petite* cell to be of positive suppressiveness was confirmed by demonstrations that, when suppressive *petite* cells are treated with EthBr, a reduction in the overall suppressiveness of the culture occurs together with the loss of mtDNA from the cells (Nagley et al., 1973). This treatment results in the generation of rho° cells. It has also been reported that ultraviolet irradiation or acriflavin treatment of suppressive *petite* cells leads to a reduction in the suppressiveness of the cultures (Uchida, 1972), which is consistent with these agents acting on the *petite* mtDNA.

The results of Saunders et al. (1971) suggested that particular sequences within the mtDNA may play a role in determining whether the *petite* mtDNA molecule as a whole confers a high or low suppressiveness value. These authors showed that a loss of the *ery1* locus in a subclone derived from rho^- *ery1-r* clone was generally accompanied by a change in suppressiveness either to a higher or lower value. This has also been observed for *mik1-o petites* derived from a rho^- *mik1-r petite* clone (Gingold, 1971). A similar phenomenon was seen in some of the *petites* described by Michels et al. (1974) initially carrying erythromycin- or oligomycin-resistance loci. A further important illustration of the idea that the suppressiveness of a *petite* is the result of contributions from many sequences within the mtDNA of the *petite* cells comes from the study by Saunders et al. (1971) of sublones of a *petite* K5 which was rho^- *ery1-r* and had a suppressiveness of 25%. In one subclone, K5272, derived in three subcloning steps from K5, the *ery1* locus was found to be still present but the suppressiveness had changed to 4%. A further subclone (K52729) of clone K5272 was found to be deleted in the *eryl* region and had a suppressiveness value of 25%, the same as the original clone K5. If all these changes result from successive deletions (and possible re-arrangements) in the mtDNA of clone K5, it is evident that various sequences of mtDNA play a role in contributing to a lower or higher value of suppressiveness.

Support for the organization of sequences within *petite* mtDNA contributing to the suppressiveness value comes from the work of Nagley *et al.* (1974a) who studied the same series of *petites* previously investigated by Saunders *et al.* (1971). The genome of *petite* K527 was compared with its parent clone K52; both K52 and K527 are *rho⁻ eryl-r petites*, but they have different suppressiveness values (53% and 4%, respectively). The genomes of *petites* K52 and K527 are apparently very differently organized because the *eryl* loci in *petite* K527 were twenty times less susceptible to elimination by EthBr than in the case of *petite* K52. This implies a great amplification of *eryl* loci in *petite* K527 as compared with K52 (Nagley *et al.*, 1974a; see also Section IV.A.3, p. 196). This could have arisen by deletion of sequences from *petite* K52, or by a sequence re-arrangement leading to differential amplification of *eryl* loci.

Whatever the manner in which *petite* mtDNA sequences may determine suppressiveness (discussed on p. 262), measurement of suppressiveness is a very useful tool in monitoring *petite* populations. This applies particularly in subcloning experiments where the stability of a *petite* can be tested, and also in the selection of *petites* containing further alterations in mtDNA. The changes in mtDNA manifest themselves genetically by suppressiveness changes.

3. *Environmental and Genetic Influence on Suppressiveness*

One of the earliest investigations into the mode of action of the postulated suppressive factor in suppressing the activity and replication of the *rho* factor was made by Ephrussi *et al.* (1966). These authors examined the properties of zygotic clones resulting from the crosses of a number of suppressive *petites* with several *grande* tester strains. Three classes of zygotes were found, which gave rise to almost exclusively *grande* colonies, wholly *petite* colonies, and a third type of colony which was clearly mixed for *petites* and *grandes* (termed *petites abcédées* by Ephrussi *et al.*, 1966). The relative frequencies of these three colony types depended mainly on the parent *petite* but the relative proportions of the colony types could be influenced by the presence of glucose or glycerol in the crossing and plating media.

The mixed clones arise when commitment of the zygote to the *petite* or *grande* state is not immediately made, and they could provide a useful approach to studying interactions between *petite* and *grande* mtDNA in suppressiveness crosses. Where mixed zygotic clones arise, we classify them as *grandes* for the purpose of the calculation of the suppressiveness value. An analysis of zygote pedigrees in crosses between suppressive *petites* and *grandes* has been made by Forster and Kleese (1975a).

Shannon *et al.* (1972) studied a *petite* which was highly suppressive (>90%) if crossed in a medium containing 2% glucose. If the glucose concentration was increased to 10%, suppressiveness of this *petite* was less than 10%. Clearly, the physiological aspects of suppressiveness need much further investigation.

Although it has briefly been mentioned by some authors (e.g. Michaelis *et al.*, 1971) that the suppressiveness of a *petite* depends upon the *grande* tester used, until recently few systematic analyses of these differences have been carried out. A nuclear gene that profoundly affects the suppressiveness of *petites* has been described by Waxman and Eaton (1974). Recently, in our laboratory, the suppressiveness of four chromosomally isogenic stable *petite* clones has been analysed. Two of the *petites* showed suppressiveness values which did not vary against all *grande* testers used. Two other *petites* showed considerable variation, and it was concluded from crosses with specially constructed *grande* strains that the differences in suppressiveness values resulted primarily from the influences of nuclear genes in the various testers, and only limited evidence for the *grande* mtDNA itself being a determinant was obtained (P. Nagley, R. Devenish and R. S. Criddle, unpublished data). These studies therefore demonstrate that nuclear factors influence the outcome of interactions between some *petites* (but not others) with particular *grande* mtDNAs. This conclusion re-inforces some of the very complex problems that must be unravelled before a proper understanding of the suppressiveness phenomenon is obtained.

4. *Molecular Models of Suppressiveness*

The various molecular models that have been proposed to explain suppressiveness all suffer from an absence of independent supporting evidence. Hence, all should be regarded as speculative until more is known about *petite* mtDNA and the suppressiveness phenomenon itself. Indeed, it is by no means certain that all *petites* which show suppressiveness follow the same pathway of interaction with the *grande* mtDNA. There may be different explanations for suppressiveness in the case of different *petites*.

Several interesting suggestions have been made to explain the phenomenon of suppressiveness. Carnevali *et al.* (1969) proposed that the replication rates of *petite* mtDNA compared with *grande* mtDNA determined the preponderance of one or the other in the zygote. This idea is attractive because it might explain the apparent lack of immediate commitment to the *grande* or *petite* state in some situations (Ephrussi *et al.*, 1966; Rank and Person, 1969). Moreover the ability of a highly suppressive *petite* to superimpose its character upon a low suppressive factor in the same cytoplasm is conveniently explained by relative replication rates of the two kinds of *petite* mtDNA (Rank, 1970), but this is not the only explanation of Rank's findings. In spite of the obvious attractions of the idea of relative replication rates of mtDNA molecules, it is not certain that this is sufficient to explain why some suppressive *petites* do give rise to such a high frequency of pure *petite* zygotic clones after crossing with a *grande* strain. It would appear that the rho^+ factor is irretrievably lost from such clones, soon after zygote formation.

Another interesting suggestion to explain suppressiveness was made by Coen *et al.* (1970) who proposed that recombination between *petite* and *grande* mtDNA molecules is involved in the suppressiveness phenomenon. Michaelis *et al.* (1973) extended this idea to suggest that the location of errors (secondary deletions or re-arrangements) in *petite* mtDNAs would determine the probability that the molecules arising from recombination between *petite* mtDNA and *grande* tester mtDNA are able to act as intact *rho* factors and give rise to *grande* zygotes. However, although it is probable that suppressiveness is somehow related to phenomena of transmission and recombination of mitochondrial genes, many pieces of information are lacking, such as knowledge of the genotypes in *petite* molecules issuing from the cross. Explanation of the total and immediate commitment of many zygotes to the *petite* state is rarely attempted.

Shannon *et al.* (1972) reported that a highly suppressive *petite* (mtDNA of density 1.677 g/ml) on crossing with a *grande* (1.683 g/ml) gave rise to *petite* diploids which appeared to all contain recombinant mtDNA molecules of density 1.679 g/ml. However, in our laboratory we have studied a suppressive *petite* with mtDNA of density 1.674 g/ml. The *petite* diploids issuing from crosses with *grande* tester strains contain mtDNA of density 1.674 g/ml, not appreciably different from the parent *petite*. A more complex pattern of behaviour was reported by Blamire *et al.* (1976). It is clear that further detailed analyses of both *petite* and *grande* mtDNA molecules resulting from such crosses (e.g. using restriction enzymes or DNA–DNA hybridization) are required in order to assess the extent of base-sequence exchanges between *petite* and *grande* mtDNA molecules entering the cross.

A third proposal has been made by Williamson (1970) to account for suppressiveness in terms of a disturbance in regulatory circuits associated with maintenance of mtDNA by defective gene products of rho^+ mtDNA. There is however at present little evidence to support this view.

Before concluding, it is worthwhile pointing out that no trend in the physical properties of mtDNA, particularly buoyant density, size or genes retained, have ever been associated with very high, intermediate or low (but not zero) suppressiveness (Bernardi *et al.*, 1970; Michaelis *et al.*, 1971; Nagley and Linnane, 1972a; Michels *et al.*, 1974). The long-known phenomenon of suppressiveness is still poorly understood in molecular terms, but it may well lie at the heart of many aspects of mitochondrial genetics.

REFERENCES

Allen, N. E. and MacQuillan, A. M. (1969). *Journal of Bacteriology* **97**, 1142.
Ashwell, M. and Work, T. S. (1970). *Annual Review of Biochemistry* **39**, 251.
Aufderheide, K. J. (1975). *Molecular and General Genetics* **140**, 231.
Avers, C. J., Billheimer, F. E., Hoffman, H. and Pauli, R. M. (1968). *Proceedings of the National Academy of Sciences of the United States of America* **61**, 90.
Avner, P. R. and Griffiths, D. E. (1970). *Federation of European Biochemical Societies Letters* **10**, 202.
Avner, P. R. and Griffiths, D. E. (1973a). *European Journal of Biochemistry* **32**, 301.
Avner, P. R. and Griffiths, D. E. (1973b). *European Journal of Biochemistry* **32**, 312.
Avner, P. R., Coen, D., Dujon, B. and Slonimski, P. P. (1973). *Molecular and General Genetics* **125**, 9.
Bak, A. L., Christiansen, C. and Stenderup, A. (1969). *Nature, London* **224**, 270.
Bak, A. L., Christiansen, C. and Christiansen, G. (1972). *Biochimica et Biophysica Acta* **269**, 527.
Baldacci, G., Carnevali, F., Frontali, L., Leoni, L., Macino, G. and Palleschi, C. (1975). *Nucleic Acids Research* **2**, 1777.
Bandlow, W. and Kaudewitz, F. (1974). *Molecular and General Genetics* **131**, 333.

Bandlow, W., Wolf, K., Kaudewitz, F. and Slater, E. C. (1974). *Biochimica et Biophysica Acta* **333**, 446.
Banks, G. R. (1973). *Nature New Biology, London* **245**, 196.
Bastos, R. N. (1975). *Journal of Biological Chemistry* **250**, 7739.
Bastos, R. N. and Mahler, H. R. (1974a). *Archives of Biochemistry and Biophysics* **160**, 643.
Bastos, R. N. and Mahler, H. R. (1974b). *Journal of Biological Chemistry* **249**, 6617.
Bech-Hansen, N. T. and Rank, G. H. (1972). *Canadian Journal of Genetics and Cytology* **14**, 681.
Bech-Hansen, N. T. and Rank, G. H. (1973a). *Canadian Journal of Genetics and Cytology* **15**, 381.
Bech-Hansen, N. T. and Rank, G. H. (1973b). *Molecular and General Genetics* **120**, 115.
Bernardi, G. (1975). *In* "Organization and Expression of the Eukaryotic Genome; Biological Mechanisms of Differentiation in Prokaryotes and Eukaryotes", (G. Bernardi and F. Gros, eds.). Proceedings of the 10th FEBS Meeting, vol. 38, pp. 41–56. North-Holland Publishing Co., Amsterdam.
Bernardi, G. and Timasheff, S. N. (1970). *Journal of Molecular Biology* **48**, 43.
Bernardi, G., Carnevali, F., Nicolaieff, A., Piperno, G. and Tecce, G. (1968). *Journal of Molecular Biology* **37**, 493.
Bernardi, G., Faures, M., Piperno, G. and Slonimski, P. P. (1970). *Journal of Molecular Biology* **48**, 23.
Bernardi, G., Piperno, G. and Fonty, G. (1972). *Journal of Molecular Biology* **65**, 173.
Bernardi, G., Prunell, A. and Kopecka, H. (1975). *In* "Molecular Biology of Nucleocytoplasmic Relationships", (S. Puiseux-Dao, ed.), pp. 85–90. Elsevier, Amsterdam.
Bernardi, G., Prunell, A., Fonty, G., Kopecka, H. and Strauss, F. (1976). *In* "Genetic Function of Mitochondrial DNA", (C. Saccone and A. M. Kroon, eds.), pp. 185–198. North-Holland Publishing Co., Amsterdam.
Bicknell, J. N. and Douglas, H. C. (1970). *Journal of Bacteriology* **101**, 505.
Billheimer, F. E. and Avers, C. J. (1969). *Proceedings of the National Academy of Sciences of the United States of America* **64**, 739.
Birky, C. W. (1975a). *Genetics* **80**, 695.
Birky, C. W. (1975b). *Molecular and General Genetics* **141**, 41.
Blamire, J., Cryer, D. R., Finkelstein, D. B. and Marmur, J. (1972a). *Journal of Molecular Biology* **67**, 11.
Blamire, J., Finkelstein, D. B. and Marmur, J. (1972b). *Biochemistry, New York* **11**, 4848.
Blamire, J., Michels, C. A., Walsh, J. M. and Friedenberg, D. L. (1976). *Molecular and General Genetics* **143**, 253.
Bleeg, H. S., Bak, A. L., Christiansen, C., Smith, K. E. and Stenderup, A. (1972). *Biochemical and Biophysical Research Communications* **47**, 524.
Bolotin, M., Coen, D., Deutsch, J., Dujon, B., Netter, P., Petrochilo, E. and Slonimski, P. P. (1971). *Bulletin de l'Institut Pasteur* **69**, 215.
Bolotin-Fukuhara, M., Faye, G. and Fukuhara, H. (1976). *In* "Genetic Function of Mitochondrial DNA", (C. Saccone and A. M. Kroon, eds.), pp. 243–250. North-Holland Publishing Co., Amsterdam.
Borst, P. (1971). *In* "Autonomy and Biogenesis of Mitochondria and Chloroplasts", (N. K. Boardman, A. W. Linnane and R. M. Smillie, eds.), pp. 260–266. North-Holland Publishing Co., Amsterdam.
Borst, P. (1972). *Annual Review of Biochemistry* **41**, 333.
Borst, P. (1974). *In* "Biogenesis of Mitochondria: Transcriptional, Translational and Genetic Aspects", (A. M. Kroon and C. Saccone, eds.), pp. 147–156. Academic Press, New York.

Borst, P. and Flavell, R. A. (1972). *In* "Mitochondria/Biomembranes", (S. G. van den Bergh, P. Borst, L. L. M. van Deenen, J. C. Riemersma, E. C. Slater and J. M. Tager, eds.), pp. 1–19. North-Holland Publishing Co., Amsterdam.
Borst, P. and Grivell, L. A. (1971). *Federation of European Biochemical Societies Letters* **13**, 73.
Borst, P. and Kroon, A. M. (1969). *International Review of Cytology* **26**, 107.
Borst, P., Kroon, A. M. and Ruttenberg, G. J. C. M. (1967). *In* "Genetic Elements: Properties and Function", (D. Shugar, ed.), pp. 81–116. Academic Press, London.
Bostock, C. J. (1969). *Biochimica et Biophysica Acta* **195**, 579.
Bucking-Throm, E., Duntze, W., Hartwell, L. H. and Manney, T. R. (1973). *Experimental Cell Research* **76**, 99.
Bulder, C. J. E. A. (1964). *Antonie van Leeuwenhoek* **30**, 1.
Bunn, C. L., Mitchell, C. H., Lukins, H. B. and Linnane, A. W. (1970). *Proceedings of the National Academy of Sciences of the United States of America* **67**, 1233.
Callen, D. F. (1974a). *Molecular and General Genetics* **128**, 321.
Callen, D. F. (1974b). *Molecular and General Genetics* **134**, 49.
Callen, D. F. (1974c). *Molecular and General Genetics* **134**, 65.
Carnevali, F. and Leoni, L. (1972). *Biochemical and Biophysical Research Communications* **47**, 1322.
Carnevali, F., Piperno, G. and Tecce, G. (1966). *Academia Nazionale dei Lincei, Rome. Rendiconti: Classe di Scienze, Fisiche, Matematichi e Naturali* **41**, 194.
Carnevali, F., Morpurgo, G. and Tecce, G. (1969). *Science, New York* **163**, 1331.
Carnevali, F., Leoni, L., Morpurgo, G. and Conti, G. (1971). *Mutation Research* **12**, 357.
Casey, J., Gordon, P. and Rabinowitz, M. (1974a). *Biochemistry, New York* **13**, 1059.
Casey, J. W., Hsu, H.-J., Rabinowitz, M., Getz, G. S. and Fukuhara, H. (1974b). *Journal of Molecular Biology* **88**, 717.
Casey, J. W., Hsu, H.-J., Getz, G. S.; Rabinowitz, M. and Fukuhara, H. (1974c). *Journal of Molecular Biology* **88**, 735.
Chanet, R., Williamson, D. H. and Moustacchi, E. (1973). *Biochimica et Biophysica Acta* **324**, 290.
Choo, K. B., Nagley, P., Lukins, H. B. and Linnane, A. W. (1977). *Molecular and General Genetics* in press.
Christiansen, C., Bak, A. L., Stenderup, A. and Christiansen, G. (1971). *Nature New Biology, London* **231**, 176.
Christiansen, C., Christiansen, G. and Bak, A. L. (1974). *Journal of Molecular Biology* **84**, 65.
Christiansen, G., Christiansen, C. and Bak, A. L. (1975). *Nucleic Acids Research* **2**, 197.
Ciferri, O., Sora, S. and Tiboni, O. (1969). *Genetics* **61**, 567.
Clark-Walker, G. D. (1972). *Proceedings of the National Academy of Sciences of the United States of America* **69**, 388.
Clark-Walker, G. D. (1973). *European Journal of Biochemistry* **32**, 263.
Clark-Walker, G. D. and Linnane, A. W. (1967). *Journal of Cell Biology* **34**, 1.
Clark-Walker, G. D. and Miklos, G. L. G. (1974a). *European Journal of Biochemistry* **41**, 359.
Clark-Walker, G. D. and Miklos, G. L. G. (1974b). *Genetical Research, Cambridge* **24**, 43.
Clark-Walker, G. D. and Miklos, G. L. G. (1975). *Proceedings of the National Academy of Sciences of the United States of America* **72**, 372.
Cobon, G. S., Groot Obbink, D. J., Hall, R. M., Maxwell, R. J., Murphy, M., Rytka, J. and Linnane, A. W. (1976). *In* "Genetics and Biogenesis of Chloroplasts and Mitochondria", (T. Bucher, W. Neupert, W. Sebald and S. Werner, eds.), pp. 453–460. North-Holland Publishing Co., Amsterdam.

Coen, D., Deutsch, J., Netter, P., Petrochilo, E. and Slonimski, P. P. (1970). In "Control of Organelle Development", (P. L. Miller, ed.), pp. 447–496. Cambridge University Press, Cambridge.
Cohen, L. M., Casey, J., Rabinowitz, M. and Getz, G. S. (1972). Journal of Molecular Biology 63, 441.
Colson, A.-M., Goffeau, A., Briquet, M., Weigel, P. and Mattoon, J. R. (1974). Molecular and General Genetics 135, 309.
Cottrell, S. F. and Avers, C. J. (1970). Biochemical and Biophysical Research Communications 38, 973.
Cottrell, S., Rabinowitz, M. and Getz, G. S. (1973). Biochemistry, New York 12, 4374.
Crandall, M. (1973a). Journal of General Microbiology 75, 363.
Crandall, M. (1973b). Journal of General Microbiology 75, 377.
Criddle, R. S., Wheelis, L., Trembath, M. K. and Linnane, A. W. (1976). Molecular and General Genetics 144, 263.
Cryer, D. R., Goldthwaite, C. D., Zinker, Z., Lam, K. B., Storm, E., Hirschberg, R., Blamire, J., Finkelstein, D. B. and Marmur, J. (1973). Cold Spring Harbor Symposium on Quantitative Biology 38, 17.
Davey, P. J., Haslam, J. M. and Linnane, A. W. (1970). Archives of Biochemistry and Biophysics 136, 54.
Dawes, I. W. and Carter, B. L. A. (1974). Nature, London 250, 709.
Deutsch, J., Dujon, B., Netter, P., Petrochilo, E., Slonimski, P. P., Bolotin-Fukuhara, M. and Coen, D. (1974). Genetics 76, 195.
Di Franco, A., Sanders, J. P. M., Heyting, C., Borst, P. and Slonimski, P. P. (1976). In "Genetic Function of Mitochondrial DNA", (C. Saccone and A. M. Kroon, eds.), pp. 291–304. North-Holland Publishing Co., Amsterdam.
Dujon, B. and Michel, F. (1976). In "Genetic Function of Mitochondrial DNA", (C. Saccone and A. M. Kroon, eds.), pp. 175–184. North-Holland Publishing Co., Amsterdam.
Dujon, B., Slonimski, P. P. and Weill, L. (1974). Genetics 78, 415.
Dujon, B., Kruszewska, A., Slonimski, P. P., Bolotin-Fukuhara, M., Coen, D., Deutsch, J., Netter, P. and Weill, L. (1975). Molecular and General Genetics 137, 29.
Dujon, B., Bolotin-Fukuhara, M., Coen, D., Deutsch, J., Netter, P., Slonimski, P. P. and Weill, L. (1976). Molecular and General Genetics 143, 131.
Ehrlich, S. D., Thiery, J. and Bernardi, G. (1972). Journal of Molecular Biology 65, 207.
Elliott, J. J. and Ball, A. J. S. (1975). Biochemical and Biophysical Research Communications 64, 277.
Ephrussi, B. (1953). "Nucleocytoplasmic Relations in Micro-Organisms", 127 pp. Clarendon Press, Oxford.
Ephrussi, B. and Grandchamp, S. (1965). Heredity 20, 1.
Ephrussi, B., Margerie-Hottinguer, H. and Roman, H. (1955). Proceedings of the National Academy of Sciences of the United States of America 41, 1065.
Ephrussi, B., Jakob, H. and Grandchamp, S. (1966). Genetics 54, 1.
Fath, W. W., Brendel, M., Laskowski, W. and Lehmann-Brauns, E. (1974). Molecular and General Genetics 132, 335.
Fauman, M. and Rabinowitz, M. (1972). Federation of European Biochemical Societies Letters 28, 317.
Fauman, M. and Rabinowitz, M. (1974). European Journal of Biochemistry 42, 67.
Fauman, M., Rabinowitz, M. and Swift, H. H. (1973). Biochemistry, New York 12, 124.
Faures-Renot, M., Faye, G., Michel, F. and Fukuhara, H. (1974). Biochimie 56, 681.
Faye, G., Fukuhara, H., Grandchamp, C., Lazowska, J., Michel, F., Casey, J., Getz, G. S., Locker, J., Rabinowitz, M., Bolotin-Fukuhara, M., Coen, D., Deutsch, J., Dujon, B., Netter, P. and Slonimski, P. P. (1973). Biochimie 55, 779.

Faye, G., Kujawa, C. and Fukuhara, H. (1974). *Journal of Molecular Biology* **88**, 185.
Faye, G., Kujawa, C., Dujon, B., Bolotin-Fukuhara, M., Wolf, K., Fukuhara, H. and Slonimski, P. P. (1975). *Journal of Molecular Biology* **99**, 203.
Faye, G., Kujawa, C., Fukuhara, M. and Rabinowitz, M. (1976). *Biochemical and Biophysical Research Communications* **68**, 476.
Finkelstein, D. B., Blamire, J. and Marmur, J. (1972). *Biochemistry, New York* **11**, 4853.
Flavell, R. A., Borst, P. and Birfelder, E. J. (1974). *European Journal of Biochemistry* **47**, 545.
Flury, U., Mahler, H. R. and Feldman, F. (1974). *Journal of Biological Chemistry* **249**, 6130.
Forrester, I. T., Nagley, P. and Linnane, A. W. (1970). *Federation of European Biochemical Societies Letters* **11**, 59.
Forster, J. L. and Kleese, R. A. (1975a). *Molecular and General Genetics* **139**, 329.
Forster, J. L. and Kleese, R. A. (1975b). *Molecular and General Genetics* **139**, 341.
Fukuhara, H. (1969). *European Journal of Biochemistry* **11**, 135.
Fukuhara, H. and Kujawa, C. (1970). *Biochemical and Biophysical Research Communications* **41**, 1002.
Fukuhara, H., Faures, M. and Genin, C. (1969). *Molecular and General Genetics* **104**, 264.
Fukuhara, H., Faye, G., Michel, F., Lazowska, J., Deutsch, J., Bolotin-Fukuhara, M. and Slonimski, P. P. (1974). *Molecular and General Genetics* **130**, 215.
Fukuhara, H., Bolotin-Fukuhara, M., Hsu, H.-J. and Rabinowitz, M. (1976). *Molecular and General Genetics* **145**, 7.
Game, J. C. (1976). *Molecular and General Genetics* **146**, 313.
Gefter, M. L. (1975). *Annual Review of Biochemistry* **44**, 45.
Gillham, N. W. (1974). *Annual Review of Genetics* **8**, 347.
Gingold, E. B. (1971). Ph.D. Thesis: Monash University.
Gingold, E. B., Saunders, G. W., Lukins, H. B. and Linnane, A. W. (1969). *Genetics* **62**, 735.
Goffeau, A., Colson, A. M., Landry, Y. and Foury, F. (1972). *Biochemical and Biophysical Research Communications* **48**, 1448.
Goffeau, A., Landry, Y., Foury, F., Briquet, M. and Colson, A. (1973). *Journal of Biological Chemistry* **248**, 7097.
Goldring, E. S., Grossman, L. I., Krupnick, D., Cryer, D. R. and Marmur, J. (1970). *Journal of Molecular Biology* **52**, 323.
Goldring, E. S., Grossman, L. I. and Marmur, J. (1971). *Journal of Bacteriology* **107**, 377.
Goldthwaite, C. D., Cryer, D. R. and Marmur, J. (1974). *Molecular and General Genetics* **133**, 87.
Gordon, P. and Rabinowitz, M. (1973). *Biochemistry, New York* **12**, 116.
Gordon, P., Casey, J. and Rabinowitz, M. (1974). *Biochemistry, New York* **13**, 1067.
Gouhier, M. and Mounolou, J. (1973). *Molecular and General Genetics* **122**, 149.
Gouhier-Monnerot, M. (1974). *Molecular and General Genetics* **130**, 65.
Granick, S. and Gibor, A. (1967). *Progress in Nucleic Acid Research and Molecular Biology* **6**, 143.
Griffiths, D. E. and Houghton, R. L. (1974). *European Journal of Biochemistry* **46**, 157.
Griffiths, D. E., Houghton, R. L. and Lancashire, W. E. (1974). *In* "Biogenesis of Mitochondria: Transcriptional, Translational and Genetic Aspects", (A. M. Kroon and C. Saccone, eds.), pp. 215–223. Academic Press, New York.
Griffiths, D. E., Lancashire, W. E. and Zanders, E. D. (1975a). *Federation of European Biochemical Societies Letters* **53**, 126.
Griffiths, D. E., Houghton, R. L., Lancashire, W. E. and Meadows, P. A. (1975b). *European Journal of Biochemistry* **51**, 393.
Grimes, G. W., Mahler, H. R. and Perlman, P. S. (1974). *Journal of Cell Biology* **61**, 565.

Grivell, L. A., Reijnders, L. and De Vries, H. (1971). *Federation of European Biochemical Societies Letters* **16**, 159.
Grivell, L. A., Netter, P., Borst, P. and Slonimski, P. P. (1973). *Biochimica et Biophysica Acta* **312**, 358.
Groot, G. S. P., Flavell, R. A., Van Ommen, G. J. B. and Grivell, L. A. (1974). *Nature, London* **252**, 167.
Groot, G. S. P., Flavell, R. A. and Sanders, J. P. M. (1975). *Biochimica et Biophysica Acta* **378**, 186.
Groot Obbink, D. J., Hall, R. M., Linnane, A. W., Lukins, H. B., Monk, B. C., Spithill, T. W. and Trembath, M. K. (1976). *In* "Genetic Function of Mitochondrial DNA", (C. Saccone and A. M. Kroon, eds.), pp. 163–173. North-Holland Publishing Co., Amsterdam.
Groot Obbink, D. J., Spithill, D. W., Maxwell, R. J. and Linnane, A. W. (1977). *Molecular and General Genetics* **151**, 127.
Grossman, L. I., Goldring, E. S. and Marmur, J. (1969). *Journal of Molecular Biology* **46**, 367.
Grossman, L. I., Cryer, D. R., Goldring, E. S. and Marmur, J. (1971). *Journal of Molecular Biology* **62**, 565.
Guerineau, M., Grandchamp, C., Paoletti, C. and Slonimski, P. P. (1971). *Biochemical and Biophysical Research Communications* **42**, 550.
Guerineau, M., Slonimski, P. P. and Avner, P. R. (1974). *Biochemical and Biophysical Research Communications* **61**, 412.
Gunge, N. (1975). *Molecular and General Genetics* **139**, 189.
Hall, R. M. and Linnane, A. W. (1977). *In* "Cell Biology: A Comprehensive Treatise", (L. Goldstein and D. M. Prescott, eds.). Academic Press, New York (in press).
Hall, R. M., Mattick, J. S., Marzuki, S. and Linnane, A. W. (1975). *Molecular Biology Reports* **2**, 101.
Hall, R. M., Nagley, P. and Linnane, A. W. (1976a). *Molecular and General Genetics* **145**, 169.
Hall, R. M., Trembath, M. K., Linnane, A. W., Wheelis, L. and Criddle, R. S. (1976b). *Molecular and General Genetics* **144**, 253.
Handwerker, A., Schweyen, R. J., Wolf, K. and Kaudewitz, F. (1973). *Journal of Bacteriology* **113**, 1307.
Hartwell, L. H. (1970). *Annual Review of Genetics* **4**, 373.
Hartwell, L. H. (1971). *Journal of Molecular Biology* **59**, 183.
Hartwell, L. H. (1973). *Experimental Cell Research* **76**, 111.
Hartwell, L. H. (1974). *Bacteriological Reviews* **38**, 164.
Hartwell, L. H., Culotti, J., Pringle, J. R. and Reid, B. J. (1974). *Science, New York* **183**, 46.
Hatzfeld, J. (1973). *Biochimica et Biophysica Acta* **299**, 34.
Hereford, L. M. and Hartwell, L. H. (1971). *Nature New Biology, London* **234**, 171.
Hereford, L. M. and Hartwell, L. H. (1973). *Nature New Biology, London* **244**, 129.
Heslot, H., Louis, C. and Goffeau, A. (1970). *Journal of Bacteriology* **104**, 482.
Heude, M. and Chanet, R. (1975). *Mutation Research* **28**, 47.
Heude, M. and Moustacchi, E. (1973). *Académie des Sciences, Paris. Comptes Rendus Hebdomadaires de Séances* **277**, 1561.
Heude, M., Chanet, R. and Moustacchi, E. (1975). *Mutation Research* **28**, 37.
Heyting, C. and Sanders, J. P. M. (1976). *In* "Genetic Function of Mitochondrial DNA", (C. Saccone and A. M. Kroon, eds.), pp. 273–280. North-Holland Publishing Co., Amsterdam.
Hixon, S. C. and Yielding, K. L. (1975). *Mutation Research* **29**, 159.

Hixon, S. C., White, W. E. and Yielding, K. L. (1975). *Journal of Molecular Biology* **92**, 319.
Hoffman, H. and Avers, C. J. (1973). *Science, New York* **181**, 749.
Hollenberg, C. P. and Borst, P. (1971). *Biochemical and Biophysical Research Communications* **45**, 1250.
Hollenberg, C. P., Borst, P., Thuring, R. W. J. and Van Bruggen, E. F. J. (1969). *Biochimica et Biophysica Acta* **186**, 417.
Hollenberg, C. P., Borst, P. and Van Bruggen, E. F. J. (1970). *Biochimica et Biophysica Acta* **209**, 1.
Hollenberg, C. P., Borst, P. and Van Bruggen, E. F. J. (1972a). *Biochimica et Biophysica Acta* **277**, 35.
Hollenberg, C. P., Borst, P., Flavell, R. A., Van Kreijl, C. F., Van Bruggen, E. F. J. and Arnberg, A. C. (1972b). *Biochimica et Biophysica Acta* **277**, 44.
Howell, N., Trembath, M. K., Linnane, A. W. and Lukins, H. B. (1973). *Molecular and General Genetics* **122**, 37.
Howell, N., Hall, R. M., Linnane, A. W. and Lukins, H. B. (1974a). *Journal of Bacteriology* **119**, 1063.
Howell, N., Molloy, P. L., Linnane, A. W. and Lukins, H. B. (1974b). *Molecular and General Genetics* **128**, 43.
Huang, M., Biggs, D. R., Clark-Walker, G. D. and Linnane, A. W. (1966). *Biochimica et Biophysica Acta* **114**, 434.
Iwashima, A. and Rabinowitz, M. (1969). *Biochimica et Biophysica Acta* **179**, 283.
Juliani, M. H., Cost, S. O. P. and Bacila, M. (1973). *Biochemical and Biophysical Research Communications* **53**, 531.
Kellerman, G. M., Biggs, D. R. and Linnane, A. W. (1969). *Journal of Cell Biology* **42**, 378.
Kellerman, G. M., Griffiths, D. E., Hansby, J. E., Lamb, A. J. and Linnane, A. W. (1971). In "Autonomy and Biogenesis of Mitochondria and Chloroplasts", (N. K. Boardman, A. W. Linnane and R. M. Smillie, eds.), pp. 346–359. North-Holland Publishing Co., Amsterdam.
Kleese, R. A., Grotbeck, R. C. and Snyder, J. R. (1972a). *Canadian Journal of Genetics and Cytology* **14**, 713.
Kleese, R. A., Grotbeck, R. C. and Snyder, J. R. (1972b). *Journal of Bacteriology* **112**, 1023.
Kotylak, Z. and Slonimski, P. P. (1976). In "Genetic Function of Mitochondrial DNA", (C. Saccone and A. M. Kroon, eds.), pp. 143–154. North-Holland Publishing Co., Amsterdam.
Kuenzi, M. T. and Roth, R. (1974). *Experimental Cell Research* **85**, 377.
Kuntzel, H. (1971). *Current Topics in Microbiology and Immunology* **54**, 94.
Kuriyama, Y. and Luck, D. J. L. (1973). *Journal of Molecular Biology* **73**, 425.
Kutzleb, R., Schweyen, R. J. and Kaudewitz, F. (1973). *Molecular and General Genetics* **125**, 91.
Kuzela, S., Smigan, P. and Kovac, L. (1969). *Experientia* **25**, 1042.
Lamb, A. J. and Rajanapo, W. (1973). *Biochemical and Biophysical Research Communications* **55**, 765.
Lamb, A. J., Clark-Walker, G. D. and Linnane, A. W. (1968). *Biochimica et Biophysica Acta* **161**, 415.
Lancashire, W. E. and Griffiths, D. E. (1975a). *European Journal of Biochemistry* **51**, 377.
Lancashire, W. E. and Griffiths, D. E. (1975b). *European Journal of Biochemistry* **51**, 403.
Lang, D. (1970). *Journal of Molecular Biology* **54**, 557.
Lazowska, J., Michel, F., Faye, G., Fukuhara, H. and Slonimski, P. P. (1974). *Journal of Molecular Biology* **85**, 393.

Linnane, A. W. (1968). In "Biochemical Aspects of Biogenesis of Mitochondria", (E. C. Slater, J. M. Tager, S. Papa and E. Quagliariello, eds.), pp. 333–353. Adriatica Editrice, Bari.
Linnane, A. W. (1971). Acta Cientifica Venezolana **22** (Suppl. 2), 51.
Linnane, A. W. and Crowfoot, P. D. (1975). In "Membrane Biogenesis: Mitochondria, Chloroplasts and Bacteria", (A. Tzagoloff, ed.), pp. 99–124. Plenum Press, New York.
Linnane, A. W. and Haslam, J. M. (1970). In "Current Topics in Cellular Regulation", (B. L. Horecker and E. R. Stadtman, eds.), vol. 2, pp. 101–172. Academic Press, New York.
Linnane, A. W., Saunders, G. W., Gingold, E. B. and Lukins, H. B. (1968a). *Proceedings of the National Academy of Sciences of the United States of America* **59**, 903.
Linnane, A. W., Lamb, A. J., Christodoulou, C. and Lukins, H. B. (1968b). *Proceedings of the National Academy of Sciences of the United States of America* **59**, 1288.
Linnane, A. W., Haslam, J. M. and Forrester, I. T. (1972a). In "The Biochemistry and Biophysics of Mitochondrial Membranes", (G. F. Azzone, E. Carafoli, A. L. Lehninger, E. Quagliarello and V. Siliprandi, eds.), pp. 523–539. Academic Press, New York.
Linnane, A. W., Haslam, J. M., Lukins, H. B. and Nagley, P. (1972b). *Annual Review of Microbiology* **26**, 163.
Linnane, A. W., Mitchell, C. H., Trembath, M. K. and Lukins, H. B. (1972c). In "Fermentation Technology Today", (G. Terui, ed.), pp. 841–846. Society of Fermentation Technology, Japan.
Linnane, A. W., Bunn, C. L., Howell, N., Molloy, P. L. and Lukins, H. B. (1973a). In "Biochemistry of Gene Expression in Higher Organisms", (J. W. Lee and J. K. Pollack, eds.), pp. 425–442. Australia and New Zealand Book Co., Sydney.
Linnane, A. W., Cobon, G. and Marzuki, S. (1973b). In "Proceedings of the Third International Specialized Symposium on Yeasts", (H. Soumalainen and C. Waller, eds.), Part II, pp. 349–367. Aiko Foundation, Helsinki.
Linnane, A. W., Howell, N. and Lukins, H. B. (1974). In "Biogenesis of Mitochondria; Transcriptional, Translational and Genetic Aspects", (A. M. Kroon and C. Saccone, eds.), pp. 193–213. Academic Press, New York.
Linnane, A. W., Kellerman, G. M. and Lukins, H. B. (1975). In "Techniques of Biochemical and Biophysical Morphology", (D. Glick and R. Rosenbaum, eds.), vol. 2, pp. 1–98. John Wiley and Sons, New York.
Linnane, A. W., Lukins, H. B., Molloy, P. L., Nagley, P., Rytka, J., Sriprakash, K. S. and Trembath, M. K. (1976). *Proceedings of the National Academy of Sciences of the United States of America* **73**, 2082.
Locker, J., Rabinowitz, M. and Getz, G. S. (1974a). *Proceedings of the National Academy of Sciences of the United States of America* **71**, 1366.
Locker, J., Rabinowitz, M. and Getz, G. S. (1974b). *Journal of Molecular Biology* **88**, 489.
Luha, A. A., Sarcoe, L. E. and Whittaker, P. A. (1971). *Biochemical and Biophysical Research Communications* **44**, 396.
Luha, A. A., Whittaker, P. A. and Hammond, R. C. (1974). *Molecular and General Genetics* **129**, 311.
Lukins, H. B., Tate, J. R., Saunders, G. W. and Linnane, A. W. (1973). *Molecular and General Genetics* **120**, 17.
Lusena, C. V. and James, A. P. (1976). *Molecular and General Genetics* **144**, 120.
MacHattie, L. A. and Thomas, C. A. (1964). *Science, New York* **144**, 1142.
Mahler, H. R. (1973a). In "Molecular Cytogenetics", (B. A. Hamkalo and J. Papaconstantinou, eds.), pp. 181–208. Plenum Press, New York.

Mahler, H. R. (1973b). *Journal of Supramolecular Structure* **1**, 449.
Mahler, H. R. (1973c). *Critical Reviews in Biochemistry* **1**, 381.
Mahler, H. R. and Bastos, R. N. (1974a). *Federation of European Biochemical Societies Letters* **39**, 27.
Mahler, H. R. and Bastos, R. N. (1974b). *Proceedings of the National Acedemy of Sciences of the United States of America* **71**, 2241.
Mahler, H. R. and Perlman, P. S. (1972a). *Archives of Biochemistry and Biophysics* **148**, 115.
Mahler, H. R. and Perlman, P. S. (1972b). *Journal of Supramolecular Structure* **1**, 105.
Mahler, H. R. and Perlman, P. S. (1973). *Molecular and General Genetics* **121**, 285.
Mahler, H. R., Perlman, P. S. and Mehrotra, B. D. (1971). *In* "Autonomy and Biogenesis of Mitochondria and Chloroplasts", (N. K. Boardman, A. W. Linnane and R. M. Smillie, eds.), pp. 429–511. North-Holland Publishing Co., Amsterdam.
Marmur, J. and Doty, P. (1962). *Journal of Molecular Biology* **5**, 109.
Maroudas, N. G. and Wilkie, D. (1968). *Biochimica et Biophysica Acta* **166**, 681.
Martin, N., Rabinowitz, M. and Fukuhara, H. (1976). *Journal of Molecular Biology* **101**, 285.
Marzuki, S., Hall, R. M. and Linnane, A. W. (1974). *Biochemical and Biophysical Research Communications* **57**, 372.
Marzuki, S., Cobon, G. S., Haslam, J. M. and Linnane, A. W. (1975). *Archives of Biochemistry and Biophysics* **169**, 577.
Mattick, J. S. and Hall, R. M. (1977). *Journal of Bacteriology*, in press.
Mattick, J. S. and Nagley, P. (1977). *Molecular and General Genetics* **152**, 267.
Mehrotra, B. D. and Mahler, H. R. (1968). *Archives of Biochemistry and Biophysics* **128**, 685.
Meselson, M. and Stahl, F. W. (1958). *Proceedings of the National Academy of Sciences of the United States of America* **44**, 671.
Mian, F. A., Kuenzi, M. T. and Halvorson, H. O. (1973). *Journal of Bacteriology* **115**, 876.
Michaelis, G., Douglass, S., Tsai, M. and Criddle, R. S. (1971). *Biochemical Genetics* **5**, 487.
Michaelis, G., Douglass, S., Tsai, M., Burchiel, K. and Criddle, R. S. (1972). *Biochemistry, New York* **11**, 2026.
Michaelis, G., Petrochilo, E. and Slonimski, P. P. (1973). *Molecular and General Genetics* **123**, 51.
Michel, F. (1974). *Journal of Molecular Biology* **89**, 305.
Michel, F., Lazowska, J., Faye, G., Fukuhara, H. and Slonimski, P. P. (1974). *Journal of Molecular Biology* **85**, 411.
Michels, C. A., Blamire, J., Goldfinger, B. and Marmur, J. (1974). *Journal of Molecular Biology* **89**, 431.
Mitchell, C. A., Bunn, C. L., Lukins, H. B. and Linnane, A. W. (1972). *Journal of Bioenergetics* **4**, 363.
Mol, J. N. M., Borst, P., Grosveld, F. G. and Spencer, J. H. (1974). *Biochimica et Biophysica Acta* **374**, 115.
Molloy, P. L., Howell, N., Plummer, D. T., Linnane, A. W. and Lukins, H. B. (1973). *Biochemical and Biophysical Research Communications* **52**, 9.
Molloy, P. L., Linnane, A. W. and Lukins, H. B. (1975). *Journal of Bacteriology* **122**, 7.
Molloy, P. L., Linnane, A. W. and Lukins, H. B. (1976). *Genetical Research, Cambridge* **26**, 319.
Morimoto, H., Scragg, A. H., Nekhorocheff, J., Villa, V. and Halvorson, H. O. (1971). *In* "Autonomy and Biogenesis of Mitochondria and Chloroplasts", (N. K. Boardman, A. W. Linnane and R. M. Smillie, eds.), pp. 282–292. North Holland Publishing Co., Amsterdam.

Morimoto, R., Lewin, A., Hsu, H.-J., Rabinowitz, M. and Fukuhara, H. (1975). *Proceedings of the National Academy of Sciences of the United States of America* **72**, 3868.
Mortimer, R. E. and Hawthorne, D. G. (1966). *Annual Review of Microbiology* **20**, 151.
Mounolou, J. C., Jakob, H. and Slonimski, P. P. (1966). *Biochemical and Biophysical Research Communications* **24**, 218.
Mounolou, J. C., Perrodin, G. and Slonimski, P. P. (1968). *In* "Biochemical Aspects of Biogenesis of Mitochondria", (E. C. Slater, J. M. Tager, S. Papa and E. Quagliariello, eds.), pp. 133–148. Adriatica Editrice, Bari.
Moustacchi, E. (1969). *Mutation Research* **7**, 171.
Moustacchi, E. (1972a). *Molecular and General Genetics* **114**, 50.
Moustacchi, E. (1972b). *Biochimica et Biophysica Acta* **277**, 59.
Moustacchi, E. (1973). *Journal of Bacteriology* **115**, 805.
Moustacchi, E. and Enteric, S. (1970). *Molecular and General Genetics* **109**, 69.
Moustacchi, E. and Williamson, D. H. (1966). *Biochemical and Biophysical Research Communications* **23**, 56.
Nagai, S., Yanagashima, N. and Nagai, H. (1961). *Bacteriological Reviews* **25**, 404.
Nagley, P. and Linnane, A. W. (1970). *Biochemical and Biophysical Research Communications* **39**, 989.
Nagley, P. and Linnane, A. W. (1972a). *Journal of Molecular Biology* **66**, 181.
Nagley, P. and Linnane, A. W. (1972b). *Cell Differentiation* **1**, 143.
Nagley, P. and Mattick, J. S. (1977). *Molecular and General Genetics* **152**, 277.
Nagley, P., Gingold, E. B., Lukins, H. B. and Linnane, A. W. (1973). *Journal of Molecular Biology* **78**, 335.
Nagley, P., Gingold, E. B. and Linnane, A. W. (1974a). *In* "Biogenesis of Mitochondria: Transcriptional, Translational and Genetic Aspects", (A. M. Kroon and C. Saccone, eds.), pp. 157–168. Academic Press, New York.
Nagley, P., Molloy, P. L., Lukins, H. B. and Linnane, A. W. (1974b). *Biochemical and Biophysical Research Communications* **57**, 232.
Nagley, P., Mattick, J. S., Hall, R. M. and Linnane, A. W. (1975a). *Molecular and General Genetics* **141**, 291.
Nagley, P., Molloy, P. L., Lukins, H. B. and Linnane, A. W. (1975b). *In* "The Eukaryote Chromosome", (W. J. Peacock and R. D. Brock, eds.), pp. 155–167. Australian National University Press, Canberra.
Nagley, P., Sriprakash, K. S., Rytka, J., Choo, K. B., Trembath, M. K., Lukins, H. B. and Linnane, A. W. (1976). *In* "Genetic Function of Mitochondrial DNA", (C. Saccone and A. M. Kroon, eds.), pp. 231–242. North-Holland Publishing Co., Amsterdam.
Nass, N. M. K. (1969). *Science, New York* **165**, 25.
Nathans, D. and Smith, H. O. (1975). *Annual Review of Biochemistry* **44**, 273.
Negrotti, T. and Wilkie, D. (1968). *Biochimica et Biophysica Acta* **153**, 341.
Netter, P., Petrochilo, E., Slonimski, P. P., Bolotin-Fukuhara, M., Coen, D., Deutsch, J. and Dujon, B. (1974). *Genetics* **78**, 1063.
Newlon, C. S. and Fangman, W. L. (1975). *Cell* **5**, 423.
O'Connor, R. M., McArthur, C. R. and Clark-Walker, G. D. (1975). *European Journal of Biochemistry* **53**, 137.
O'Connor, R. M., McArthur, C. R. and Clark-Walker, G. D. (1976). *Journal of Bacteriology* **126**, 959.
Ogur, M., Minckler, S., Lindegren, G. and Lindegren, C. C. (1952). *Archives of Biochemsitry and Biophysics* **40**, 175.
Ojala, D. and Attardi, G. (1974). *Journal of Molecular Biology* **88**, 205.
Padmanabam, G., Hendler, F., Patzer, J., Ryan, R. and Rabinowitz, M. (1975). *Proceedings of the National Academy of Sciences of the United States of America* **72**, 4293.

Paoletti, C., Couder, H. and Guerineau, M. (1972). *Biochemical and Biophysical Research Communication* **48**, 950.
Parks, L. W. and Starr, P. R. (1963). *Journal of Cellular and Comparative Physiology* **61**, 61.
Perlman, P. S. (1976). *Genetics* **82**, 645.
Perlman, P. S. and Birky, C. W. (1974). *Proceedings of the National Academy of Sciences of the United States of America* **71**, 4612.
Perlman, P. S. and Mahler, H. R. (1971a). *Nature New Biology, London* **231**, 12.
Perlman, P. S. and Mahler, H. R. (1971b). *Biochemical and Biophysical Research Communications* **44**, 261.
Perlman, P. S. and Mahler, H. R. (1973). *Molecular and General Genetics* **121**, 295.
Petes, T. D. and Fangman, W. L. (1973). *Biochemical and Biophysical Research Communications* **55**, 603.
Petes, T. D. and Williamson, D. H. (1975). *Cell* **4**, 249.
Petes, T. D., Byers, B. and Fangman, W. L. (1973). *Proceedings of the National Academy of Sciences of the United States of America* **70**, 3072.
Pinon, R., Salts, Y. and Simchen, G. (1974). *Experimental Cell Research* **83**, 231.
Piperno, G., Fonty, G. and Bernardi, G. (1972). *Journal of Molecular Biology* **65**, 191.
Pittman, D., Ranganathan, B. and Wilson, F. (1959). *Experimental Cell Research* **17**, 368.
Prakash, L. (1975). *Journal of Molecular Biology* **98**, 781.
Prunell, A. and Bernardi, G. (1974). *Journal of Molecular Biology* **86**, 825.
Puglisi, P. P. and Algeri, A. (1971). *Molecular and General Genetics* **110**, 110.
Putrament, A., Baranowska, H. and Prazno, W. (1973). *Molecular and General Genetics* **126**, 357.
Rabinowitz, M. (1968). *Bulletin de la Société Chimie Biologique* **50**, 311.
Rabinowitz, M. and Swift, H. (1970). *Physiological Reviews* **50**, 376.
Rabinowitz, M. Getz, G. S., Casey, J. and Swift, H. (1969). *Journal of Molecular Biology* **41**, 381.
Rabinowitz, M., Casey, J., Gordon, P., Locker, J., Hsu, H. and Getz, G. S. (1974). *In* "Biogenesis of Mitochondria: Transcriptional, Translational and Genetic Aspects", (A. M. Kroon and C. Saccone, eds.), pp. 89–105. Academic Press, New York.
Rabinowitz, M., Jakovcic, S., Martin, N., Hendler, F., Halbreich, A., Lewin, A. and Morimoto, R. (1976). *In* "Genetic Function of Mitochondrial DNA", (C. Saccone and A. M. Kroon, eds.), pp. 219–230. North-Holland Publishing Co., Amsterdam.
Radloff, R., Bauer, W. and Vinograd, J. (1967). *Proceedings of the National Academy of Sciences of the United States of America* **57**, 1514.
Rank, G. H. (1970). *Canadian Journal of Genetics and Cytology* **12**, 340.
Rank, G. H. (1973). *Heredity* **30**, 265.
Rank, G. H. and Bech-Hansen, N. T. (1972a). *Canadian Journal of Microbiology* **18**, 1.
Rank, G. H. and Bech-Hansen, N. T. (1972b). *Genetics* **72**, 1.
Rank, G. H. and Martin, R. (1972). *Canadian Journal of Genetics and Cytology* **14**, 197.
Rank, G. H. and Person, C. (1969). *Canadian Journal of Genetics and Cytology* **11**, 716.
Reijnders, L. and Borst, P. (1972). *Biochemical and Biophysical Research Communications* **47**, 126.
Reijnders, L., Kleisen, C. M., Grivell, L. A. and Borst, P. (1972). *Biochimica et Biophysica Acta* **272**, 396.
Reijnders, L., Sloof, P. and Borst, P. (1973). *European Journal of Biochemistry* **35**, 266.
Robberson, D., Aloni, Y., Attardi, G. and Davidson, N. (1972). *Journal of Molecular Biology* **64**, 313.
Roodyn, D. B. and Wilkie, D. (1968). "The Biogenesis of Mitochondria", 123 pp. Methuen, London.
Rose, A. H. and Harrison, J. S. (1969). "The Yeasts", vol. 1, 508 pp. Academic Press, New York.

Rytka, J., English, K. J., Hall, R. M., Linnane, A. W. and Lukins, H. B. (1976). *In* "Genetics and Biogenesis of Chloroplasts and Mitochondria", (T. Bucher, W. Neupert, W. Sebald and S. Werner, eds.), pp. 427–434. North-Holland Publishing Co., Amsterdam.
Sager, R. (1972). "Cytoplasmic Genes and Organelles", 405 pp. Academic Press, New York.
Sanders, J. P. M., Flavell, R. A., Borst, P. and Mol. J. N. M. (1973). *Biochimica et Biophysica Acta* **312**, 441.
Sanders, J. P. M., Weijers, P. J, Groot, G. S. P. and Borst, P. (1974). *Biochimica et Biophysica Acta* **374**, 136.
Sanders, J. P. M., Borst, P. and Weijers, P. J. (1975a). *Molecular and General Genetics* **143**, 53.
Sanders, J. P. M., Heyting, C. and Borst, P. (1975b). *Biochemical and Biophysical Research Communications* **65**, 699.
Sanders, J. P. M., Heyting, C., Di Franco, A., Borst, P. and Slonimski, P. P. (1976). *In* "Genetic Function of Mitochondrial DNA", (C. Saccone and A. M. Kroon, eds.), pp. 259–272. North-Holland Publishing Co., Amsterdam.
Saunders, G. W., Gingold, E. B., Trembath, M. K., Lukins, H. B. and Linnane, A. W. (1971). *In* "Autonomy and Biogenesis of Mitochondria and Chloroplasts", (N. K. Boardman, A. W. Linnane and R. M. Smillie, eds.), pp. 185–193. North-Holland Publishing Co., Amsterdam.
Schafer, K. P. and Kuntzel, H. (1972). *Biochemical and Biophysical Research Communications* **46**, 1312.
Schatz, G. (1970). *In* "Membranes of Mitochondria and Chloroplasts", (E. Racker, ed.), pp. 251–314. Van Nojstrand-Reinhold, New York.
Schatz, G. and Mason, T. L. (1974). *Annual Review of Biochemistry* **43**, 51.
Schatz, G. and Saltzgaber, J. (1969). *Biochemical and Biophysical Research Communications* **37**, 996.
Schatz, G., Haslbrunner, E. and Tuppy, H. (1964). *Biochemical and Biophysical Research Communications* **15**, 127.
Schildkraut, C. L., Marmur, J. and Doty, P. (1962). *Journal of Molecular Biology* **4**, 430.
Schneller, J. M., Stahl, A. and Fukuhara, H. (1975). *Biochimie* **57**, 1051.
Schwab, R., Sebald, W. and Kaudewitz, F. (1971). *Molecular and General Genetics* **110**, 361.
Schweizer, E. and Halvorson, H. O. (1969). *Experimental Cell Research* **56**, 239.
Schweyen, R. J., Steyrer, U., Kaudewitz, F., Dujon, B. and Slonimski, P. P. (1976a). *Molecular and General Genetics* **146**, 117.
Schweyen, R. J., Weiss-Brummer, B., Backhaus, B. and Kaudewitz, F. (1976b). *In* "Genetic Function of Mitochondrial DNA", (C. Saccone and A. M. Kroon, eds.), pp. 251–258. North-Holland Publishing Co., Amsterdam.
Sena, E. P., Welch, J. W., Halvorson, H. O. and Fogel, S. (1975). *Journal of Bacteriology* **123**, 497.
Shannon, C., Rao, A., Douglass, S. and Criddle, R. S. (1972). *Journal of Supramolecular Structure* **1**, 145.
Shannon, C., Enns, R., Wheelis, L., Burchiel, K. and Criddle, R. S. (1973). *Journal of Biological Chemistry* **248**, 3004.
Shearman, C. W. and Kalf, G. F. (1975). *Biochemical and Biophysical Research Communications* **63**, 712.
Sherman, F. (1959). *Journal of Cellular and Comparative Physiology* **54**, 37.
Sherman, F. (1963). *Genetics* **48**, 375.
Sherman, F. and Ephrussi, B. (1962). *Genetics* **47**, 695.

Sherman, F. and Lawrence, C. W. (1974). *In* "Handbook of Genetics", (R. C. King, ed.), vol. 1, pp. 359–393. Plenum Press, New York.
Sherman, F. and Slonimski, P. P. (1964). *Biochimica et Biophysica Acta* **90**, 1.
Simpson, L. (1972). *International Review of Cytology* **32**, 139.
Sinclair, J. H., Stevens, B. J., Sanghavi, P. and Rabinowitz, M. (1956). *Science, New York* **156**, 1234.
Slonimski, P. P. and Tzagoloff, A. (1976). *European Journal of Biochemistry* **61**, 27.
Slonimski, P. P., Perrodin, G. and Croft, J. H. (1968). *Biochemical and Biophysical Research Communications* **30**, 232.
Smith, D., Tauro, P., Schweizer, E. and Halvorson, H. O. (1968). *Proceedings of the National Academy of Sciences of the United States of America* **60**, 936.
Smith, D. G., Wilkie, D. and Srivastava, K. C. (1973). *Microbios* **6**, 231.
Somlo, M., Avner, P. R., Cosson, J., Dujon, B. and Krupa, M. (1974). *European Journal of Biochemistry* **42**, 439.
Sriprakash, K. S., Choo, K. B., Nagley, P. and Linnane, A. W. (1976a). *Biochemical and Biophysical Research Communications* **69**, 85.
Sriprakash, K. S., Molloy, P. L., Nagley, P., Lukins, H. B. and Linnane, A. W. (1976b). *Journal of Molecular Biology* **104**, 485.
Stevens, B. J. and Moustacchi, E. (1971). *Experimental Cell Research* **64**, 259.
Storm, E. M. and Marmur, J. (1975). *Biochemical and Biophysical Research Communications* **64**, 752.
Stuart, K. D. (1970). *Biochemical and Biophysical Research Communications* **39**, 1045.
Subik, J., Kolarov, J. and Kovac, L. (1974). *Federation of European Biochemical Societies Letters* **45**, 263.
Suda, K. and Uchida, A. (1974). *Molecular and General Genetics* **128**, 331.
Sigimura, T., Okabe, K. and Imamura, A. (1966). *Nature, London* **212**, 304.
Swift, H. and Wolstenholme, D. R. (1969). *In* "Handbook of Molecular Cytology", (A. Lima-de-Faria, ed.), pp. 972–1046. North-Holland Publishing Co., Amsterdam.
Swift, H. and Wolstenholme, D. R. (1969). *In* "Handbook of Molecular Cytology", (A. Biogenesis of Mitochondria", (E. C. Slater, J. M. Tager, S. Papa and E. Quagliariello, eds.), pp. 3–19. Adriatica Editrice, Bari.
Tabak, H. F. and Weijers, P. J. (1976). *Federation of European Biochemical Societies Letters* **69**, 211.
Tewari, K. K., Jayaraman, J. and Mahler, H. R. (1965). *Biochemical and Biophysical Research Communications* **21**, 141.
Thomas, D. Y. and Wilkie, D., (1968). *Biochemical and Biophysical Research Communications* **30**, 368.
Tingle, M. A., Kuenzi, M. T. and Halvorson, H. O. (1974). *Journal of Bacteriology* **117**, 89.
Trembath, M. K., Bunn, C. L., Lukins, H. B. and Linnane, A. W. (1973). *Molecular and General Genetics* **121**, 35.
Trembath, M. K., Monk, B. C., Kellerman, G. M. and Linnane, A. W. (1975). *Molecular and General Genetics* **141**, 9.
Trembath, M. K., Molloy, P. L., Sriprakash, M. S., Cutting, G. J., Linnane, A. W. and Lukins, H. B. (1976). *Molecular and General Genetics* **145**, 43.
Tzagoloff, A., Akai, A. and Needleman, R. B. (1975a). *Journal of Bacteriology* **122**, 826.
Tzagoloff, A., Akai, A. and Needleman, R. B. (1975b). *Proceedings of the National Academy of Sciences of the United States of America* **72**, 2054.
Tzagoloff, A., Akai, A., Needleman, R. B. and Zulch, G. (1975c). *Journal of Biological Chemistry* **250**, 8236.

Tzagoloff, A., Akai, A, and Foury, F. (1976a). *Federation of European Biochemical Societies Letters* **65**, 391.
Tzagoloff, A., Foury, F. and Akai, A. (1976b). *In* "Genetic Function of Mitochondrial DNA",(C. Saccone and A. M. Kroon, eds.), pp. 155–161. North-Holland Publishing Co., Amsterdam.
Uchida, A. (1972). *Japanese Journal of Genetics* **47**, 159.
Uchida, A. and Suda, K. (1973). *Mutation Research* **19**, 57.
Upholt, W. B. and Borst, P. (1974). *Journal of Cell Biology* **61**, 383.
Van Kreijl, C. F., Borst, P., Flavell, R. A. and Hollenberg, C. P. (1972). *Biochimica et Biophysica Acta* **272**, 61.
Vidova, M. and Kovac, L. (1972). *Federation of European Biochemical Societies Letters* **22**, 347.
Wakabayashi, K. (1974a). *Journal of Antibiotics, Tokyo* **27**, 373.
Wakabayashi, K. (1974b). *Proceedings of the Japanese Academy* **50**, 396.
Wakabayashi, K. and Gunge, N. (1970). *Federation of European Biochemical Societies Letters* **6**, 302.
Wakabayashi, K. and Kamei, S. (1973). *Federation of European Biochemical Societies Letters* **33**, 263.
Wallis, O. C., Ottolenghi, P. and Whittaker, P. A. (1972). *Biochemical Journal* **127**, 46P.
Waters, R. and Moustacchi, E. (1974). *Biochimica et Biophysica Acta* **366**, 241.
Waxman, M. F. and Eaton, N. R. (1974). *Molecular and General Genetics* **133**, 37.
Waxman, M. F., Eaton, N. and Wilkie, D. (1973). *Molecular and General Genetics* **127**, 277.
Weislogel, P. O. and Butow, R. A. (1970). *Proceedings of the National Academy of Sciences of the United States of America* **67**, 52.
Weislogel, P. O. and Butow, R. A. (1971). *Journal of Biological Chemistry* **246**, 5113.
Wells, J. R. M. (1974). *Experimental Cell Research* **85**, 278.
Werkheiser, D. and Pittman, D. (1972). *Genetics* **72**, 177.
Weth, G. and Michaelis, G. (1974). *Molecular and General Genetics* **135**, 269.
Wheelis, L., Trembath, M. K. and Criddle, R. S. (1975). *Biochemical and Biophysical Research Communications* **65**, 838.
Whittaker, P. A. and Wright, M. (1972). *Biochemical and Biophysical Research Communications* **48**, 1455.
Whittaker, P. A., Hammond, R. C. and Luha, A. A. (1972). *Nature New Biology, London* **238**, 266.
Wickner, R. B. (1974). *Journal of Bacteriology* **117**, 252.
Wilkie, D. (1963). *Journal of Molecular Biology* **7**, 527.
Wilkie, D. (1970). *In* "Control of Organelle Development", (P. L. Miller, ed.), pp. 71–83. Cambridge University Press, Cambridge.
Wilkie, D. and Thomas, D. Y. (1973). *Genetics* **73**, 367.
Wilkie, D., Saunders, G. and Linnane, A. W. (1967). *Genetical Research, Cambridge* **10**, 199.
Williamson, D. H. (1970). *In* "Control of Organelle Development", (P. L. Miller, ed.), pp. 247–276. Cambridge University Press, Cambridge.
Williamson, D. H. (1973). *Biochemical and Biophysical Research Communications* **52**, 731.
Williamson, D. H. and Fennell, D. J. (1974). *Molecular and General Genetics* **131**, 193.
Williamson, D. H. and Moustacchi, E. (1971). *Biochemical and Biophysical Research Communications* **42**, 195.
Williamson, D. H., Maroudas, N. G. and Wilkie, D. (1971a). *Molecular and General Genetics* **111**, 209.
Williamson, D. H., Moustacchi, E. and Fennell, P. (1971b). *Biochimica et Biophysica Acta* **238**, 369.

Wintersberger, E. (1966). *Biochemical and Biophysical Research Communications* **25**, 1.
Wintersberger, E. (1967). *Hoppe-Zeyler's Zeitschrift für Physiologische Chemie* **348**, 1701.
Wintersberger, E. (1968). *In* "Biochemical Aspects of the Biogenesis of Mitochondria", (E. C. Slater, J. M. Tager, S. Papa and E. Quagliariello, eds.), pp. 189–201. Adriatica Editrice, Bari.
Wintersberger, E. and Viehhauser, G. L. (1968). *Nature, London* **220**, 699.
Wintersberger, U. and Hirsch, J. (1973). *Molecular and General Genetics* **126**, 71.
Wintersberger, U. and Wintersberger, E. (1970a). *European Journal of Biochemistry* **13**, 11.
Wintersberger, U. and Wintersberger, E. (1970b). *European Journal of Biochemistry* **13**, 20.
Wintersberger, U., Hirsch, J. and Fink, A. M. (1974). *Molecular and General Genetics* **131**, 291.
Wolf, K., Sebald-Althaus, M., Schweyen, R. J. and Kaudewitz, F. (1971). *Molecular and General Genetics* **110**, 101.
Wolf, K., Dujon, B. and Slonimski, P. P. (1973). *Molecular and General Genetics* **125**, 53.
Wolf, K., Burger, G., Lang, B. and Kaudewitz, F. (1976a). *Molecular and General Genetics* **144**, 67.
Wolf, K., Lang, B., Burger, G. and Kaudewitz, F. (1976b). *Molecular and General Genetics* **144**, 75.
Wright, R. E. and Lederberg, J. (1957). *Proceedings of the National Academy of Sciences of the United States of America* **43**, 919.
Yotsuyanagi, Y. (1966). *Académie des Sciences, Paris, Comptes Rendus Hebdomadaires des Séances* **262**, 1348.
Zeman, L. and Lusena, C. V. (1974a). *Federation of European Biochemical Societies Letters* **38**, 171.
Zeman, L. and Lusena, C. V. (1974b). *Federation of European Biochemical Societies Letters* **40**, 84.
Zeman, L. and Lusena, C. V. (1975). *European Journal of Biochemistry* **57**, 561.

Disruption of Micro-organisms

W. T. COAKLEY, A. J. BATER AND D. LLOYD

Department of Microbiology, University College, Newport Road, Cardiff CF2 1TA, Wales

I. Introduction	. . . .	279
II. Strength of Cell Walls in Relation to Structure	. . . . .	281
A. Cell Outer Layers	. . . .	281
B. Relationship Between Strength of Cell Wall and Breakage Method Employed	. . . .	286
III. Principles of Cell Breakage	. . . .	288
A. Osmotic Lysis	. . . .	288
B. Basic Hydrodynamics	. . . .	289
C. Cells in Flow Systems	. . . .	291
D. Mechanical Methods for Cell Disruption	. . . . .	295
IV. Lability of Cell Extract Components	. . . .	305
A. Thermal Lability	. . . .	305
B. Chemical Inactivation	. . . .	306
C. Mechanical Comminution	. . . .	307
D. General Approach to Optimization of the Activity of Extracts	. .	324
V. Controlled Breakage of Eukaryotic Micro-organisms	. . . .	328
A. Controlled Mechanical Breakage	. . . .	328
B. Disruption after Pretreatment of Organisms	. . . . .	331
VI. Concluding Remarks	. . . .	334
References	. . . .	335

I. Introduction

Techniques for the disruption of cells have evolved over a period of nearly eighty years, and for the most part suitable methods for the solution of individual needs have been arrived at by a pragmatic approach. It is clear, however, that from the viewpoint of future developments much is to be gained from a rigorous physical and biochemical understanding of the processes involved in cell breakage. In recent years a requirement for a greater degree of sophistication has become evident as the emphasis has in many areas shifted from the aim of quantitative enzyme release to that of the preservation of as much information content as possible within the disrupted system. This modification intentionally places a high priority on a thorough understanding of the disrupting techniques and their controlling parameters. Thus the experimentalist should ideally appreciate the following components of the system:

1. The nature of the boundary layers of the organism, their tensile strength, the chemical constitution of those structural elements which make the greatest contribution to this strength, and the intermolecular interactions which preserve this structure.

2. The nature of the forces generated by the breakage technique and the ability to adjust applied forces so as to optimize the formation of the desired product.

The desired product is, in the simplest cases, a highly stable homogeneous soluble molecular species; for this, any technique which ensures complete removal of permeability barriers will suffice to provide maximum yield of the product, and difficulties arise only with those cell-types which are only partially disrupted by the most fierce mechanical devices at present in use, or with cells which have boundary layers attacked by none of the commercially-available enzymes. In recent years the extension of both mechanical and enzymic capabilities has reduced the number of intractable species almost to vanishing point, although some micro-organisms still pose problems. If the component is soluble but unstable to high hydrodynamic shearing forces, or to attack by free radicals or by cellular components, then care is necessary to optimize processes for its release in the active form.

At higher levels of organization, i.e. where the molecular species is membrane associated, two different approaches are possible: (a) to disperse all the membranes of the organisms and then purify the individual component or, (b) to first obtain the membrane fraction or organelle-enriched fraction of the extract as a starting point for purification of a molecular species or for the study of its function *in situ*. The philosophy behind cell breakage is evidently quite different in case (b) and it is the needs of this approach which have led to the development of techniques which can release subcellular particles, still possessing some degree of structural and functional integrity, which may provide information on their physiological role *in vivo*. The requirements of this approach which is now employed in the majority of fundamental biochemical investigations have provided the challenge for the development of what may be termed "minimal disruptive conditions", i.e. refined procedures which minimize damage to intracellular organelles. Thus many of the methods used for quantitative release of enzymes from micro-organisms, especially those which rely on the development of high liquid or solid shearing forces are not suitable for the release of "intact" membranes or organelles. Indeed, as further information on the complexity of intracellular organization becomes available, it becomes increasingly evident that the extraction of "intact" organelles may often be an unattainable goal. Thus many organelles are extensively branched, coiled or reticulate; one may completely envelop a second class of organelles; similarly, organelles previously thought to be structurally and functionally separate, may be interconnected. The whole process of subcellular fractionation is therefore fraught with the probability of disruptive artefact, and by definition, the most critical choice in any proposed requirement for subcellular fractionation is the method of cell disruption employed.

If we cannot study completely undamaged organelles *in vitro,* then even a pessimistic observer would have to admit that carefully prepared vesicular fragments can provide a wealth of information, data which may with caution be related to events *in vivo.* Here again choice of a method for producing vesicles is the most important preparative step. Now that many independent criteria are available for assessment of the orientation of the surfaces relative to that obtaining *in situ*, a whole family of disruptive procedures remains to be critically reevaluated.

Evidently the procedures employed for the controlled release of intracellular components and for the comminution of fractions, assume an even greater importance as progress is made in the understanding of the intimate details of structure-function interrelationships. This review updates that of Hughes *et al.* (1971), surveying recently developed methods and the increasing appreciation of the principles underlying some older techniques.

II. Strength of Cell Walls in Relation to Structure

A. CELL OUTER LAYERS

The widely varying strengths of the envelopes of biological cells is apparent from the differences in resistance to disruption by various methods. Mechanical strength is lowest in cells which are completely lacking in cell walls and reaches its peak in cells possessing highly ordered, multilayered or fibrous cell walls. In the following sections these will be dealt with in order of increasing strength.

1. Cells Lacking Cell Walls

The primary boundary layer of any cell is the cytoplasmic membrane and in many cases no other layer is present. As the cytoplasmic membrane is an osmotic barrier, such cells have to be protected from the hypotonic lysis to which they are susceptible. Various protective measures have evolved in different organisms, each suited to its own particular environment.

The cells comprising tissues of higher animals have no protection from osmotic damage, as under physiological conditions they are constantly bathed in isotonic fluids. If mammalian cells, for example, are removed from this environment and suspended in a hypotonic buffer, the cells swell rapidly until the cytoplasmic membrane can no longer stretch; at this point lysis occurs due to pores opening in the membrane. This is the basis of methods for preparation of membranes from a variety of cell types, e.g. adipocytes (Rodbell and Krishna, 1974) and erythrocytes (Steck and Kant, 1974; Hanahan and Ekholm, 1974).

The protozoa, most of which are found naturally in constantly hypotonic environments, are able to survive because they possess a contractile vacuole. This is an intracellular body which collects excess water from the cytoplasm and periodically expels it from the cell. Thus these cells are protected by the elimination of osmotic stress but are still very fragile. They are easily disrupted by gentle mechanical treatments such as hand homogenization, although the use of the Chaikoff Press is preferred due to its greater reproducibility.

Another class of cells which are sensitive to osmotic lysis, and must therefore be classified as structurally weak, are the halophilic bacteria. Although they are morphologically very similar to the Gram-negative bacteria, they appear to lack the rigid peptidoglycan layer which confers strength upon the cell wall of the latter. Halophilic bacteria are adapted for survival in the conditions of high osmotic strength encountered in salt water and require high concentrations of inorganic salts for growth and maintenance of structural integrity. If, for example, cells of *Halobacterium halobium* are subjected to conditions of low ionic strength, the membranes disintegrate into fragments

(Blaurock and Stockenius, 1971). Similarly, the *Mycoplasma* group of bacteria, which are found as pathogens and parasites of man and a number of other animals, and the mycoplasma-like bodies found in plants are devoid of cell walls. They also are adapted for growth in a specialized environment and they rapidly lyse if placed in a hypotonic medium.

2. *Bacterial Cell Walls*

With the exception of the halophilic bacteria and *Mycoplasma* spp., bacteria are not normally susceptible to osmotic damage. They possess cell walls of varying complexity which confer mechanical strength and rigidity, and are reponsible for the maintenance of cell shape.

The basis of bacterial cell walls is a macromolecular network of peptidoglycan. In Gram-negative bacteria approximately 10% of the cell envelope is peptidoglycan but in Gram-positive bacteria it can account for 80% of the cell wall. The peptidoglycan is similar in all bacteria and normally consists of a β 1,4-linked polysaccharide backbone of alternating N-acetylmuramic acid and N-acetylglucosamine residues cross-linked by peptides, which vary greatly between genera. The degree of cross-linking can be very high, especially in Gram-positive cells, and the peptidoglycan has often been considered to form a single macromolecule encompassing the whole cell. Current evidence, however, favours an amorphous, elastic model of cell wall composition rather than a highly condensed structure (Ou and Marquis, 1970, 1972). Ou and Marquis (1970) suggest that the peptidoglycan of Gram-positive cocci can be viewed as a three-dimensional rope ladder with rigid polysaccharide rungs and flexible polypeptide ropes; this implies an elastic, restraining role for the structure rather than the common concept of it acting as a rigid shell.

Gram-negative bacteria possess a single, two-dimensional layer of peptidoglycan bounded on the inside by the cytoplasmic membrane and on the outside by a lipopolysaccharide-rich outer membrane. The latter may be linked to the peptidoglycan *via* glycolipids and lipopolysaccharides. These bacteria are more susceptible to disintegration by mechanical stress than are their Gram-positive counterparts, where the cell wall comprises a three-dimensional matrix of peptidoglycan and teichoic acid which may be up to 80 nm thick (Glauert, 1962). For instance, in a comparison of total nitrogen release from various organisms subjected to hydrodynamic shear in a French Press, Hughes et al., (1971) noted that 70 MN/m^2 was sufficient to release 65% of total nitrogen from *Escherichia coli* while 200 MN/m^2 only released 57% from *Bacillus megaterium*. Gram-positive cocci are even more resistant than Gram-positive rods; in the above study, 200 MN/m^2 released 31% and 25% of the total nitrogen from *Staphylococcus aureus* and *Sarcina lutea* respectively. This may be due to a denser, more compact wall in cocci than in rods. For example, in *Arthrobacter crystalopoietes*, which can exist as a rod or a coccus, the peptidoglycan of the coccoid form has been found to have shorter glycan strands (19 versus 60 disaccharide units) and longer cross-linking peptides than that of the rod-form (Krulwich et al., 1967a, b). Marquis (1973) interpreted this as evidence for a closer association of polymer strands in coccal walls than in rod walls, resulting in a greater compactness. The extra compactness and higher density may account for the greater mechanical strength of coccal walls.

Although bacteria exhibit a similar response to ultrasound treatment as to French Pressing, disruption by the more extreme forces produced in the Hughes Press does not differentiate between bacterial types. This is because the solid-shear forces set up are sufficient to overcome the mechanical strength of even the most resistant bacterial cell.

An interesting comparison of the strengths of Gram-negative and Gram-positive bacteria was presented by Isaac and Ware (1974) who embedded bacteria in a glycerol/gelatin gel and by direct microscopic observation estimated their resistance to stretching. The Gram-negative bacteria tested, *Spirillum* sp. and *Escherichia coli*, would stretch to twice or three times their original lengths respectively, although in the first case this was due to uncoiling of the normally spiral cell as well as to true elongation. In contrast, a Gram-positive rod, *Bacillus megaterium,* could only be extended by 0.25 of its original length, and no deformation or elongation of a Gram-positive coccus, *Staphylococcus aureus,* was observed.

Although, as mentioned previously, bacteria are not normally sensitive to osmotic pressure, this is due only to the presence of the peptidoglycan layer. The latter can be digested by the enzyme lysozyme, which attacks the β 1,4-linkages in the polysaccharide backbone, although with Gram-negative cells EDTA treatment is usually required to allow the lysozyme to penetrate the outer membrane. Lysozyme treatment renders bacteria sensitive to osmotic pressure and must be carried out in isotonic sucrose solutions if lysis is to be avoided. The effects of such treatment on the morphology of bacterial membranes is discussed in the section on bacterial membrane vesicles.

3. Fungal Cell Walls

Among the basidiomycetes, the ascomycetes, and the Fungi Imperfecti, the characteristic morphology is that of a coenocytic mycelium. However, several groups in these classes have largely lost the mycelial habit of growth and have become free-living unicellular organisms, termed yeasts. Both yeast and mycelial forms of fungi are characterized by the presence of rigid cell walls with a high polysaccharide content. Most fungi also contain chitin in their cell walls, the exceptions having cellulose as their main component (Rogers and Perkins, 1968). A great variety of other polysaccharide types are present, including glucans and mannans; for a review of fungal cell-wall polysaccharides see Gander (1974). These cell walls all exhibit great strength, and methods such as freeze-pressing, or treatment with vibratory ball mills or Mickle-type shakers, are commonly used to disrupt fungi. Chemical methods, long used for the isolation of intracellular components of yeasts, such as digestion of the wall polymers with enzyme extracted from the gut of the snail, *Helix pomatia,* followed by osmotic lysis, are now being employed in studies with filamentous fungi (Aguirre and Villanueva, 1962).

4. Algal Cell Walls

With the exception of the euglenoid flagellates, the algae possess complex cell walls containing polysaccharides arranged in microfibrillar arrays. Four groups can be recognized on the basis of their predominant sugar (Rogers and Perkins, 1968); these contain respectively, cellulose, xylans, β 1,4-linked mannose, or a combination of these, as the major cell-wall polymer. All have very strong cell walls; for example, in the comparative study of resistance to disruption by French Pressing described earlier (p. 283), only 35% of cells of *Chlorella pyrenoidosa* were disrupted by hydrodynamic stresses sufficient to break 57% of *Bacillus megaterium,* 70% of *Saccharomyces cerevisiae,* and more than 90% of *Escherichia coli* (Hughes *et al.,* 1971).

In contrast to the other algae, the euglenoid flagellates do not possess a polysaccharide cell wall, but are protected from osmotic lysis by a pellicle consisting of thin strips of proteinaceous material wound around like a number of bandages which fuse in pairs at the base and apex of the cell (Leedale, 1967). These pellicular strips, being internal to the cytoplasmic membrane, form a pliable, extensible, endoskeleton which provides mechanical strength while allowing the cell to change its shape.

B. RELATIONSHIP BETWEEN STRENGTH OF CELL WALL AND BREAKAGE METHOD EMPLOYED

The foregoing sections have demonstrated the dependence of cell envelope strength upon cell-wall structure. As the strength of the cell wall increases, it becomes necessary to exert higher mechanical stresses upon the cells before disruption occurs. The result of a survey of the mechanical breakage methods employed in studies reported in thirty issues of the *Journal of General Microbiology* is presented in Table 1; in this Table, organisms are grouped according to similarities in cell-wall structure and are listed in order of increasing mechanical strength; the disruptive potential of the breakage methods listed, increases from left to right. A good correlation is found between increasing cell strength and increasing violence of breakage methods. The one exception is with the filamentous fungi, where 50% of the studies examined, reported the use of less rigorous disruptive procedures. There are two reasons for this. First, for any breakage technique, there is a limiting size for the treated particle below which the method is ineffective; the mycelial nature of the filamentous fungi means that in one dimension the cell is very large and therefore more susceptible to disruption. The second reason for the high frequency of gentle techniques is that these fungi are coenocytic and if the aim of disruption is the isolation of sub-cellular organelles or cytoplasmic constituents, it is sufficient to produce breaks in the mycelium at relatively large intervals, after which the internal components may be extruded by gentle homogenization.

At first glance the Table indicates the proneness of workers to use more rigorous treatment than is really necessary for cell disruption. Such a tendency should be guarded against as increasing comminution

TABLE 1. Survey of mechanical breakage methods used in studies reported in volumes 80–94 of the *Journal of General Microbiology* (1974–1976)

	Frequency of use of each method or group of methods[a]								
	Total number of studies reported	Osmotic shock, freeze thawing, and hand homogenizers	Motor-driven homogenizers and blenders	Hand grinding with abrasives	Sonication	Liquid extrusion presses	Bead mills	Freeze presses	Combinations of above methods
Protozoa and mycoplasmas	5	40			60				
Gram-negative bacteria and blue green algae	75	3	1	3	62	19	4	8	
Gram-positive rods and actinomycetes	16	6		6	44	25	13	6	
Gram-positive cocci	10			10	20	10	40	20	
Yeasts	20		5		5	10	30	50	
Filamentous fungi	20	5	15	30			35	5	10

[a] The frequency is expressed as percentage of the total number of studies reported for the particular group of organisms.

of cellular material enhances the production of artefacts and renders the debris more and more unlike the starting material. It must be remembered, however, that the gentlest technique which will disrupt a given cell is not necessarily the most efficient for producing the desired extract. This variation in subcellular material yielded by different techniques will be discussed in later sections.

III. Principles of Cell Breakage

A. OSMOTIC LYSIS

When a wall surrounds the cytoplasmic membrane of a microorganism it protects the cell against osmotic lysis. Removal of, or weakening of the wall to produce protoplasts or sphaeroplasts can lead to cell lysis if the cell is suspended in hypotonic solution, or can at least make the cell more vulnerable to mechanical stress. Osmotic stress is also exerted on the outer layers of cells which normally inhabit salt rich environments, e.g. the halobacteria, marine micro-organisms and mammalian cells, when they are placed in hypotonic environments.

The response of protoplasts of *Bacillus megaterium* to slowly applied osmotic stress consists mainly of elastic yield and development of tautness, while the response to rapidly applied stress includes brittle fracture of membranes which are not fully extended (Marquis, 1967). The response of cells to a gradual reduction in the osmolarity of the suspending medium can in general be described as a three stage process. Initially the cell volume increases without an increase in the surface area of the membrane, i.e. no tension develops in the membrane and the cell obeys the Boyle-Van't Hoff law during swelling, as when the surface area of protoplasts lose their invaginations (Marquis, 1967), the convoluted surfaces of lymphocytes unfold (Corner and Marquis, 1969) or when an erythrocyte changes from a biconcave disc to a sphere (Burton, 1970). With further reduction of the external osmolarity, tension develops in the membrane. Corner and Marquis (1969) concluded from their observations of *B. megaterium* protoplasts that the membrane becomes permeable to previously impermeant solute molecules due to dilation of membrane pores. The resultant inflow of solute lowers the cytoplasmic water activity so that additional water enters the cell, more swelling ensues, with further membrane stretching, further dilation of pores and consequent increase in permeability. The process cascades and finally leads to the third stage of response, i.e. cell bursting, to leave cell ghosts with a single torn or damaged area. The resistance of bacterial protoplasts to osmotic shock decreases as the incubation temperature is increased (Eisenberg and Corner, 1973).

In very violent rapid osmotic lysis of erythrocytes, the membrane is irreversibly torn and the haemoglobin leaves in "blobs" (Burton,

1970). Where the haemolysis is rapid but results in cell ghosts with a restored permeability barrier, Seeman (1967) found that transient holes of the order of 20–50 nm wide were produced in the membrane. These holes existed only for 15–25 seconds after the onset of rapid haemolysis, i.e. the holes shrank leaving an intact membrane. Burton (1970) proposed a mechanical treatment for the production of holes in the erythrocyte membrane and concluded that, whether the model used is of an elastic structure with cylindrical pores, or at the other extreme a dynamic double surface layer of oriented molecules, a small increase in area of the membrane under stress would result in a great increase in the size of either static or dynamic pores.

When Gram-negative bacteria are subjected to osmotic shock, periplasmic binding proteins are selectively released. A considerable number of different proteins is released amounting to about 5% of the total cell protein. These proteins largely fall into two functional categories: (i) degradative enzymes such as alkaline phosphatase, 5′-nucleotidase, ribonuclease 1, DNA endonuclease 1; and (ii) binding proteins, each of which is capable of specifically binding some small ligand such as sulphate ion, individual amino acid, or certain sugars. None of the binding proteins has been shown to have any enzymic activity, but many of them have been implicated in the transport of their specific ligands (Oxender, 1972; Singer, 1974).

Osmotic lysis is frequently used in the production of membrane vesicles for transport studies and other studies of membrane proteins. These aspects of osmotic lysis are discussed later (p. 310).

B. BASIC HYDRODYNAMICS

Cells in suspension are stressed when the suspension flows. Many mechanical devices have been developed to increase this stress to a level at which the cell will break. The shear stress, in a liquid where laminar (stream-lined) flow conditions obtain, is given by

$$\tau = \eta G \tag{1}$$

where η is the co-efficient of viscosity and G is the velocity gradient or shear rate. The velocity gradient is the velocity change with distance, perpendicular to the direction of flow. The velocity gradient can be written as dv/dx where v is the laminar flow velocity and x is measured in a direction normal to the direction of flow. If a cell suspension is considered to be similar to a suspension of non-ionized spheres, then the viscosity, η_s, of a dilute cell suspension will be little different from the viscosity, η, of the liquid. The viscosity of a suspension of spheres is given by:

$$\eta_s = \eta(1 + 2.5c) \tag{2}$$

where c is the volume concentration of the spheres (Einstein, 1906). The above relationship holds well when $c < 0.1$. If the model of the cell is a fluid droplet then the viscosity of a dilute droplet suspension is given by:

$$\eta_s = \eta\{1 + [(5\lambda + 2)/2(\lambda + 1)]c\} \tag{3}$$

where λ is the ratio of the internal viscosity of the droplet to that of the suspending liquid (Taylor, 1932). The above expression reduces to Eq. (2) for the case of a solid when $\lambda = \infty$.

In a flow system of a given geometry, the flow conditions may be characterized by the non-dimensional Reynolds number Re where:

$$Re = \rho \bar{v} d / \eta \tag{4}$$

ρ is the density of the liquid, $\bar{v}$ is the average velocity of the liquid and d is a lateral dimension of the channel. For a given flow system there is an upper limit to the Reynolds number above which laminar flow breaks down and turbulent flow sets in. From the above definition of Re we see that flow is more likely to be streamlined if the viscosity is high and the flow rate and channel width are small. In a straight tube when d in Eq. (4) is the inner diameter, experimental results generally show that fully developed flow will be laminar when the Reynolds number is less than about 2300 and will be turbulent at greater values of Re. However when exceptional precautions are taken, the laminar condition can be maintained in a straight pipe at values of Re up to 100,000 (Pfenniger, 1961). Turbulence can set in at comparatively low values of Re if there is a sudden alteration in the lateral dimensions of the channel at entry or exit, or if there are regions of high curvature in the channel.

In discussing turbulence we will follow the terminology developed by Hinze (1975). Real turbulent motion is an irregular condition of flow in which the various quantities, such as velocity, pressure and temperature, show a random variation with time and space coordinates, so that statistically distinct average values can be discerned. When considering cells forced through narrow channels in turbulent flow we will be dealing with "wall turbulence", i.e. turbulence whose structure is directly influenced by the presence of a solid boundary, when the cells travel through the channel, and with "free turbulent jet flow" when the cell suspension emerges rapidly from the orifice into a container. In wall turbulence, at any value of Re there is an extremely thin layer at the wall where the flow is predominantly viscous. Outside the viscous layer is a "buffer region" where inertial effects become of increasing importance. A region of fully developed turbulence completes the "wall region" flows (Hinze, 1975). There is in the tube core a further turbulent region, unaffected by the wall, but which interacts with the "wall region" turbulence. The "buffer" region near the wall is the most active as far as production and dissipation of turbulence energy is concerned. The turbulence shear stress is directly proportional to distance from the tube axis over almost the whole cross section of the tube (Laufer, 1954). Turbulence consists of the superposition of eddies of various size, or scales, and intensities. The scales of the larger eddies are determined by the dimension of the flow channel. As the eddy scale decreases, in general the velocity gradient in the eddy increases, and the viscous shear stresses which tend to counteract the eddy, increase. Thus there is a minimum eddy scale of size s given by:

$$s = (\nu^3/\epsilon)^{1/4} \qquad (5)$$

below which viscosity controls the flow. In the above equation, ν is the kinematic viscosity and ϵ is the energy input to the system per unit mass (Hinze, 1975).

C. CELLS IN FLOW SYSTEMS

Theoretical treatments of the behaviour of cells in flow have been influenced by the understanding available of the microrheology of dispersions. The movements of solid spheres and of rods, which might serve as models for cells such as cocci or bacterial rods, in flow systems have been reviewed by Goldsmith and Mason (1967). In a concentric cylinder system with one cylinder rotating relative to the other (Couette flow), a moving solid sphere in suspension will rotate about a vertical axis with a constant angular velocity of $G/2$. Deformable drops migrate in laminar tube-flow (Poiseuille flow) across the streamlines towards the tube axis. Such drops serve as good models for erythrocytes in flow (Goldsmith, 1971). When the particle Reynolds number Re_p is low, solid spheres do not exhibit radial migration, but at higher values of Re_p a tubular pinching effect is observed in which the spheres move away from wall and from the axis to an equilibrium radial position (Goldsmith, 1971). The deformations of droplets are similar in Couette and in Poiseuille flows and are functions of the shear stress, the interfacial tension, and the ratio λ of the viscosities of the suspended to the suspending phases. The deformations observed when comminution of the droplets occurs at different values of λ are shown in Fig. 1 which is compiled from illustrations of Goldsmith and Mason (1967). When λ is less than 0.2, the drops assume a sigmoid shape with

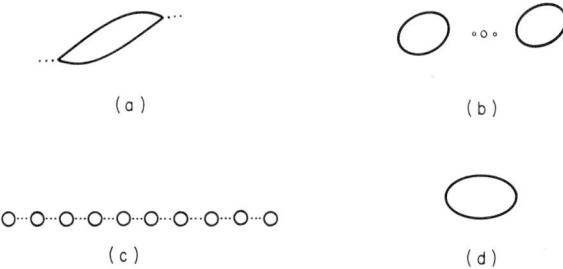

FIG. 1. Limiting deformations of droplets in laminar Couette or Poiseuille flow for different values of the ratio λ of droplet to suspending phase viscosities: (a) $\lambda < 0.2$; (b) $2.0 > \lambda > 0.2$; (c) $2.0 > \lambda > 0.2$ slow growth of deformation; (d) $\lambda > 2.0$. [(a), (b), (d), compiled from, and (c) described by Goldsmith and Mason, 1967.]

pointed ends from which fragments of the disperse phase are released (Fig. 1a). When λ exceeds 0.2, the central portion of the drop suddenly begins to extend into a cylinder which progressively necks off in the middle until two nearly identical daughter drops are formed separated by three satellite droplets (Fig. 1b). If the rate of growth of the disturbance is too slow for necking to occur, then the drop extends into a thread which progressively increases in length until at a sufficiently small diameter, it breaks up into tiny droplets. If λ is greater than 2.0 the drop may elongate to a limiting deformation and align itself along the streamline (Fig. 1d). Taylor (1934) has suggested that, when the difference in normal stresses at the interface which tend to disrupt a drop, exceed the forces due to surface tension which tend to hold the drop together, the drop will burst. The threshold stress at which disruption will occur is a weak function of λ and is given to a good approximation by $T/2a$ where T is the interfacial tension and a is the radius of the drop. Nyborg (1968) pointed out that the critical flow stress necessary to disrupt erythrocytes is of the order which would be expected on calculating $T/2a$, using published values for the erythrocyte interfacial tension. The droplet model for the disruption of weak-walled cells received further support when "microspheres" (haemoglobin filled vesicles) were observed in erythrocyte suspensions subjected to high hydrodynamic shear stress (Williams et al., 1970; Champion et al., 1971), although we will see later (p. 294) that it may also be possible under certain conditions for vesicles to arise from interactions between a cell and a solid boundary wall. Richardson (1974, 1975) has extended the treatment of erythrocyte lysis in laminar flow to include the effects of cell membrane visco-elasticity and has proposed that, if the cell survives the first few rotations without disrupting, then the critical time t_c to haemolysis is given by:

$$t_c \approx \alpha/\eta G^2 \qquad (6)$$

where α is a constant which is a property of the cell outer layer. Recent attempts to verify the above relationship for erythrocyte disruption will be discussed later (p. 295). Hinze (1953) considered droplet deformation in turbulent flow and concluded that the drop would undergo "bulgy" deformations in which parts of the surface would protrude at some length into the suspending phase. If a cell behaved similarly in turbulent flow, then surface tension might act to form vesicles from the tips of such protrusions. Dumb-bell behaviour of droplets, i.e. an elongation into two product droplets, similar to that described in Fig. 1(b), has been observed in turbulent flow (Collins and Knudsen, 1970). Ellipsoidal and dumb-bell shapes in erythrocytes in turbulent flow have been demonstrated by Sutera and Mehrjardi (1975).

Critical studies of cell disruption in streamlined flow have largely been concerned with erythrocyte breakage. Human erythrocytes are more resistant to disruption than *Acanthamoeba castellanii* and have the same strength as mouse ascites carcinoma cells (Williams et al., 1970; Williams, 1972). When the results of experiments to disrupt erythrocytes in flows of different geometries were compared by Leverett et al. (1972), a trend similar to that in Fig. 2 was discerned which it was claimed separated two regions of response of the cells to stress. At high stresses, cells were disrupted by shear stress in the bulk of the fluid. When cells were stressed in low stress flows for a long time, the disruption observed generally involved a low proportion of the cells and was considered to arise from interactions between the cells and the walls of the shearing devices (cell-surface interactions). The latter mechanism has received considerable attention in recent years because of the occurrence of erythrocyte lysis in artificial blood-circulating devices. The problems associated with a theoretical approach to the general problem of cell-surface interactions have been pointed out by Weiss (1971). Blackshear et al. (1971) reported that, when erythrocytes contacted a wall, they left behind a residue. The cells

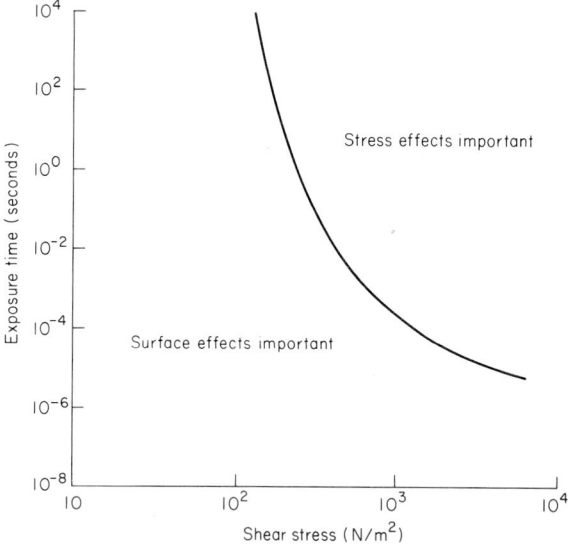

FIG. 2. The separate stress-time regimes suggested by Leverett et al. (1972) for haemolysis by cell-surface interaction and by flow stress in the bulk of the liquid.

sometimes stuck, often at a small site on the wall, and were observed to fly like a kite suspended by a strand of membraneous material with the main body of the cell extending into the adjacent flowing stream. Both haemolysis and cell fragmentation were related to drastic shape changes in the attached cells. A membrane tension develops in the stressed main body of a cell which is tethered to a wall. This stress can become large enough for pores to open, which allow haemoglobin to escape, or the tether can collapse to form a microsphere (vesicle). Threshold conditions for both phenomena have been described. The presence of vibrations in the system favour microsphere production rather than haemolysis (Blackshear et al., 1971).

It has been suggested by Richardson (1974, 1975) that, if an erythrocyte survives the first few rotations in a high stress flow, then the critical time to disruption is controlled by the viscoelasticity of the outer layer as described by Eq. (6). Recent results obtained by Coakley et al. (1977) show that, at least for erythrocytes in dextran solution, disruption in high stress-flow occurs in times less than 1.0 ms, and imply that when the stress exceeds 3 kN/m² erythrocyte disruption is an immediate effect.

D. MECHANICAL METHODS FOR CELL DISRUPTION

1. *Low Stress Devices*

When laminar flow conditions exist in a tube, the shear stress τ_r at a point in the liquid which is a radial distance r from the centre of the tube is given by:

$$\tau_r = [4\eta V(r/a)]/\pi a^3 \quad (7)$$

where V is the volume flow rate, and a is the radius of the tube. The stress is greatest at the wall $(r = a)$ and is zero on the tube axis. The residence time t_r for a particle travelling along a streamline at a distance r from the axis is:

$$t_r = L\pi a^2/2V[1 - (r/a)^2] \quad (8)$$

where L is the length of the tube. It can be seen from Eq. (8) that cells near the wall are exposed to stress for longer times than cells near the tube axis. In cell suspension flow situations, concentration effects can act to distort the parabolic velocity profile on which Eqs. (7) and (8) are based and lead to a blunting of the velocity profile (Goldsmith, 1971).

In the case of the rapid flow of a cell suspension along the annulus between concentric cylinders of hand homogenizers or of the Chaikoff Press, the shear stress τ_z, for laminar conditions, at a distance z from the centre of the annulus is:

$$\tau_z = 6\eta Vz/\pi r_m b^3 \quad (9)$$

where r_m is the mean radius of the cylinders and b is the width of the annulus (Coakley, 1974). It can be seen that the stress is strongly dependent on the width of the annulus. As with the case of flow in a tube, the stress and exposure time vary with distance from the cylinder walls.

When the inner cylinder rotates, as with motor-driven homogenizers of the Potter–Elvehjem type, the flow is complicated by the onset of a pseudoturbulent flow, i.e. regular arrays of vortices around the annulus which were first described by Taylor (1935, 1936). The vortices set in at a critical value of Taylor number Ta where Ta is given by:

$$Ta = 2\pi f r_m^{1/2} b^{3/2}/\nu \qquad (10)$$

where f is the frequency of rotation of the cylinder. When the rotation is accompanied by axial flow, as when the outer cylinder of the

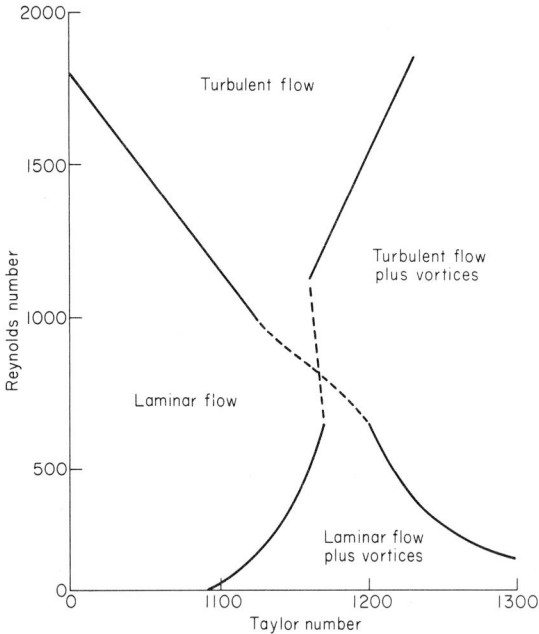

FIG. 3. Four flow patterns in the narrow gap between horizontal concentric cylinders with axial flow and the inner cylinder rotating, as a function of the non-dimensional Reynolds and Taylor numbers. (Kaye and Elgar, 1958.)

homogenizer is moved up and down to facilitate mixing and cooling during operation, different combinations of flow can exist. Kaye and Elgar (1958) examined a problem similar to the above, i.e. the flow of cooling air through a motor annulus and identified the four flow regimes which are outlined in Fig. 3 as functions of Reynolds and Taylor numbers. A dilute 15 ml suspension of the amoeba *A. castellanii* can be completely disrupted in a motor-driven homogenizer of the

Potter–Elvehjem type with a gap of 100 μm and an inner cylinder of diameter 25 mm rotating at a frequency of 83 Hz (Hughes et al., 1971). The sample was mixed by four, even upwards and downwards movements of the outer cylinder in 30 seconds. The Reynolds and Taylor numbers were 102 and 58, respectively, which would place the flow in the laminar region of Fig. 3. Vectorial addition of the axial and radial stresses (Coakley, 1974) shows that the stress on cells near the walls was 70 N/m^2. Laminar-flow conditions in the annulus are facilitated if the outer cylinder is lightly held so that the inner cylinder can be self centering as it rotates in the homogenizer.

In the Chaikoff Press (Emanuel and Chaikoff, 1957a, 1957b) a cell suspension is forced through a narrow orifice with the aid of a hydraulic press. The method has been successfully used to disrupt *Tetrahymena pyriformis*, and tissue cells, e.g. liver, brain, spleen, kidney, fish gonads and ascites cells. Intact nuclei can be recovered from disrupted liver cells. Homogenization of liver cells processed once by passage through the Chaikoff Press was superior to that obtained by 200 strokes in a glass hand-homogenizer. In this laboratory we found complete disruption of suspensions of *Crithidia fasciculata* pressed from a reservoir of 22 mm diameter through a 17 μm annulus of 8 mm diameter at a plunger speed of approximately 0.1 m/s. The Reynolds number of 1500 is close to the laminar/turbulent transition value. A calculation (Coakley, 1974) based on the assumption of laminar flow in the annulus gives 34 kN/m^2 as the stress to which cells would be exposed near the wall. Disruption in the Chaikoff Press is more reproducible than with repeated strokes with a glass homogenizer, either hand or motor driven. The reproducibility of breakage with the press might be further increased by driving the plunger with a constant-speed motor, instead of depending on reproducing driving pressures with a hydraulic press.

The pressures exerted on a cell suspension during processing with the Chaikoff Press are usually around 6.9 MN/m^2 (1000 p.s.i.) but can range up to 70 MN/m^2. Pressures of these orders and higher are also exerted on cell suspensions disrupted in the Yeda Press (Shneyour and Avron, 1970), the Kontes mini-bomb, which is based on the work of Hunter and Commerford (1961), the French Press (Milner et al., 1950), the APV homogenizer (Hetherington et al., 1971), various combinations of valves, outlet tubes, and high pressure pumps (Brookman and Davies, 1973; Brookman 1974, 1975; Sharpe, 1976), and Freeze Presses (Hughes et al., 1971; Magnusson and Edebo, 1976a).

2. Explosive Decompression and Shock-Tube Techniques

The explosive decompression (nitrogen bomb or nitrogen cavitation) technique is based on the work of Fraser (1951). A cell suspension is placed in a pressure vessel and the gas pressure over the suspension is increased to levels ranging from 3.5 to 35 MN/m^2. After a time delay, chosen to equilibrate the gas content in the atmosphere and in the liquid, a needle valve is opened to allow the cell suspension to flow out at the rate of a few drops per second. It is generally assumed (Fraser, 1951; Hunter and Commerford, 1961) that a high gas pressure builds up in the cells before the needle valve is opened, and that the sudden removal of the balancing external pressure ruptures the cells. The observation by Fraser (1951) that the extent of disruption of *E. coli* depended on the gas used for pressurization suggested that the gas played an important role in the mechanism of disruption. Few critical studies have been made of the importance of the duration of the period for which the cells should be under pressure. *Escherichia coli* can be disrupted at 10 MN/m^2 in the Yeda Press (a variation on the explosive decompression technique) where no waiting period is stipulated (Shneyour and Avron, 1970) and at a pressure of 8.3 MN/m^2 in the pneumatic disrupter described by Sharpe (1976). Both of these pressures are comparable to the pressure of 3.5 MN/m^2 required by Fraser (1951) to disrupt *E. coli* after shaking under pressure for three minutes, and suggest that mechanical stress on the cells passing through the outlet valve influences disruption in some situations. Higher pressures of 70 MN/m^2 are required for *E. coli* disruption in the piston-driven French Press (Hughes *et al.*, 1971).

Wallach (1972) considered explosive decompression to be the method of choice for disruption of mammalian cells with minimum comminution of subcellular structures. Application of pressures of 4.0 MN/m^2 to L-cells results in complete disruption of the cells with negligible damage to organelles (Dowben *et al.*, 1969). The yield of polyribosomes is higher and more reproducible in cell extracts prepared by explosive decompression than in those prepared by a Potter–Elvehjem homogenizer (Dowben *et al.*, 1968).

Foster *et al.* (1962) described an apparatus specially designed to disrupt pathogens using a rapid decompression technique which depended on the rupture of a disc which separated a cell suspension under high pressure from a gas phase at atmospheric pressure. There was extensive disruption of 50 ml suspensions of *Mycoplasma gallinarium* and *Leptospira pomona* at pressures in excess of 12 MN/m^2, and 30–60% breakage of *Serratia marcescens* and 10–25% disruption of *Brucella abortus* and *Staphylococcus aureus*. For optimum cell breakage it was required that the cell suspension be maintained under pressure for 75 min before rupturing the disc. The theory and operation of a similar shock-tube device have been described by Edwards and Wiseman (1971). They found that yeasts could be disrupted easily at pressures of less than 7 MN/m^2. They concluded that the passage of a rarefaction wave through the suspension in a time less than 1.0 ms results in cell rupture. They contrasted the rapid decompression thus achieved with the slower rate of pressure release through the valve system used in conventional explosive decompression devices, and concluded that the stress on the cell wall is greater in the shock tube system. The role of explosive decompression in cell disruption in piston-driven extrusion devices will be discussed later (p. 300).

3. Piston-Pressure Extrusion

The devices grouped in the piston-pressure extrusion category operate with the cell suspension in contact with a piston rather than with high pressure gas. The most widely used device of this type is the French Press in which cell suspensions are forced through a needle valve at pressures as high as 280 MN/m^2 (Milner et al., 1950; French and Milner, 1955). The piston is driven by a hydraulic press capable of applying a force of 10–20 tonnes. In a comparative study, Hughes et al. (1971) found that *E. coli* and *Saccharomyces cerevisiae* were easily disrupted. Over 50% of a suspension of the latter cells was disrupted on applying a pressure of 120 MN/m^2. Almost 70% of a suspension of *Mycobacterium tuberculosis* BCG, which is a difficult cell to disrupt, was broken in the French Press at a pressure of 200 MN/m^2. Less than 40% of suspensions of *Chlorella pyrenoidosa*, *Sarcina lutea* and *Staphylococcus aureus* were disrupted in a single pass at 200 MN/m^2.

Brookman and Davies (1973) used an outlet needle valve system, as in the French Press, but replaced the batch hydraulic press by a continuous-flow high pressure pump. They observed 90% disruption of *Sacch. cerevisiae* in a single pass on applying a pressure of approximately 300 MN/m^2. Similar results were obtained with a specially designed valve of simple geometry. Earlier Hetherington *et* al. (1971) had demonstrated that Sacch. cerevisiae could be disrupted using a Manton–Gaulin A.P.V. reciprocating pump and a discharge valve. They found the following relationship between protein release and the applied pressure, P:

$$\log[R_m/(R_m - R)] = KP^{2.9} \qquad (11)$$

where K is a constant at a fixed temperature and R_m is the total protein available for release. About 35% of the cells were disrupted at 45 MN/m^2 compared with a pressure of 130 MN/m^2 for similar breakage in the needle valve system of Brookman and Davies (1973). The flow rate in the latter system was significantly slower than in the homogenizer described by Hetherington et al. (1971). Brookman (1974) reviewed suggested mechanisms for particle disruption in valves. These mechanisms include shear in the liquid, impact against the valve walls, turbulence, cavitation, and the effects of a sudden pressure drop on the cell. Brookman's own (1974) careful measurements of pressure and valve displacement suggested that pressure drop on the extruded material was the main cause of cell disruption. This conclusion was supported by further experiments (Brookman, 1975) with a reciprocating pump and narrow-bore outlet tubes. Brookman (1975) went on to suggest that disruption in the needle valve of the French Press may also arise because of the sudden pressure drop on the extruded cells. A disruption mechanism which depends on the rate of pressure release would explain the anomaly noted above where Brookman and Davies (1973) required a higher pressure at a slower flow rate to disrupt Sacch. cerevisiae compared with the results of Hetherington et al. (1971).

A further class of piston-pressure extrusion devices, developed from the work of Hughes (1951), consists of systems which force frozen suspensions or pastes of cells through a small gap into a receiving chamber at pressures ranging from 70 to 600 MN/m^2. Subsequent developments of the Hughes Press and the "X press" have been described by Hughes et al. (1971) and by Magnusson and Edebo (1976a, b, c). The Hughes Press ruptures material ranging from whole animal tissues to the highly refractory staphylococci, green algae, yeast and fungal mycelia. Over 80% of a frozen paste of Streptococcus faecalis can be disrupted in five passes through a Hughes Press (Hughes et al., 1971). The various mechanisms suggested to explain the way in which the Hughes Press disrupts cells are discussed by Magnusson and Edebo (1976, a, b), and by Scully and Wimpenny (1974). The latter authors constructed a model for the plastic flow of the cell preparation in the press. If *Sacch. cerevisiae* is suspended in high salt concentrations or in 20% (w/v) gelatin before freeze pressing, in an X-press, the percentage of cells disrupted increases significantly (Magnusson and Edebo, 1976a). The increased disruption in high salt concentrations has been attributed to changes in the explosive pressure fluctuations which occur when the frozen paste liquefies as it is pressed through the central hole in the X-press, while the effect of the gelatin is to lower the pressure fluctuations but increase the viscous shear stress on the cells. The effects of salt and gelatin are not additive.

4. *Summary of Disruption of Cells in Suspensions under High Pressure*

With the exception of the shock tube devices, disruption of cells in suspensions under high pressures is based on controlled release of the cell suspension through a narrow orifice. The extent of the disruption depends on the applied pressure, the geometry of the outlet, and in some cases on the gas concentration in the suspension, although the effect of this variable has received little critical attention. From the data available, it appears that cells such as *E. coli* extruded through a needle valve of a nitrogen bomb are disrupted at significantly lower driving pressures than is the case with the piston-driven French Press. The shock-tube technique of cell disruption operates at pressures of the order of 7 MN/m^2 and can disrupt cells such as *Sacch. cerevisiae* and to a lesser extent *Staph. aureus,* which would require disruption pressures in the range 48–280 MN/m^2 in piston-driven extrusion devices such as the French or Hughes Press. In extrusion devices, the applied pressure and the outlet valve geometry interact to determine the rate at which the suspension emerges from the outlet. The outlet flow rate in turn influences the shear stress, turbulence, rate of pressure release on cells, and the length of time that a cell spends travelling through the high stress outlet region.

5. *Colloid Mills*

The development and mechanism of operation of crushing, grinding, and wet milling techniques (colloid mills) for cell disruption have been reviewed at length by Hughes *et al.* (1971). Modern vibration mills, e.g. the Braun homogenizer, provide a widely used, efficient, rapid method for the disruption of cells over a wide range of wall strengths. Batches of 10 g wet weight of *Sacch. cerevisiae* can be extensively disrupted in the Braun homogenizer in one min, so that some of the disadvantages of earlier vibration mills (Rodgers and Hughes, 1960) have been overcome. Larger cell volumes may be disrupted in industrial agitator mills such as the Netzsch–Molinex KE5 mill (Currie *et al.*, 1972) or the Dyno-mill (Mogren *et al.*, 1974; Deters *et al.*, 1976). Like the Braun homogenizer, these mills disrupt cells in suspensions agitated in the presence of glass beads. The effects of bead size, agitator speed, bead concentration, cell concentration and disruption temperature, on cell breakage have been examined (Currie *et al.*, 1972). Efficient disruption of large quantities of *Sacch. cerevisiae*, *Candida utilis*, *Sacch. carlsbergensis* and *Scenedesmus obliquus* have been obtained with a Dyno-mill (Mogren *et al.*, 1974).

6. *Ultrasonics*

Ultrasound has been widely used to disrupt cells since the early experiments of Harvey and Loomis (1929, 1932). Much of the early work has been reviewed by Él'piner (1964). Many of these studies were carried out with megahertz ultrasonic waves, while most commercially available devices manufactured in the last twenty years operate at 20 kHz.

The passage of ultrasonic waves through a liquid may give rise to cavitation bubbles (Flynn, 1964). Cell killing during such irradiations is mainly connected with cavitation activity (Thacker, 1973). If the amplitude of vibration of a bubble is carefully controlled, as in experiments with bubbles trapped against a solid surface (Hughes and Nyborg, 1962) or bubbles trapped in a standing wave field in a liquid (Eller and Crum, 1970; Gould, 1974), a number of states of bubble activity may be recognized. A laminar streaming motion (microsteaming) is set up about an air bubble trapped against a solid boundary when the bubble amplitude vibration is low. When the bubble is vibrating in a high viscosity dextran solution, the laminar stresses in the microstreaming boundary layer are large enough to disrupt erythrocytes (Rooney, 1972). The dextran solution acts to suppress surface wave activity which would otherwise occur on bubbles driven at intermediate intensities in a medium with the viscosity of water. When surface wave activity sets in on a bubble in water, the bubble appearance becomes frosted and smaller air bubbles may be thrown off by the main surface active bubble. Nyborg and Hughes (1967), in a motion-picture study of bubbles in cavitation streamers at 20 kHz, observed that visible bubbles were continuously being formed and disappearing, leading to an apparently constant number of bubbles in the field at any time. They suggested that the bubbles were annihilated as a result of surface wave activity becoming so vigorous that bubbles of around radial resonance size disintegrated into clouds of microbubbles.

From the above we see that the mechanism of cell disintegration will depend on the type of bubble motion which is prevalent in the liquid. When glass bottles containing cell suspensions are placed in ultrasonic cleaning baths, it is likely that the sound pressure in the bottles is low and that a comparatively gentle type of bubble motion is responsible for cell breakage. When probes are placed directly into containers of cell suspensions at ultrasonic power densities of 0.2 to 3.0 W/ml, it is likely that surface activity and shock waves are mainly responsible for cell disintegration. The acoustic power in the liquid when treating an aqueous load of volume V ml for a time t s may be satisfactorily determined by measuring the temperature rise $H\,°C$ and calculating the power W as $(4.18\ V \cdot H/t)$ watts. The calculation should be limited to the irradiation time range where H/t is essentially constant, and every effort should be made to minimize heat loss to the surroundings. Reporting the power input and sample volume when publishing descriptions of cell disruption would facilitate standardization of techniques among workers in different laboratories. In our experience, it is difficult to have a power density much greater than 3.0 W/ml in a container, where the probe tip is directly coupled to an aqueous load at 20 kHz, without inducing "cavitation unloading", i.e. growth of large air bubbles on the surface of the probe. These bubbles inhibit transmission of sound into the liquid and result in a decreased rate of cell breakage. The presence of "cavitation unloading" may be detected as a sudden change in the amplitude and frequency of the noise coming from the cavitating liquid.

It has been known for some time (Kinsloe et al., 1954) that, for many micro-organisms, the number of cells surviving an irradiation period of duration t, is related to the initial population N_0 by the following equation:

$$N_t = N_0 e^{-k_1 t} \qquad (12)$$

where k_1 is a disintegration rate constant. This form of relationship would follow from a situation where there is a constant number of cavitation bubbles and a fixed probability that any cell will travel into the destructive region around a bubble. In high intensity 20 kHz cell disintegration, most of the cavitation bubbles are close to the face of the probe. The radiation pressure "quartz wind" arising from attenuation of the sound in the liquid, effects good mixing in most containers, and ensures that unbroken cells are continuously conveyed into the cell disrupting volume in front of the transducer. Thacker (1973) has discussed some situations where the survival curve departs from the exponential form of Eq. (12).

In a study of the disruption of *Sacch. cerevisiae* with a commercial 20 kHz cell disintegrator, James *et al.* (1972) confirmed that the cell population decreased exponentially with time, at a rate k_1. They found that k_1 was directly proportional to intensity above a threshold intensity level. At a fixed intensity, k_1 was inversely proportional to the treated volume V, i.e. $k_1 V$ was constant. The product $k_1 V$ is numerically equal to the volume of suspension processed in unit time, i.e. the number of cells disrupted in unit time occupy a volume $k_1 V$ of suspension. The product $k_1 V$ is independent of cell concentration up to a limit of 60 g wet weight/100 ml of suspension (James *et al.*, 1972). At higher concentrations, the viscosity of the suspension is such that bulk mixing in the container is reduced and the breakage rate falls.

Many ultrasonic disintegrators now include a flow system for continuous disintegration of cell suspensions. If the cell suspension flows at a rate y ml/min through a container of volume V for which the disintegration rate in a batch system is k_1 min^{-1}, then the fraction X_{eq} of the cell population which will be broken after the suspension has been flowing long enough for equilibrium conditions to be reached is given (James *et al.*, 1972) by:

$$X_{eq} = k_1 V/(y + k_1 V) \qquad (13)$$

If the flow system and generator are switched on at time zero, the fraction of cells disintegrated in the container increases gradually to reach the value X_{eq}. The time required for breakage to reach 95% of X_{eq} is $3/[k_1(1 + Z)]$ where $Z = y/k_1 V$ (Coakley *et al.*, 1974).

7. *Scale Up of Mechanical Methods*

The effectiveness of a number of methods of disrupting kilogram amounts of strong-walled cells in a short time has been examined by different workers, using *Sacch. cerevisiae* as a test organism. The Manton–Gaulin homogenizer may be used to obtain 90% disruption of a suspension of *Sacch. cerevisiae* at a rate exceeding 4.0 kg dry weight of yeast per hour, potentially rising to 21 kg/hr (Hetherington *et al.*, 1971). Studies of the operation of these valve homogenizer techniques, which essentially scale-up the French Press by replacing the hydraulic press by a continuously operating pump, have been reported by Brookman and Davies (1973) and Brookman (1974, 1975). Magnusson and Edebo (1976c) have recently reported the development of a semi-continuous version of the X-press in which about 2.5 kg dry weight of *Sacch. cerevisiae* can be freeze-pressed per hour. A 20 kHz sonicator, operating in a flow system, could disrupt 60% of yeast at a concentration of 60 g wet weight/100 ml of suspension at a rate of 0.35 kg dry weight of yeast per hour when the acoustic power was 185 W (James *et al.*, 1972). More high powered generators are now available which could significantly increase the above yield from ultrasonic devices. Currie *et al.* (1972) showed that a continuously operating colloid mill could give 90% disruption of *Sacch. cerevisiae* at a rate of 12 kg dry weight of yeast per hour. Mogren *et al.* (1974) had 85% disruption of yeast at a throughput of 20 kg dry weight of cells per hour in a Dyno-mill colloid mill. A smaller version of the Dyno-mill disrupted 2.5 kg (pressed, not dry) of *Sacch. cerevisiae* per hour (Deters *et al.*, 1976). In a method which is attractive for laboratory use, Tzagoloff (1969) disrupted a kilogram of pressed yeast which had been frozen in liquid nitrogen, in a Waring blendor in 3 min.

It can be seen that the problem of disruption of large quantities of cells has largely been overcome by the variety of methods described above. The selection of a best method for a particular application will depend less on the ability of the device to disrupt a large quantity of material than on the integrity of the component of interest in the extract.

IV. Lability of Cell Extract Components

A. THERMAL LABILITY

A temperature rise in a stressed cell suspension accompanies most mechanical methods of cell disruption whether by ultrasound or by pressure extrusion through narrow orifices or ball mills. An exception is the explosive decompression technique where it is claimed that adiabatic cooling of the sample follows from the abrupt release of pressure (Hunter and Commerford, 1961). The temperature rise, expressed as degrees C per 7 MN/m^2 (1000 p.s.i.) in piston-extrusion devices, has been given as 0.75 (French and Milner, 1955), 1.5 (Brookman, 1974), 0.70 and 1.3 (Hetherington et al., 1971). When cells are processed at pressures of 140 MN/m^2, the above temperature increase co-efficients result in significant temeprature rises. In practice the French Press is precooled to 0°C. A modification of the French Press, the Sorvall–Ribi fractionator, has an external needle valve which is chilled with nitrogen (Hughes et al., 1971). Cell suspensions processed in the APV homogenizer are precooled to 5°C or are collected in cooled containers (Brookman, 1974; Hetherington et al., 1971). The temperature rise in a high powered continuous flow colloid mill is a function of the flow rate and of the loading of the device, and can range from 6° to 45°C (Mogren et al., 1974).

Significant temperature rises can occur in cell suspensions disrupted in the Potter–Elvehjem homogenizer. The effects of this temperature rise may be minimized by precooling the sample and pulsing the disruption procedure, cooling the homogenizer in ice at intervals. The Chaikoff Press is usually precooled before use and must be cooled by pumping brine at −5°C through a cooling jacket if the width of the annular gap is less than 15 μm (Emanuel and Chaikoff, 1957b). Most commercially available ultrasonic generators now include a method for cooling the treatment vessel. It has proved possible to maintain the temperature of a 280 ml volume, treated at 180 W for 30 min at 8°C during treatment (Coakley et al., 1973).

B. CHEMICAL INACTIVATION

Cell extracts prepared by ultrasonic cavitation are exposed to free radicals which are generally believed to result from the breakdown of water molecules by the high temperatures produced within cavitation bubbles during the adiabatic compression of the bubble (Flynn, 1964). Ultrasonic cavitation has been shown to result in production of atomic and molecular hydrogen, hydroxyl and hydroperoxyl radicals, hydrogen peroxide and solvated electrons (Margulis, 1971). The active species are similar to those produced as a result of molecular water breakdown by ionizing radiation. Sonochemical modification of many molecules of biological interest has been reported. These reports have been comprehensively reviewed by Él'piner (1964) and are summarized here. Decarboxylation and de-amination of the side chains of aromatic compounds occur. Changes also take place in the ultraviolet adsorption spectra of nucleic acid bases. Enzymes are inactivated, oxidases being more sensitive than reductases. A recent study by Staas and Spurlock (1975) confirmed earlier findings of degradation of amino acids exposed to ultrasound. Ammonia, carbon monoxide and formaldehyde were produced in all treated samples and some new products were observed. These authors suggest a scheme for general pathways of de-amination, decarboxylation, and production of formaldehyde.

Many of the studies of chemical changes in biological molecules exposed to cavitation have been carried out with high intensity sound over long time intervals. Few attempts have been made to assess these results in the context of their implication for extract preparation by cavitation. To this end, alcohol dehydrogenase, lysozyme and catalase have been assayed after exposure to cavitating 20 kHz ultrasound (Coakley et al., 1973). Catalase was little affected but alcohol dehydrogenase and lysozyme were both inactivated at an exponential rate by a sonochemical mechanism. The most sensitive enzyme, alcohol dehydrogenase (20 μg/ml), had an inactivation rate of 0.16 min^{-1}. The rate was reduced to 0.026 min^{-1} at the higher enzyme concentration of 100 μg/ml in the presence of 500 μg of bovine serum albumin/ml., added as a competitive scavenger for the inactivating species. The rate constant for the disruption of *Sacch. cerevisiae* under similar sonication conditions (90 W) is 0.07 min^{-1} (James et al., 1972) which is nearly three times faster than the latter rate of enzyme inactivation. The following precautions were suggested to minimize the sonochemical degradation of enzymes: (i) sonicating as high a cell concentration as possible to make maximum use of the observed protein concentration protection effect; (ii) sonicating the cells in growth medium so that radical scavengers (e.g. sugars) in the medium will protect the extract; (iii) during prolonged sonication, gassing the cell suspension with an appropriate gas such as hydrogen may avoid all sonochemical effects.

C. MECHANICAL COMMINUTION

1. *Molecules*

Since the bacterial chromosome is the largest molecule in the cell, studies of the susceptibility of DNA to breakage under mechanical stress give some insights into the effects of extraction methods on the size of extract components. Purified DNA in salt solution has a conformational state intermediate between that of a rigid rod and a random coil. The molecule has a persistence length of 0.1 μm, i.e. bases need to be separated by a distance of 0.1 μm before it can be said that the bases are randomly placed relative to each other. A large DNA molecule such as T_2 bacteriophage DNA has a molecular weight of 1.2×10^8 daltons, a length of 60 μm and a mean end-to-end distance in solution of 2.5 μm (Harrington, 1970). When a solution flows, the dissolved molecules tend to rotate in the flow, rotating most slowly at a

TABLE 2. Molecular weights of DNA in solution following degradation by different methods

Degradation method	Molecular weight daltons ($\times 10^{-6}$)	References
Extrusion through a narrow annulus at a piston pressure of 21 MN/m²	5.0	Harrington and Zimm (1965)
DNA in 50% glycerol-BPES buffer in a "Virtis" homogenizer	0.1	Harrington and Zimm (1965)
Pumped through glass atomizer	1.0	Cavalieri and Rosenberg (1959)
Cavitating ultrasound (9 kHz; 60 min)	0.3	Doty et al. (1958)
Extrusion through syringe needle	5.0	Richards and Boyer (1965)
French Press, one pass, 62 MN/m²	0.38	Richards and Boyer (1965)
French Press, five passes, 62 MN/m²	0.20	Richards and Boyer (1965)
Cavitating ultrasound (20 kHz; 15 min)	0.23	Richards and Boyer (1965)

preferred angle to the streamlines. This preferred angle, which is a function of the shear rate, may be measured as the flow birefringence extinction angle. It has been assumed that the molecule elongates to its full length during flow (Levinthal and Davison, 1961), but time-dependent changes in the extinction angle at high shear rates (Champion and Coakley, 1969; Champion and North, 1971) may indicate a degree of recoiling of the stretched molecule. That stressed polymers break at a central point has been demonstrated for T_2 bacteriophage DNA in laminar tube flow (Levinthal and Davison,

1961), inferred for DNA disrupted in a "Virtis" homogenizer with a rotating blade or rod (Burgi and Hershey, 1961; Green, 1966) and inferred for polymer breakdown by cavitating ultrasound (Ovenall *et al.*, 1958). The shear stress necessary to halve T_2 DNA molecules in laminar flow is approximately 100 N/m² (Levinthal and Davison, 1961). The molecular weights of DNA following degradation by different methods are shown in Table 2. The size of product following treatment by those methods most used in rigid-walled microbe disruption, i.e. the French Press and ultrasound, give comparable degradation of the DNA.

Él'piner (1964) concluded that, in general, the mechanical forces associated with ultrasonic cavitation will not degrade a globular protein of molecular weight less than 20,000–25,000 daltons. However he observed that, on prolonged irradiation, individual amino acids or low molecular-weight polypeptides may be broken off protein molecules, by chemical attack. Follows *et al.* (1971) examined extraction of seven enzymes from *Sacch. cerevisiae* in the piston-pressure extrusion APV homogenizer when operated at 42 MN/m². The extraction procedure was carried out over 90 min in a recycle system. Most of the protein released, had been extracted in the first 30 min of operation. Subsequent recycling of the enzymes had little effect on their activity. The activity of a multimeric enzyme consisting of a carrier protein of molecular weight 30,000 daltons and four subunits of molecular weight about 70,000 daltons decreased on recycling through an APV homogenizer at pressures of 28 MN/m² (Augenstein *et al.*, 1974). Inactivation of pure alcohol dehydrogenase was not observed on subjecting the solution to increasing pressures in the French Press (Hughes *et al.*, 1971). The same authors found similar yields of four enzymes of the tricarboxylic acid cycle in *E. coli* extracts prepared in the French Press, Hughes Press and by ultrasound. Enzyme inactivation in the French Press and ultrasonic extracts became significant if cells were processed at a concentration of less than 0.3 g wet weight/ml. Applying a pressure of 280 MN/m² to a suspension of *E. coli* in the French Press for 20 min before disruption led to a loss of about half of the activity in the four enzymes. Augenstein *et al.* (1974) argue that mechanically labile enzymes may be degraded intracellularly during homogenization in the APV homogenizer at 77 MN/m². Reports of high-pressure enzyme inactivation, in the absence of shear stress, have been reviewed by Joly (1965). The pressures required for enzyme inactivation are in the higher part of the range of pressures employed in piston-pressure extrusion devices. The extent of enzyme inactivation is dependent on the time for which the pressure is applied.

2. Membrane Vesicles

a. *Definition of Vesicles.* When a cell is subjected to a force sufficient to overcome the mechanical strength of the cell wall, the outcome is eventually cell breakage. However, the exact response to stress depends both upon the nature of the applied force and upon the structure of the cell envelope. For example, when bacteria are freeze-pressed in a Hughes Press, the solid-shear forces set up tend to cause disruption by transverse cleavage of the cell, causing little comminution of the cell envelope, which can subsequently be recovered as large open-ended fragments (Hughes *et al.,* 1971). In contrast, disruptive procedures such as the French Press or ultrasound produce a much higher degree of fragmentation of cell envelopes. Much of the membranous material recovered after such treatment is in the form of vesicles.

At this point, a definition of membrane vesicles is in order. They are topologically continuous fragments of membrane, of variable size and sidedness, which may or may not be permeable to molecules in solution. Production of vesicles by different processes and the variation in their properties will be discussed in the following sections. Vesicularization of a membrane preparation is a very important consideration, all too frequently overlooked, because, as will become apparent, this phenomenon can greatly affect the interpretation of results, especially when the system under investigation is distributed asymmetrically across the bilayer of the membrane.

b. *Membrane Asymmetry.* The currently accepted view of membrane structure is of a two-dimensional solution of proteins and glycoproteins in a viscous phospholipid solvent (Singer and Nicholson, 1972). Much evidence is available which suggests that both the protein/glycoprotein and phospholipid are distributed anisotropically across the bilayer. Many studies of membrane asymmetry, or sidedness, have utilized erythrocytes as model systems as they have a plasma membrane of relatively simple composition, and plasma-membrane isolation is not hindered by contamination from membranous intracellular organelles.

The two major techniques which have been used to study distribution of erythrocyte membrane proteins are selective digestion of exposed polypeptides with proteases, and specific labelling of exposed proteins with membrane-impermeable radio-active or fluorescent markers. In both cases, a comparison can be made of membranes from treated intact cells, where only the outer membrane face is exposed, and of membranes from treated erythrocyte ghosts, in which both inner and outer membrane faces are exposed to treatment. Comparisons can either be made by direct microscopic examination (fluorescent antibody labelling; Maddy, 1964) or, more commonly, by an examination of the electrophoretic profiles of the membrane preparations on SDS-polyacrylamide gels. The latter approach is used to detect protease digestion, indicated by an increase in the electrophoretic mobility of the affected polypeptide (Bender et al., 1971), and to identify isotopically labelled proteins by comparing the optical density profile of a gel stained for total protein with the distribution of the isotopic label. Commonly used radio-active markers are ^{35}S-diazobenzene sulphonate (Berg, 1969), ^{35}S-formylmethionylsulphone methyl phosphate (Bretscher, 1971), and an ^{125}I-label produced by treating the sample with a mixture of ^{125}I-iodide, lactoperoxidase and hydrogen peroxide (Phillips and Morrison, 1970); the last specifically labels exposed tyrosine and histidine residues. An analogous method for specifically labelling exposed carbohydrates, utilizes enzymic oxidation of galactose residues with galactose oxidase followed by reduction with ^{3}H-sodium borohydride (Morell et al., 1966). These techniques have combined to show that all of the glycosyl residues on glycoproteins and glycolipids are exposed on the outer membrane face only, that the major erythrocyte glycoprotein, glycophorin, in inserted through the membrane in such a way that it is exposed on both inner and outer faces, and that probably all of the remaining non-glycosylated proteins are exposed on the inner membrane face only (Bretscher, 1973; Marchesi, 1972; Steck, 1972). Enzyme activities have also been studied topologically in erythrocytes; acetylcholinesterase activity is only detectable on the outer membrane face, whereas NADH oxidase and glyceraldehyde 3-phosphate dehydrogenase activities are only detectable on the inner membrane face (Steck, 1974).

Analogous methods have been used to investigate anisotropy in the distribution of phospholipids. Digestion of intact and of ghosted erythrocytes with phospholipases, followed by analysis of the breakdown products, has led to the suggestion that phosphatidylcholine is present in the outer layer of the bilayer while phosphatidylethanolamine occurs in the inner layer (Bretscher, 1973). This is supported by fluorescent labelling studies with acetamido-4'-thiocyanodisulphonate which reacts with exposed amino groups (Knauf and Rothstein, 1971). Although the erythrocyte system is probably the most intensively studied, much evidence is available which suggests an asymmetric distribution of components in membranes of most cells, whether plasma or organelle membranes of eukaryotic cells, or cytoplasmic or outer membranes of prokaryotic cells.

c. *Membrane Vesicles from Erythrocytes.* Formation of vesicles from erythrocytes can be induced by various treatments; they are produced under conditions of stress set up in velocity gradients, presumably by shear-dependent deformation of the visco-elastic membrane, or they will form spontaneously after suitable manipulations of the temperature or ionic nature of the suspending medium.

If a suspension of erythrocytes in isotonic saline is subjected to a velocity gradient, either by treatment in a cone and plate viscometer (Champion *et al.,* 1971) or in conditions of acoustic microstreaming near a wire transversely oscillating at 20 kHz (Williams *et al.,* 1970), a range of shear rates can be obtained in which erythrocytes do not undergo haemolysis but fragment into perfect spheres accompanied by many small haemoglobin-filled vesicles of about 1 μm diameter (called microspheres in the original publications). Champion *et al.* (1971) suggested that this occurs by a stripping off of fragments from the ends of a distorted cell as in Fig. 1a (p. 292); erythrocytes swollen in hypotonic (0.54%) saline, where this deformation of the cell was not possible, were disrupted under stress conditions which caused vesicle formation in isotonic (0.9%) saline. Although determination of vesicle sidedness was not made in these studies, the fact that the vesicles are haemoglobin-filled is indicative of a right-side-out orientation. This, together with the absence of observed haemoglobin release during vesicle formation, suggests a pinching-off mechanism of vesicle production rather than release of open fragments which subsequently round off and seal their torn edges. Such a pinching-off also occurs when erythrocytes are heat treated. Williams (1970) heated suspensions of erythrocytes in isotonic saline to 49°–50°C and found that, after ten minutes exposure, many cells showed one or two protuberances while the surrounding medium contained many small (about 1 µm diameter) vesicles.

Effects of ionic manipulations upon red cell membranes are to be found in the work of Steck (1974) who incubated haemoglobin-free erythrocyte ghosts at 37°C in the presence or absence of magnesium ions. Unsealed ghosts incubated in 0.5 mM phosphate buffer (pH 8.6) for one to two hours yield a population of sealed, inside-out vesicles on pelleting and homogenization of the resuspended membranes through a hypodermic needle. If, however, 0.1 mM magnesium sulphate is introduced into the vesicularization medium immediately before centrifugation, and if all subsequent steps are carried out in magnesium-containing buffers, the resulting vesicle population has the normal right-side-out orientation. It is pertinent to note that electrophoretic profiles of the inside-out vesicle membranes are lacking in a set of polypeptides known as spectrin (Steck, 1972), which are thought to constitute a re-inforcing network across the inner surface of the erythrocyte membrane (Singer, 1975). A similar absence of the spectrin polypeptides from membranes of the vesicle population induced by elevation of the intracellular calcium level of erythrocytes was reported by Allan et al. (1976). Spectrin is a peripheral protein and, as such, may be solubilized by low ionic strength, divalent cation-free media (Singer, 1974) such as Steck's vesicularization medium. If incipient endocytosis (budding into the cell interior, as in phagocytosis) occurred during the one to two hour incubation period, subsequent homogenization would be expected to yield inverted vesicles. If addition of magnesium ions promoted re-attachment of spectrin to the membranes, as has been demonstrated for other peripheral proteins (e.g. ATPase of *Micrococcus lysodeikticus* and of *Bacillus megaterium*; Munoz et al., 1969; Mirsky and Barlow, 1971; hexokinase of inner mitochondrial membranes; Rose and Warms, 1967), the shape and mechanical properties of the incipient endocytotic vesicles might be altered in such a way that subsequent homogenization yielded right-side-out vesicles (Fig. 4).

Spectrin has previously been implicated in calcium-induced morphological changes of erythrocytes by Palek et al. (1974) who observed the change from biconcave discs to echinocytes (spiculed spheres). Spectrin is a rod-like protein closely related to myosin and is associated with actin or an actin-like molecule of the cytoplasmic face of the erythrocyte membrane (Singer, 1974). Like moysin, addition of calcium ions results in aggregation of spectrin molecules in solution, and causes spectrin-actin complexes to contract (Guidotti, 1972). Palek et al. (1974) proposed that such a contraction of the spectrin-network was responsible for the observed evagination of erythrocytes which leads to echinocyte formation. However, Allan and Michell (1975a, b)

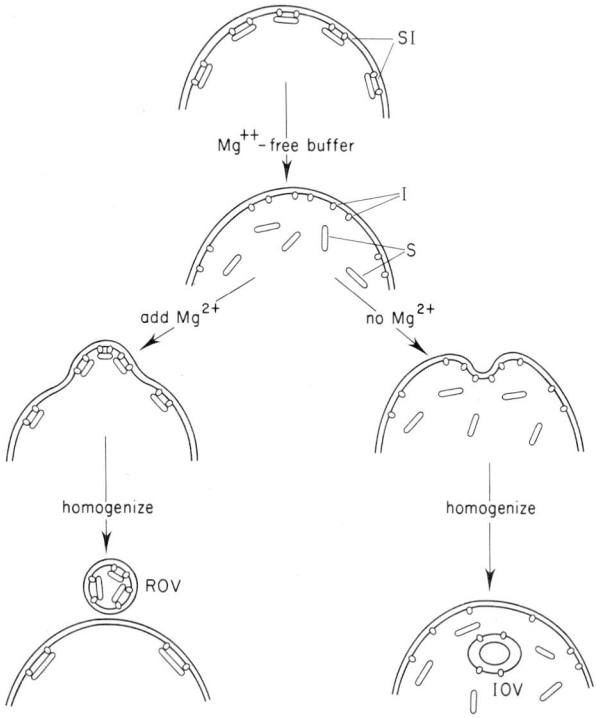

FIG. 4. Possible mechanism by which magnesium ions reverse the polarity of the erythrocyte vesicularization process described by Steck (1974). SI indicates spectrin-integral protein complexes; S, uncomplexed spectrin; I, uncomplexed integral proteins; ROV, right-side-out vesicles; IOV, inside-out vesicles.

found that elevation of the intracellular Ca^{2+} concentration by incubation of erythrocytes with Ca^{2+} and the divalent cation iono-phore A23187 (Reed and Lardy, 1972) leads to enhanced production of 1,2-diacylglycerol and suggested that Ca^{2+}-dependent change in membrane lipid composition, rather than any direct effect on protein conformation, might be responsible for the change in shape and for the outward vesicularization.

The latter hypothesis is supported by the observation that the vesicles themselves are markedly enriched in diacylglycerol. Exocytosis involves membrane fusion through interactions at the cytoplasmic surface of the microvillus-like protrusions formed initially. Diacylglycerol, which has been shown to be fusogenic (Ahkong et al., 1973; DeBoer and Loyter, 1971) is probably produced in the inner surface of the membrane bilayer in response to calcium treatment, and it would therefore be able to mediate fusion of the cytoplasmic surfaces (Allan et al., 1976). In contrast, treatment of erythrocytes with exogenous phospholipase C, which produces diacylglycerol in the outer surface of the bilayer, promotes stomatocyte formation and endocytosis (Allan et al., 1975). Asymmetry of the vesicularization process is emphasized by the failure of ionophore treatment to promote endocytosis or cell–cell fusion, both of which involve membrane fusion through interactions at the outer membrane face. Although the mechanism of membrane fusion induced by diacylglycerol and similar fusogenic lipids is not known, it may be due to localized changes in membrane curvature resulting from alterations in the packing of the phospholipids in either face of the bilayer (Allan et al., 1975; Sheetz and Singer, 1974).

The work of Allan et al. cited above suggests that 1,2-diacylglycerol, even in relatively small concentrations, may influence membrane morphology and fusibility in erythrocytes. The process may be regulated *in vivo* by diacylglycerol kinase, a very active erythrocyte enzyme which phosphorylates diacylglycerol to non-fusogenic phosphatidate (Allan et al., 1976); as the latter activity requires ATP, elevation of the intracellular Ca^{2+} concentration will shift the equilibrium away from phosphatidate formation due to a lowering of the intracellular ATP levels by calcium-stimulated ATPase. Observations of Allan and his colleagues raise the possibility that, in physiological situations where both a rise in intracellular Ca^{2+} concentration and changes in the morphology, fluidity or permeability of the plasma membrane occur, changes in membrane properties may be brought about by similar Ca^{2+}-dependent effects on membrane lipid composition. Another observation that, under conditions of energy starvation, diacylglycerol accumulates as a result of phosphatidate phosphohydrolase activity which cannot be balanced by diacylglycerol kinase activity (Allan et al., 1976), emphasizes the inadvisability of extrapolating results obtained with stored cell or membrane preparations to situations where the cells are metabolizing normally.

Disruption of mammalian tissues other than blood for the purpose of membrane isolation is most commonly done with coaxial pestle homogenizers of the Dounce or Potter–Elvehjem type. Wallach (1967) has suggested using explosive decompression which leads to fragmentation of plasma membrane and endoplasmic reticulum into small vesicles of unknown orientation while preserving the integrity of other membranous organelles. A method analogous to that used for erythrocyte membrane isolation has also been used to prepare ghosts from adipocytes (Rodbell and Krishna, 1974). The use of high-shear techniques or of freeze-thawing is not recommended for preparation of membrane isolates as lysosomal damage is frequently concomitant with cellular fragmentation and resulting release of autolytic enzymes may cause severe damage to membrane fragments.

d. *Bacterial Membrane Vesicles.* Interest in bacterial membrane vesicles increased significantly when it was demonstrated that, under certain circumstances, they could be made actively to concentrate solutes against a concentration gradient. The use of vesicle preparations for transport studies has many advantages, not the least of which is the enhanced recovery of transported markers which are not normally metabolized in the absence of cytoplasmic enzymes.

Among the first to describe preparation of transport active vesicles from bacteria was Kaback (Kaback and Deuel, 1969) whose experimental procedure with *E. coli* has now become accepted as a standard technique (Kaback, 1971). The basis of the Kaback preparation is formation of sphaeroplasts by treatment with EDTA/lysozyme in isotonic or hypertonic sucrose solution, followed by rapid dilution of the sphaeroplast suspension into 300 to 500 volumes of dilute buffer.

Many workers have studied vesicles produced in this way and some controversy exists about their morphological properties. Kaback maintains that his technique generates sealed vesicles which are almost completely right-side-out, and he has published several studies supporting his claim. For example, one study utilized high-resolution autoradiography to examine vesicle preparations labelled with ^{3}H-vinylglycollate (Short *et al.*, 1974); the labelled vesicle preparation was dried onto an electron microscope grid, overlaid with a gel containing photographic emulsion, incubated for a suitable period, and examined under an electron microscope. Transport of vinylglycollate is the limiting step in covalent labelling of membrane proteins with ^{3}H-vinylglycollate, and hence only those vesicles which had actively transported the isotope would be overlaid with silver grains. Up to 95%

of the vesicles were so labelled, indicating that they were transport active and so could not be unsealed or inverted (Short et al., 1974). A study of D-lactate dehydrogenase response to antibodies prepared against the purified enzyme showed that the dehydrogenase, present only on the cytoplasmic face of the E. coli inner membrane, was not inactivated when Kaback vesicles were incubated with the antibodies, providing further evidence that the vesicles are not inverted or damaged sufficiently to allow penetration of antibodies (Short et al., 1975). Adenosine triphosphatase is also inaccessible to antibodies unless the vesicles are disrupted with ultrasound or by vigorous homogenization.

Independent support comes from direct examination of E. coli vesicles by freeze-fracture electron microscopy. Application of this technique to intact cells of E. coli has revealed that the particles exposed on the resulting two inner faces of the cytoplasmic membrane are distributed asymmetrically; Altendorf and Staehelin (1974) were therefore able to assess the orientation of E. coli vesicles by comparing freeze-fractured preparations with similarly treated samples of intact cells. Their results show that vesicles prepared exactly as described by Kaback (1971) consist almost entirely of right-side-out vesicles; their size (0.8–1.1 μm diameter) suggests that each cell gives rise to a single vesicle. As biochemical investigations frequently employ membrane preparations which have been stored deep frozen, Altendorf and Staehelin (1974) assessed the effect of one freeze-thaw cycle on the architecture of their vesicles. The number of inside-out vesicles rose to 25% as judged by distribution of particles on the convex and concave fracture faces. However, the inverted vesicles were much smaller in size than the right-side-out vesicles, and were estimated to contribute only 2–3% of the surface area of the preparation. Thus, a single freeze-thawing would have minimal effects on measurements of transport activity, especially as nearly half of the inverted vesicles are enclosed within vesicles of normal orientation. This indicates that inside-out vesicles arise by endocytosis into right-side-out vesicles; free inverted vesicles were commonly observed to be clustered in groups between the larger right-side-out vesicles, thereby suggesting that they could have arisen via endocytosis into a normal vesicle which subsequently disintegrated (Altendorf and Staehelin, 1974).

To be weighted against the above evidence are several studies of vesicle sidedness which are based upon measurements of the distribution of activity of a number of marker enzymes which are known to be located on the cytoplasmic face of the *E. coli* inner membrane. For example, Weiner (1974) assayed dehydrogenases for glycerol 3-phosphate, succinate, and D-lactate, in *E. coli* vesicles prepared according to Kaback, using the membrane impermeant dye, potassium ferricyanide, as an electron acceptor. When toluene or Triton X100 was added to destroy the permeability barrier, a two-fold increase in ferricyanide reduction was obtained with all three substrates. In contrast, with intact cells or unlysed sphaeroplasts, no activity could be detected for any of the three dehydrogenases tested. Therefore, approximately half of each dehydrogenase activity present in the Kaback vesicle preparation is accessible to the electron acceptor; Weiner (1974) proposed that this was caused by the presence of two vesicle populations in the Kaback preparation, one sealed and right-side-out, the other unsealed or inverted.

Similar results were obtained by Futai (1974) who tested the accessibility of NADH-ferricyanide oxidoreductase and of ATPase to their membrane impermeant substrates, and of ATPase to a specific antibody raised against the purified enzyme. In all cases, the enzymes were completely inaccessible in sphaeroplasts unless the permeability barrier was destroyed with toluene, Triton-X100, or sodium cholate, while in Kaback vesicles approximately half of each enzyme was accessible in the absence of detergent or toluene.

Several explanations have been put forward in attempts to reconcile these results which at first glance seem quite incompatible. Suggestions that Kaback vesicle preparations consist of a mixture of right-side-out and inside-out vesicles do not fit the data from freeze-fracturing (Kaback, 1972; Altendorf and Staehelin, 1974) or the fact that the majority of vesicles in the preparation are transport-active (Short *et al.*, 1974) and so from the vectorial nature of the transport process (Harold, 1972) must have the normal orientation. One explanation is strain differences among the bacteria used, as Kaback's work has been done on *E. coli* ML, rather than the more commonly used K12 or B strains. Another possibility is that various groups of workers, e.g. Thienen and Postma (1973) who state that in their hands Kaback's procedure yields a heterogeneous population of variously oriented vesicles, have used slight modifications of the original procedure. As has been noted earlier, a minor alteration in the constitution of an erythrocyte vesicularization medium is sufficient completely to reverse orientation of the resulting vesicle preparation (Steck, 1972).

However, assuming that the preparative techniques are identical, other explanations are required. It has been suggested that some vesicles may be a mosaic of differently oriented patches of membrane (Harold, 1972; Weiner, 1974). Such hybrid vesicles have been detected after sonication of a mixture of vesicles of different orientation (Tsukagoshi and Fox, 1971). However it is difficult to envisage a mechanism which would produce such vesicles in the absence of high shear forces such as those encountered in a cavitating ultrasonic field.

Another way of reconciling the conflicting evidence is to assume that, although the overall orientation of the membrane does not alter during vesicle formation, certain marker proteins change their location (Futai, 1974). Several alternative mechanisms for translocation of membrane proteins from the inner to the outer surface of the cytoplasmic membrane have been discussed by Altendorf and Staehelin (1974): (i) During the lysis process enzymes may be released from the inner surface of the membrane into the medium, where re-adsorption to the outer surface could occur. However, the extent of dilution during lysis is such that re-adsorption is unlikely, although not impossible. (ii) Stretching of the membrane during the lysis process could exert sufficient mechanical stress upon membrane components such as the ATPase complex to cause them to become dislocated and effectively "flip" from one surface to the other. (iii) During lysis numerous small, transient pores (20–50 nm diameter) open in the cytoplasmic membrane through which the cell contents are released (Corner and Marquis, 1969). Rapid outward flow of cytoplasmic material may induce a streaming of membrane material in the region of the pore, causing some proteins to be swept onto the outer membrane face (Fig. 5).

Whereas orientation of vesicles produced from *E. coli* by osmotic lysis of sphaeroplasts is still uncertain, there is general agreement that treatment of whole cells with high shear techniques such as the French Press or sonication yields populations of vesicles which are predominantly inverted. This has been demonstrated by sidedness studies using marker enzyme assays (Futai, 1974). In contrast to the large size (1 μm diameter) of sphaeroplast-derived vesicles, French Press vesicles are much smaller (40–110 nm diameter; Altendorf and Staehelin, 1974). Tsuchiya and Rosen (1975) noted that increasing the treatment pressure of an *E. coli* cell suspension in a French Press from 28 MN/m^2 to 100 MN/m^2 did not alter the orientation of the resulting vesicle

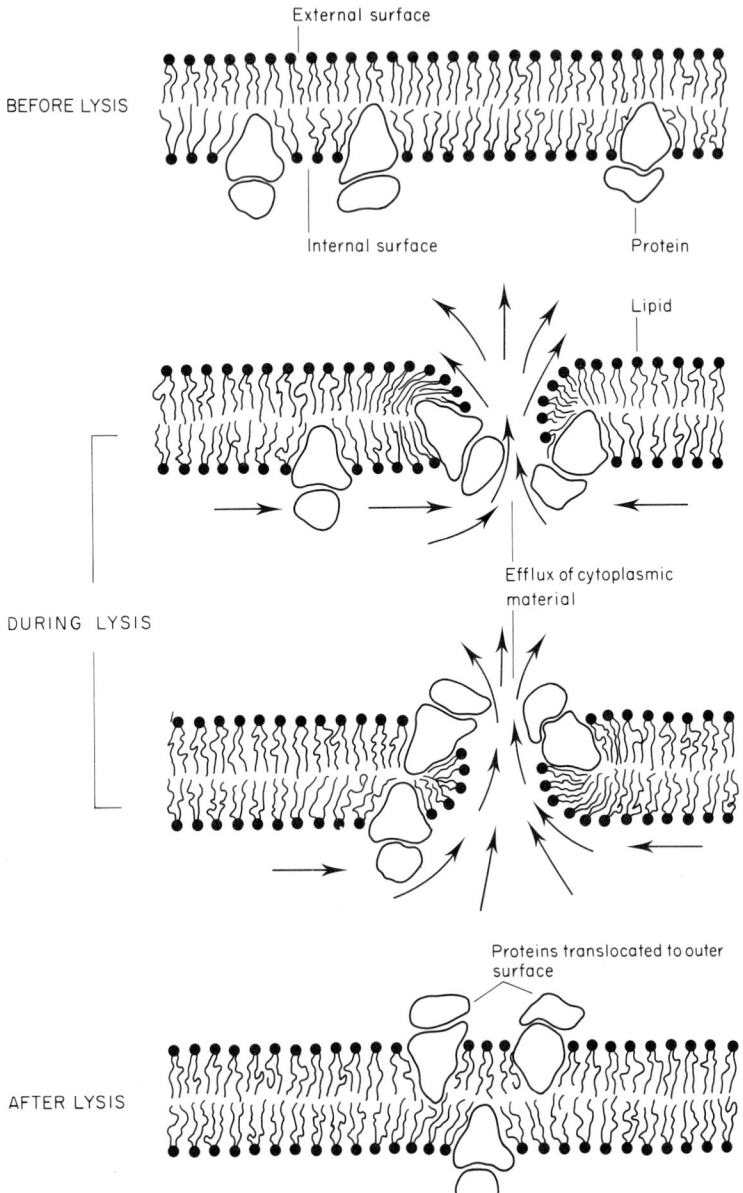

FIG. 5. Possible mechanism for translocation of membrane proteins from the inner to the outer membrane surface during lysis. The rapid efflux of cytoplasmic material during lysis may induce a flow of membrane proteins and lipids in the region of the small transient pores. Adapted from Altendorf and Staehelin (1974).

preparation but merely served to decrease the average size of the vesicles. Such inverted vesicles are not normally transport-active as they are of the wrong polarity. However, Rosen and McClees (1974) have described active concentration of calcium ions by French Press vesicles of *E. coli*; this suggests that the calcium pump acts *in vivo* to extrude calcium from the cytoplasm. These inverted vesicles may be useful in studying other reverse transport systems which would not otherwise be detectable.

Although the majority of this section of the article has dealt with membrane vesicles produced from *E. coli*, a wide variety of bacteria have been used in similar studies. The Kaback procedure has been modified to produce vesicles from *Bacillus subtilis* for study of transport of amino acids (Konings and Freese, 1972) and of dicarboxylic acids (Bisschop *et al.*, 1975); hypotonic lysis of sphaeroplasts and subsequent sonication have been used to produce right-side-out and inside-out vesicles respectively from *Mycobacterium phlei* (Asano *et al.*, 1973). Short and Kaback (1974) used the Kaback method to study amino-acid transport by *Staph. aureus* vesicles; similar methods were used by Burnell *et al.* (1975) to follow sulphate transport by vesicles of *Micrococcus denitrificans* and by Matin and Konings (1973) to study transport of lactate and succinate by vesicles of *Pseudomonas aeruginosa*, *E. coli* and *B. subtilis*. In Table 3, some of the common techniques which have been used to produce membrane vesicles from erythrocytes or bacteria are summarized, together with what is known about the morphology of the resulting vesicle preparations. Methods which have been applied to the determination of the sidedness of vesicle preparations are listed in Table 4.

e. *Interpretation of Results from Membrane Preparations.* The variability which seems to be inherent in the vesicularization process can make interpretation of results very difficult and can lead to erroneous conclusions. An example may serve to illustrate this point. Konings and Freese (1972), in a study of amino-acid transport by membrane vesicles of *B. subtilis* produced by the Kaback method, reported that oxidation of NADH was not able to drive amino-acid transport whereas a variety of other electron donors whose oxidation rate was similar to that of NADH were able to energize active transport. This could be taken as evidence against the existence of a transport system driven by a proton gradient which requires that any substrate should drive active transport if its oxidation is coupled to electron transport

TABLE 3. Methods of vesicle production

Cell	Procedure	Morphology of vesicles	Reference
Erythrocytes	Hypotonic lysis followed by incubation for two hours at 37°C and homogenization	Inside-out, sealed	Steck (1972)
	As above but add magnesium sulphate prior to homogenization	Right-side out, sealed	Steck (1972)
	Incubate at 49°–50°C for 10 minutes	Right-side out, impermeable to haemoglobin, 1 μm diameter	Williams (1970)
	Incubate with Ca^{2+} and ionophore A23187	Right-side out	Allan and Michell (1975a, b)
	Incubate with phospholipase-C	Inside-out	Allan et al. (1975)
Bacteria	Hypotonic lysis of sphaeroplasts	Right-side out, sealed ca 1.1 μm diameter	Kaback (1971); Altendorf and Staehelin (1974)
	French press disruption of intact cells	Predominantly inside-out 40–110 nm diameter, sealed	Altendorf and Staehelin (1974); Rosen and McClees (1974)
	Sonication of subcellular particles	Inside-out	Asano et al. (1973)

TABLE 4. Methods for determination of vesicle sidedness

Cell	Method	Orientation indicated by positive result[a]	References
Erythrocyte	NADH: Cytochrome c reductase activity	Inside-out	Steck (1974)
	Glyceraldehyde 3-phosphate dehydrogenase activity	Inside-out	Steck (1974)
	Acetylcholinesterase activity	Right-side out	Steck (1974)
	Accessibility of sialic acid to sialidase	Right-side out	Steck (1974)
	Isotopic labelling of glycoproteins	Right-side out	Steck (1972)
Escherichia coli	^{3}H-Vinylglycollate labelling	Right-side out	Short et al. (1974)
	Accessibility of D-lactate dehydrogenase and ATPase to antibodies	Inside-out	Short et al. (1975)
	Activity of dehydrogenases for succinate, D-lactate, and glycerol 3-phosphate	Inside-out	Weiner (1974)
	NADH: Ferricyanide oxidoreductase and ATPase activities	Inside-out	Futai (1974)
	Active transport of calcium ions	Inside-out	Rosen and McClees (1974)
Various bacteria	Active transport of sugars and amino acids	Right-side out	Many workers: see Harold (1972)
Escherichia coli	Examination by freeze-fracture microscopy	Both orientations are directly identifiable	Altendorf and Staehelin (1974)

[a] Assumes that vesicles are impermeable to reagents.

(Harold, 1972). However, a closer analysis of the vesicle population revealed a very small percentage of open or inverted vesicles which accounted for all of the NADH oxidase activity; these vesicles were incapable of active transport because of their inappropriate orientation, while the remainder of vesicle population could not oxidize NADH as the latter was unable to cross the membrane (Hampton and Freese, 1976).

In the light of this discussion, vesicle preparations which have been freeze-thawed, sonicated, or even stored for prolonged periods under conditions of energy starvation should be viewed with extreme caution. Altendorf and Staehelin (1974) describe an increase in the proportion of inverted vesicles to 25% after a single freeze-thawing cycle; although they discount this fraction as being unimportant in terms of transport activity, accounting as it does for only 2–3% of the surface area, it is sufficient to cause large errors in kinetic measurements or in sidedness determinations based upon the assay of marker enzymes. The morphology of a sonicated membrane preparation is very dependent upon the period of treatment. In a study of the effect of 20 kHz ultrasound on vesicle-free membrane preparations produced by Hughes pressing of *Ps. aeruginosa* it was found that maximum formation of right-side-out sealed vesicles was reached after 15–20 minutes sonication at 150 W; further sonication produced a decrease in this vesicle population (A. J. Bater, R. C. Brown, W. T. Coakley, unpublished observations).

In summary, it must be emphasized that, in any preparation of subcellular membranes, it must not be assumed that the measurable activity of any enzyme represents the total activity present. It is unwise to assume even that a constant proportion of the total activity is measurable, as small variations in preparative technique are sufficient to alter significantly the degree to which various membrane components are sequestered from the suspending medium by impermeable membrane barriers.

D. GENERAL APPROACH TO OPTIMIZATION OF THE ACTIVITY OF EXTRACTS

1. *Batch Systems*

Where the component of interest in a cell extract is degraded during the extraction procedure, there will be situations, as with the recycling of material disrupted by a piston-pressure extrusion device or the continued exposure of material extracted by cavitating ultrasound, where a continuation of the treatment will lead to an increase in cell disruption but to a decrease in the activity of the extract. An approach to a determination of the optimum conditions of treatment has been developed for the case of ultrasonic extraction by Coakley et al. (1974).

Cells in a cavitating sound field are disrupted mechanically with a rate constant k_1 (p. 303). Comminution of the extract in some cases,

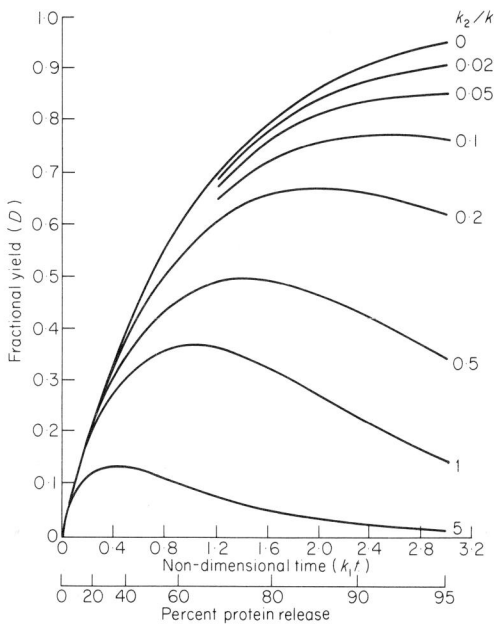

FIG. 6. A plot of the fraction (D) of the total pool of a component in the cells which is released without degradation, by batch sonication, against $k_1 t$ for several values of k_2/k_1 when k_2 is a constant. The lower numbers on the abscissa scale refer to the percentage of cell breakage associated with the corresponding value of $k_1 t$. (Coakley et al., 1974.)

e.g. polymer degradation, will also proceed to a limiting size at an exponential rate (Peacocke and Pritchard, 1968) and may be assigned a rate constant k_2. In cases where the inactivation rate is concentration dependent, as with sonochemical inactivation of enzymes, k_2 may be described as a function of concentration. From Eq. (11) it follows that the fraction, x, of a cell population disrupted after a treatment time t is given by:

$$x = 1 - e^{-k_1 t} \tag{14}$$

The fraction of the total cell contents released in a time Δt is then $k_1(1 - x)\Delta t$. Let the undegraded fraction of the total product pool in solution at time t be D, then its change ΔD in an interval Δt will be equal to the fraction of the product pool released from cells minus the fraction which has been inactivated in the time Δt. We have, therefore, if release and degradation of the product are independent events:

$$\Delta D = k_1(1 - x)\Delta t - k_2 D\Delta t \tag{15}$$

Solution of this equation when k_2 is a constant gives:

$$\Delta D = (e^{-Bk_1 t} - e^{-k_1 t})/(1 - B) \tag{16}$$

where $B = k_2/k_1$. The value of D as a function both of $k_1 t$ and of the percentage of cells disrupted is shown in Fig. 6. A family of curves is plotted for different values of the ratio k_2/k_1. The rate of ultrasonic enzyme inactivation in solution is broadly proportional to the reciprocal of protein concentration (Coakley et al., 1973). A family of curves similar to those in Fig. 6 has been derived for such a case (Coakley et al., 1974).

2. Flow Systems

Extract preparation is often carried out in flow systems. The dependence of protein release on flow rate y, cell disruption constant k_1 and container volume V has already been described (p. 304). In the case where extracted material is degraded at a constant rate, the fractional yield of undegraded material reaches a maximum value after a time t given by:

$$t = \ln(k_2/k_1)/(k_2 - k_1) \tag{17}$$

The yield then falls to an equilibrium value D_{eq} given by:

$$D_{eq} = Z/[(1 + Z)(Z + B)] \tag{18}$$

where $Z = y/k_1 V$, and will maintain such a value as flow continues. It follows that care should be taken to ensure that equilibrium conditions have been reached before determining a value of D as being appropriate for the established flow system. It can be shown, other variables being fixed, that the yield reaches the equilibrium value most rapidly for small values of the container volume V. The value of D_{eq} as a function of Z is shown in Fig. 7 for different values of the ratio k_2/k_1.

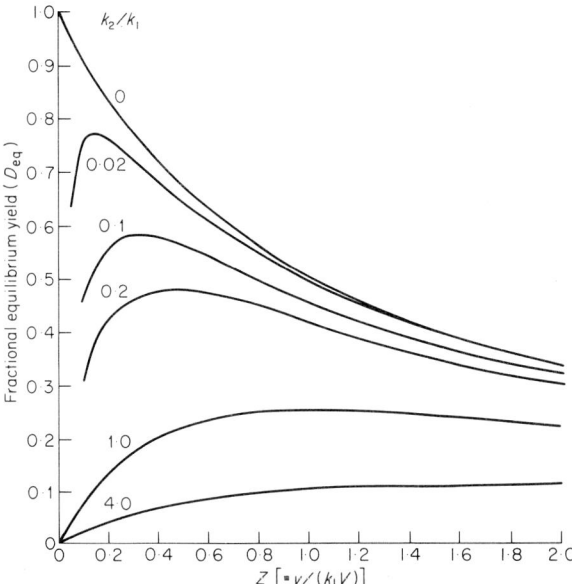

FIG. 7. A plot of the equilibrium fraction (D_{eq}) of undegraded product obtainable from a cell suspension in a flow sonication system, against $y/(k_1 V)$ for several values of k_2/k_1 when k_2 is a constant. (Coakley et al., 1974.)

If we compare the maximum value of D_{eq} for flow systems from Fig. 7, and the optimum yield for batch systems from Fig. 6 as a function of k_2/k_1 (Fig. 8), we see that the batch system is superior to a flow system if a high yield of undegraded material is required. The above treatment

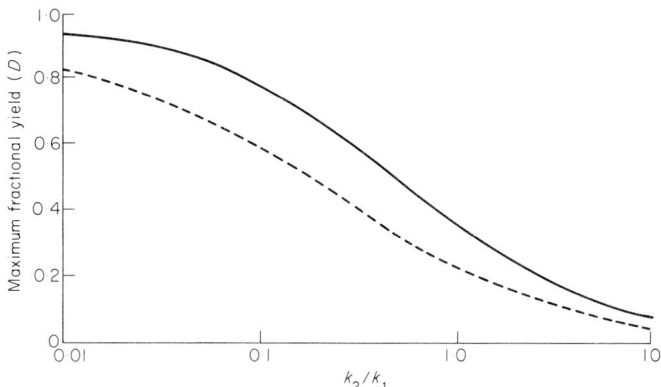

FIG. 8. The maximum fractional yield of product obtainable in batch (———) and flow systems at equilibrium (– – –) for different values of the ratio k_2/k_1 when k_2 is constant. (Coakley et al., 1974.)

may have application to devices other than sonicators. In a ball mill, the high stress region at the contact faces of the Ballotini beads resemble cavitation bubbles in that they are discrete regions of stress and they require that the cell suspension be mixed so that the new cells will flow into the high stress regions. Currie et al. (1972) and Mogren et al. (1974) have concluded that cell disruption in colloid mills (the Dyno-mill and Netzch-Moleux mill) is a first-order rate process. These mills are therefore likely to release undegraded material as outlined in Fig. 7. Recycling of a cell suspension through the Manton–Gaulin APV homogenizer valve (Augenstein et al., 1974; Hetherington et al., 1971) results in increased cell breakage. This increased breakage argues that there are localized areas of high stress in the valve and that cells may not come into contact with those regions in a single pass. The system will then have some similarities to a cavitation field. Augenstein et al. (1974) have shown that both disruption of cells and inactivation of a labile enzyme in the APV homogenizer give exponential trends when plotted against number of passes. Not all valve systems may be treated as above since Sharpe (1976) found no increase in cell disruption on repeated processing of a cell suspension.

V. Controlled Breakage of Eukaryotic Micro-organisms

A. CONTROLLED MECHANICAL BREAKAGE

Two criteria must be fulfilled before a mechanical breakage method can be routinely adopted as the initial step in isolation of functional membrane preparations: (a) the efficiency of cell disruption must be reasonably high (not less than 50%); and (b) breakage must be achieved without extensive structural and functional disorganization of intracellular organelles. It is self evident that no mechanical method can ever fully satisfy these criteria, and the uneasy compromise has to be found by control of an independent variable, usually "intensity" or time of treatment. Of the many machines currently commercially available, few have been adequately tested with a view to optimization of release of functional organelles or organelle fragments from a range of cell-types, although there may be little to choose between many of these devices as means for highly efficient disruption of even the most refractory systems. We recommend that preliminary experiments should be conducted along lines recommended by Hughes et al. (1971)

under "methods of assessment". If nuclear isolation is being attempted, the recent review of Buetow (1976) is useful, and criteria for purity and functional integrity of mitochondria-enriched subcellular fractions has been detailed by Lloyd (1974). Less attention has been paid to the integrity of other organelles released from microorganisms, although purification of peroxisomes (Müller, 1975), and lysosomes (Morgan et al., 1973) are rapidly becoming major foci of attention and presumably the principles for these procedures are similar to those for nuclear and mitochondrial preparations.

The more fragile of the protozoa (e.g. amoebae and some flagellates like *Polytomella* spp. and *Dunaliella* spp.) may be effectively disrupted by a few strokes of a hand homogenizer (e.g. Kontes or Dounce), but even this Phylum provides rather few species so amenable to easy distintegration. At the next level of resistance (e.g. flagellates like *Euglena* spp. or the trypanosomes) the Chaikoff Press (Emanuel and Chaikoff, 1957a) or the method of explosive decompression (Hunter and Commerford, 1961) become necessary; these treatments do not necessarily lead to fragmentation of organelles. Thus the Yeda Press, in which cells are equilibrated with dinitrogen or argon at pressures up to 12 MN/m^2 prior to release through a needle valve, has been used to prepare highly active chloroplasts from *Euglena gracilis* (Shneyour and Avron, 1970). For these organisms, and for those with highly refractory walls, the French Press (Milner et al., 1950) rates high in the popularity polls, and both this device and its many automated successors (Hetherington et al., 1971; Brookman, 1974) find wide application where effective breakage of organisms is required and where there is little concern for loss of intrinsic subcellular organization. However under carefully controlled conditions (<35 MN/m^2) these machines become possible candidates in the search for methods for release of partially intact organelles. The Hughes Press (Hughes, 1951) has never been critically assessed in this role; it is possible that it could be useful when operated in conjunction with cryoprotective compounds especially in view of the successful use of its modified descendant, the Eaton Press (Eaton, 1962) for the preparation of yeast nuclei (Duffus, 1969). Bhargava and Halvorson (1971) have adapted a French Press so as to extrude frozen cell pastes and suspensions by replacing the needle valve with a stainless steel tube (2 cm long and with an internal diameter of 0.95 mm). Freshly harvested yeast cells suspended in 1 M-sorbitol–20% glycerol–5%

polyvinylpyrrolidone and rapidly frozen in dry ice–ethanol mixture yielded preparations of reasonably intact nuclei after a single passage through this machine. Controlled conditions of freezing and thawing have been devised which are favourable to the continued functioning of yeast mitochondrial systems *in vitro* (Balcavage et al., 1970), and application of these regimes to treatments of cell suspensions during breakage procedures might yield useful preparative techniques.

A variety of colloid mills and high-speed shakers are now available; used in conjunction with suspensions of acid-washed Ballotini beads these provide a useful armoury of widely accepted methods for the isolation of mitochondria and submitochondrial particles from protozoa and yeasts. The successful adoption of these methods has arisen from the thoroughness with which investigation of the variables involved in the breakage process has been carried out. Thus Guarnieri et al. (1970) achieved controlled disruption of yeast (up to 30 g wet wt. cells) in a Mini-Mill (Gifford–Wood) which provided submitochondrial preparations with a high degree of functional integrity. The M.S.K. Braun Cell Homogenizer has a similar maximum capacity and is cooled with liquid carbon dioxide. A modification of the sample container of this rotary shaker enables simultaneous processing of multiple small samples (in a sixteen-place "multismash" adaptor) and is very useful when extracts from many mutants or strains are to be screened for a particular activity (Needleman and Tzagoloff, 1975). On a large scale, grinding in the frozen state in a Waring blender (at the temperature of liquid nitrogen) has been employed for preparation of mitochondrial membranes (Tzagoloff, 1969) as has the Manton–Gaulin homogenizer (Mason et al., 1973). Balcavage and Mattoon (1968) optimized conditions in an Eppenbach MV-6-3 micromill at gap setting 0.04 in and speed 7000 rev min^{-1} with 600 ml of a 66% (w/v) suspension of yeast and 500 ml of 0.2 mm glass beads. Shaking for 8 min gave a mitochondrial preparation which compares favourably with those obtained by the gentler but more expensive sphaeroplast method. This machine has also been used for disruption of *Neurospora crassa* (Hall and Greenawalt, 1964). Another large-scale procedure uses a Dyno-Mill KD-L, a continuous-flow disintegrator which has a capacity of 2.5 kg of yeast cells h^{-1}. With about 80% cell breakage, up to 7 g mitochondrial protein per kg pressed yeast have been obtained (Deters et al., 1976). In this device, a cell suspension is pumped through a cooled cylindrical glass chamber in which glass beads are rapidly being sitrred. Variables include the composition of the buffer medium, flow rate, temperature, size of glass beads, velocity of stirring, and cell concentration. Large beads (0.5 mm) were found to be more effective than small ones (0.1 mm), and 3000 rev min^{-1} proved optimal; flow rate and cell concentration were less critical.

Hand grinding with glass beads or silicon carbide at 0°–4°C is still the method of choice for disruption of *Euglena gracilis* (Buetow and Buchanan, 1965) *Crithidia fasciculata* (Toner and Weber, 1967, 1972) and *Prototheca zopfii* (Lloyd, 1965) for studies on the mitochondria of these organisms; this method is easily applied to minute samples of cell suspensions.

Methods of disruption employed in procedures for the isolation of nuclei include high speed blending for 15 seconds in a Waring Blender for *Tetrahymena pyriformis* (Grovosky et al., 1975), treatment in a glass-Teflon homogenizer or nitrogen cavitation for *Paramecium aurelia* (Cummings and Tait, 1975); protectants such as Ca^{2+}, polyamines and sugars, and disrupting agents such as deoxycholate are essential for release of intact organelles. Continuous flow homogenizers (Desjardins *et al.*, 1965) and the Super Dispax homogenizer, which have been recommended for isolation of mammalian nuclei (Taylor *et al.*, 1975) might also prove useful for preparations from micro-organisms.

Continuous-flow sonication at 20 kHz (James *et al.*, 1972) has not yet been assessed as a method for preparation of membraneous organelles or membrane fragments. In theory it should provide an excellent approach to the problem of larger scale preparations of submitochondrial particles, possibly with results better than those attainable by batch treatment (Schatz and Racker, 1966).

Preparation of fractions containing organelles from filamentous fungi is in some ways easier than from single cells, as in many species only a few breaks or tears in the hyphal walls are necessary for release of much of their contents. Kawakita (1970) has described the construction of a roller-mill which achieves this end and releases high-quality mitochondria from *Aspergillus oryzae*.

B. DISRUPTION AFTER PRETREATMENT OF ORGANISMS

1. *Preparation and Disruption of Sphaeroplasts*

Yeast protoplasts can be produced by mechanical or autolytic methods (Holden and Tracey, 1950; Nečas, 1956), by specific inhibition of cell wall synthesis (Berliner and Reca, 1970), or by enzymic treatment of the wall. The only approach of any practical importance is the last, pioneered by Eddy and Williamson (1957), who used the digestive juice obtained from the snail *Helix pomatia*. Snail enzyme can be purchased from Endo Laboratories Inc., Garden City, N.Y. (Glusulase) or from L'Industrie Biologique Francaise, 35–49 Quai du Moulin de Cage, Genevilliers, Paris. Live snails can be obtained from Gerrard and Haig Ltd., Gerrard House, East Preston, Sussex. Sphaeroplasts can also be prepared by using lytic enzymes produced by micro-organisms, but few of these are commercially available, although an endo-β-(1→3)-glucanase can be obtained from Kirin Brewery Co. Ltd., Gumna, Japan (Zymolase 5000) which has been used successfully with *Saccharomyces* (six species), *Candida* (eight species), *Pichia* (three species), *Torulopsis* (two species), *Brettanomyces anomalus, Debaromyces subglobus, Endomyces capsularis, Hansenula capsulata, Nematospora coryli* and *Saccharomycodes ludwigii* (Kitamura et al., 1971, 1972; Kaneko et al., 1973, Yamamoto et al., 1974). Other lytic preparations from fungi (e.g. *Rhizopus, Basidiomycetes* or *Chaetomium* spp.) or bacteria (e.g. *Bacillus* or *Streptomyces* spp.) have been shown to contain endo-β-(1 → 3)-, exo-β-(1 → 3)-, β-(1 → 6)-glucanases, phosphomannanase, α-mannanase, protease, chitinase, lipase, and/or phospholipases (Kuo and Yamamoto, 1975). Procedures for sphaeroplast formation require the use of osmotic stabilizers (sucrose, mannitol, sorbitol, and potassium or magnesium chloride), and dissolution of the walls is often attainable only in the presence of mercapto compounds (2-mercaptoethylamine, 2-mercaptoethanol, thioglycollate, or dithiothreitol) which rupture disulphide bridges. Evidently sensitivity to enzymic procedures provides a great deal of information on the forces necessary for the maintenance of wall structure. Variations in sensitivity to enzyme attack have been reported between one yeast strain and other of the same species and also between species. Growth conditions also influence ease of sphaeroplast production; thus both genetic and environmental factors are involved. Modification of the cell wall of *Schizosaccharomyces pombe* by growth in the presence of 2-deoxyglucose is necessary for the production of sphaeroplasts by treatment with snail enzymes (Poole and Lloyd, 1972; Foury and Goffeau, 1973).

Procedures for the preparation of sphaeroplasts of filamentous fungi have been reviewed (Villanueva et al., 1973; Peberdy, 1972); these are similar to those described for yeasts. No successful general methods for unicellular green algae such as *Chlorella* or *Scenedesmus* have been developed; in species of the Chlorococcales, an outer resistant layer of sporopollenin protects many species from lytic attack (Burczyk *et al.*, 1971; Atkinson *et al.*, 1972). One species of *Chlorella* which may lack sporopollenin is amenable to sphaeroplast production (Braun and Aach, 1975). A species-specific autolysin has been prepared from *Chlamydomonas reinhardii* culture filtrates during interaction of compatible gametes; after ammonium sulphate precipitation and dialysis, the enriched solution may be used for large-scale protoplast production (Schlösser *et al.*, 1976). Proteolytic enzymes have been used to prepare protoplasts of *Euglena gracilis* after hypotonic swelling (Kirk, 1964; Price and Bourke, 1966) but the yields are very low. However, when frozen-thawed cells are digested with protease in the presence of 2% Triton X-100, virtually all of the organisms are transformed into sphaeroplasts which can be broken by low-speed homogenization (Parenti *et al.*, 1969).

Sphaeroplasts can be gently disrupted by osmotic means, metabolic lysis (Indge, 1968), by careful hand homogenization, or by specific interaction with synthetic polybasic macromolecules, diethylamino-ethyl-dextran (mol. wt. 5×10^5) or poly-DL-lysine (mol. wt. $3-7 \times 10^4$, (Dürr *et al.*, 1975). The last-named method has been developed for gentle release of vacuoles from sphaeroplasts of *Candida utilis*. Adsorption of polybases at 0°C was followed by complete lysis on elevating the incubation temperature to 30°C. Vacuoles subsequently isolated contained large pools of amino acids, and a further indication of their functional integrity was provided by their ability to transport arginine. Judged by these criteria, polybase-induced lysis seems to be the best method available at present and may be widely applicable.

2. Disruption After Chemical or Physical Modification of Limiting Layers

Some procedures have been evolved which do not extensively destroy the structural integrity of the limiting layers of an organism, as is the case in sphaeroplast formation, but which nevertheless facilitate subsequent mechanical breakage at low stress values. An example of this approach is the removal of ergosterol from the pellicle of *Crithidia fasciculata* by titration with digitonin (Kusel and Storey, 1972) prior to mitochondrial isolation. This method has also been employed for

Tetrahymena pyriformis but is only successful if the naturally-occurring sterol, tetrahymenol, has been replaced by incorporation of ergosterol from supplemented media (Conner *et al.,* 1971).

3. *Destruction of Permeability Barriers Without Removal of Limiting Layers*

Permeability barriers can be removed by a variety of chemical procedures without necessitating removal of the walls. Thus treatment with dimethylsulphoxide (Adams, 1972), toluene (Serrano *et al.,* 1973), phenylpiperidyl-2-acetic acid methyl ester (Spoerl, 1971) or basic proteins (e.g. mammalian cytochrome *c*; Svihla *et al.,* 1969), all facilitate leakage of intracellular components from yeasts. Physical treatments which release some soluble enzymes from yeasts include drying on filters at room temperature (Mowshowitz, 1976) and freeze drying (Carter and Halvorson, 1973).

4. *Genetic Modification of Organisms to Provide Easily Disrupted Strains*

A large number of cell wall-defective mutants of *Chlamydomonas reinhardii* have been isolated (Davies and Plaskitt, 1971; Hyams and Davies, 1972). Mutant CW15 produces only a minute amount of wall, and this is not attached to the plasmalemma; intact organelles are easily obtained from this strain (Robreau and LeGall, 1974; Gould, 1975). The poor growth of the mutant can be overcome by increasing the concentration of sulphate in the growth medium (Lien and Knutsen, 1976).

A wall-less variant of *Neurospora crassa* (Gaertner and Leef, 1970) has been used for fractionation studies, and a mutant of *Aspergillus nidulans* containing 7–15% of the normal chitin complement has been described by Katz and Rosenberger (1971). Mutants of yeast which lyse at a non-permissive growth temperature may also prove useful (Cabib and Duran, 1975).

VI. Concluding Remarks

Whereas many of the problems of the efficient extraction of intracellular components have been overcome *via* a diversity of methodological developments, it is evident that the current demands made on disruption procedures on the part of microbial physiologists engaged in fundamental studies of subcellular organization can only be partially satisfied. Whilst it would seem highly inappropriate for the present reviewers to belittle the considerable achievements of the classical analytical approaches of biochemists towards the separation and purification of functional components at levels of organization from the molecular to that of the organelle, increasing awareness of the possibilities of artefactual modification adds a cautionary note to their enthusiasm. Extrapolation of data obtained with cell-free extracts to the *in vivo* situation requires, wherever possible, independent confirmation, e.g. from the ever increasingly powerful techniques of analytical cytology. Results obtained with disrupted cell preparations can only provide information on those reaction pathways and kinetics which are possible, not necessarily on those which have a functional significance. Perhaps we should follow A. N. Whitehead's advice: "Seek simplicity and then distrust it".

REFERENCES

Adams, B. G. (1972). *Analytical Biochemistry* **45**, 137.
Aguirre, M. J. and Villanueva, J. R. (1962). *Nature, London* **196**, 692.
Ahkong, Q. F., Fisher, D., Tampion, W. and Lucy, J. A. (1973). *Biochemical Journal* **136**, 147.
Allan, D., Billah, M. M., Finean, J. B. and Michell, R. H. (1976). *Nature, London* **261**, 58.
Allan, D., Low, M. G., Finean, J. B. and Michell, R. H. (1975). *Biochimica et Biophysica Acta* **413**, 308.
Allan, D. and Michell, R. H. (1975a). *Biochemical Society Transactions* **3**, 751.
Allan, D. and Michell, R. H. (1975b). *Nature, London* **258**, 348.
Allan, D., Watts, R. and Michell, R. H. (1976). *Biochemical Journal* **156**, 225.
Altendorf, K. H. and Staehelin, L. A. (1974). *Journal of Bacteriology* **117**, 888.
Asano, A., Cohen, N. S., Baker, R. F. and Brodie, A. F. (1973). *Journal of Biological Chemistry* **248**, 3386.
Atkinson, A. W., Gunning, B. E. S. and John, P. C. L. (1972). *Planta* **107**, 1.
Augenstein, D. C., Thrasher, K., Sinskey, A. J. and Wang, D. I. C. (1974). *Biotechnology and Bioengineering* **16**, 1433.
Balcavage, W. X. and Mattoon, J. R. (1968). *Biochimica et Biophysica Acta* **153**, 521.
Balcavage, W. X., Beck, J. C., Beck, D., Greenawalt, J. W., Parker, J. H. and Mattoon, J. R. (1970). *Cryobiology* **6**, 385.
Bender, W. W., Garan, H. and Berg, H. C. (1971). *Journal of Molecular Biology* **58**, 783.
Berg, H. C. (1969). *Biochimica et Biophysica Acta* **183**, 65.
Berliner, M. D. and Reca, M. E. (1970). *Science, New York* **167**, 1255.
Bhargava, M. M. and Halvorson, H. O. (1971). *Journal of Cell Biology* **49**, 423.
Bisschop, A., Doddema, H. and Konings, W. N. (1975). *Journal of Bacteriology* **124**, 613.
Blackshear, P. L. Jr., Forstrom, R. J. and Dorman, F. D. (1971). *Federation Proceedings. Federation of American Societies for Experimental Biology*, **30**, 1600.
Blaurock, A. E. and Stockenius, W. (1971). *Nature New Biology, London* **233**, 152.

Braun, E. and Aach, H. G. (1975). *Planta* **126**, 181.
Bretscher, M. S. (1971). *Journal of Molecular Biology* **58**, 775.
Bretscher, M. S. (1973). *Science, New York* **181**, 622.
Brookman, J. S. G. (1974). *Biotechnology and Bioengineering* **16**, 371.
Brookman, J. S. G. (1975). *Biotechnology and Bioengineering* **17**, 465.
Brookman, J. S. G. and Davies, M. (1973). *Biotechnology and Bioengineering* **15**, 693.
Buetow, D. E. (1976). *In* "Methods in Cell Biology" (D. M. Prescott, ed.), vol. xiii, pp. 283–311. Academic Press, New York.
Buetow, D. E. and Buchanan, P. J. (1965). *Biochimica et Biophysica Acta* **96**, 9.
Burczyk, J. H., Grzybek, H., Banas, J. and Banas, E. (1971). *Experimental Cell Research* **63**, 451.
Burgi, E. and Hershey, A. D. (1961). *Journal of Molecular Biology* **3**, 458.
Burnell, J. N., John, P. and Whatley, F. R. (1975). *Biochemical Journal* **150**, 527.
Burton, A. C. (1970). *In* "Permeability and Function of Biological Membranes" (L. Bolis, A. Katchalsky, R. D. Keynes, W. R. Lowenstein and A. Pethica, eds.), pp. 1–19. North Holland/American Elsevier, Amsterdam and New York.
Cabib, E. and Duran, A. (1975). *Journal of Bacteriology* **124**, 1604.
Carter, B. L. A. and Halvorson, H. O. (1973). *Experimental Cell Research* **76**, 152.
Cavalieri, L. F. and Rosenberg, B. H. (1959). *Journal of the American Chemical Society* **81**, 5136.
Champion, J. V. and Coakley, W. T. (1969). *Biopolymers* **7**, 815.
Champion, J. V. and North, P. F. (1971). *Journal de Chimie Physique* **11**, 1585.
Champion, J. V., North, P. F., Coakley, W. T. and Williams, A. R. (1971). *Biorheology* **8**, 23.
Coakley, W. T. (1974). *Brain Research* **70**, 281.
Coakley, W. T., Brown, R. C., James, C. J. and Gould, R. K. (1973). *Archives of Biochemistry and Biophysics* **159**, 722.
Coakley, W. T., Brown, R. C., James, C. J. and Gould, R. K. (1974). *Biotechnology and Bioengineering* **16**, 659.
Coakley, W. T., James, C. J. and Macintosh, I. J. C. (1977). *Biorheology* in press.
Collins, S. B. and Knudsen, J. G. (1970). *American Institute of Chemical Engineers Journal* **16**, 1075.
Conner, R. L., Mallory, F. B., Landrey, J. T., Ferguson, K. A., Kaneshiro, E. S. and Ray, E. (1971). *Biochemical and Biophysical Research Communications* **44**, 995.
Corner, T. R. and Marquis, R. E. (1969). *Biochimica et Biophysica Acta* **183**, 544.
Cummings, D. J. and Tait, A. (1975). *In* "Methods in Cell Biology" (D. M. Prescott, ed.), vol. IX, pp. 281–309. Academic Press, New York.
Currie, A., Dunnill, P. and Lilly, M. D. (1972). *Biotechnology and Bioengineering* **14**, 725.
Davies, D. R. and Plaskitt, A. (1971). *Genetical Research, Cambridge* **17**, 33.
DeBoer, E. and Loyter, A. (1971). *Federation of the European Biochemical Societies Letters* **15**, 325.
Desjardins, R., Smetana, K. and Busch, H. (1965). *Experimental Cell Research* **40**, 127.
Deters, D., Muller, U. and Homberger, H. (1976). *Analytical Biochemistry* **70**, 263.
Doty, P., McGill, B. B. and Rice, S. A. (1958). *Proceedings of the National Academy of Sciences of the United States of America* **44**, 432.
Dowben, R. M., Gaffey, T. A. and Lynch, P. A. (1968). *Federation of the European Biochemical Societies Letters* **2**, 1.
Dowben, R. M., Lynch, P. M., Nadler, H. C. and Hsia, D. Y. (1969). *Experimental Cell Research* **58**, 167.
Duffus, J. H. (1969). *Biochimica et Biophysica Acta* **195**, 233.
Dürr, M., Boller, T. and Weimken, A. (1975). *Archives of Microbiology* **105**, 319.

Eaton, N. R. (1962). *Journal of Bacteriology* **83**, 1359.
Eddy, A. A. and Williamson, D. H. (1957). *Nature, London* **179**, 1252.
Edwards, D. G. and Wiseman, A. (1971). *Process Biochemistry* November issue, p. 32.
Einstein, A. (1906). *Annalen der Physik* **19**, 289.
Eisenberg, A. D. and Corner, T. R. (1973). *Journal of Bacteriology* **114**, 1177.
Eller, A. I. and Crum, L. A. (1970). *Journal of the Acoustical Society of America* **47**, 762.
Él'piner, I. E. (1964). "Ultrasound; Physical, Chemical and Biological Effects", 371 pp. Consultants Bureau, New York.
Emanuel, C. F. and Chaikoff, I. L. (1957a). *Biochimica et Biophysica Acta* **24**, 254.
Emanuel, C. F. and Chaikoff, I. L. (1957b). *Biochimica et Biophysica Acta* **24**, 261.
Flynn, H. (1964). *In* "Physical Acoustics" (W. P. Mason, ed.), vol. 1B, p. 62. Academic Press, New York.
French, C. S. and Milner, H. W. (1955). *In* "Methods in Enzymology" (S. P. Colowick and N. O. Kaplan, eds.), vol. 1, pp. 64–67. Academic Press, New York.
Follows, M., Hetherington, P. J., Dunnill, P. and Lilly, M. D. (1971). *Biotechnology and Bioengineering* **13**, 549.
Foster, J. W., Cowan, R. M. and Maag, T. A. (1962). *Journal of Bacteriology* **83**, 330.
Foury, F. and Goffeau, A. (1973). *Journal of General Microbiology* **75**, 227.
Fraser, D. (1951). *Nature, London* **167**, 33.
Futai, M. (1974). *Journal of Membrane Biology* **15**, 15.
Gaertner, F. H. and Leef, J. L. (1970). *Biochemical and Biophysical Research Communications* **41**, 1192.
Gander, J. E. (1974). *Annual Review of Microbiology* **28**, 103.
Glauert, A. M. (1962). *British Medical Bulletin* **18**, 245.
Goldsmith, H. L. (1971). *Federation Proceedings. Federation of American Societies for Experimental Biology* **30**, 1578.
Goldsmith, H. L. and Mason, S. G. (1967). *In* "Rheology: Theory and Applications", (F. R. Eirich, ed.), vol. 4, pp. 85–250. Academic Press, New York.
Grovosky, M. A., Yao, M-C., Keevert, J. B. and Pleger, G. L. (1975). *In* "Methods in Cell Biology" (D. M. Prescott, ed.), vol. IX, pp. 311–327. Academic Press, New York.
Gould, R. K. (1974). *Journal of the Acoustical Society of America* **56**, 1740.
Gould, R. K. (1975). *Journal of Cell Biology* **65**, 65.
Green, D. M. (1966). *Journal of Molecular Biology* **22**, 15.
Guarnieri, M., Mattoon, J. R., Balcavage, W. X. and Payne, C. (1970). *Analytical Biochemistry* **34**, 39.
Guidotti, G. (1972). *Annual Review of Biochemsitry* **41**, 731.
Hall, D. O. and Greenawalt, J. W. (1964). *Biochemical and Biophysical Research Communications* **17**, 565.
Hampton, H. L. and Freese, E. (1974). *Journal of Bacteriology* **118**, 497.
Hanahan, D. J. and Ekholm, J. E. (1974). *In* "Methods in Enzymology" (S. P. Colowick and N. O. Kaplan, eds.), vol. 31, pp. 168–172. Academic Press, New York.
Harold, F. M. (1972). *Bacteriological Reviews* **36**, 172.
Harrington, R. E. (1970). *Biopolymers* **9**, 159.
Harrington, R. E. and Zimm, B. H. (1965). *Journal of Physical Chemistry* **69**, 161.
Harvey, E. N. and Loomis, A. L. (1929). *Journal of Bacteriology* **17**, 373.
Harvey, E. N. and Loomis, A. L. (1932). *Journal of General Physiology* **15**, 147.
Hetherington, P. J., Follows, M., Dunnill, P. and Lilly, M. D. (1971). *Transactions of the Institute of Chemical Engineers* **49**, 142.
Hinze, J. O. (1953). *American Institute of Chemical Engineering Journal* **1**, 289.
Hinze, J. O. (1975). "Turbulence", 2nd edition, 790 pp. McGraw-Hill, New York.

Holden, M. and Tracey, M. V. (1950). *Biochemical Journal* **47**, 407.
Hughes, D. E. (1951). *British Journal of Experimental Pathology* **32**, 97.
Hughes, D. E. and Nyborg, W. L. (1962). *Science, New York* **138**, 108.
Hughes, D. E., Wimpenny, J. W. T. and Lloyd, D. (1971). In "Methods in Microbiology" (J. R. Norris and D. W. Ribbons, eds.), vol. 5B, pp. 1–54. Academic Press, London.
Hunter, M. J. and Commerford, S. L. (1961). *Biochimica et Biophysica Acta* **47**, 580.
Hyams, J. and Davies, D. R. (1972). *Mutation Research* **14**, 381.
Indge, K. J. (1968). *Journal of General Microbiology* **51**, 441.
Isaac, L. and Ware, G. C. (1974). *Journal of Applied Bacteriology* **37**, 335.
James, C. J., Coakley, W. T. and Hughes, D. E. (1972). *Biotechnology and Bioengineering* **14**, 33.
Joly, M. (1965). "A Physico-Chemical Approach to the Denaturation of Proteins", 350 pp. Academic Press, London.
Kaback, H. R. (1971). In "Methods in Enzymology" (S. P. Colowick and N. O. Kaplan, eds.), vol. 22, pp. 99–120. Academic Press, New York.
Kaback, H. R. (1972). *Biochimica et Biophysica Acta* **265**, 367.
Kaback, H. R. and Deuel, T. F. (1969). *Archives of Biochemistry and Biophysics* **132**, 118.
Kaneko, T., Kitamura, K. and Yamamoto, Y. (1973). *Agricultural and Biological Chemistry* **37**, 2295.
Katz, D. and Rosenberger, R. F. (1971). *Archives of Microbiology* **80**, 284.
Kawakita, M. (1970). *Journal of Biochemistry, Tokyo* **68**, 625.
Kaye, J. and Elgar, E. C. (1958). *Transactions of the American Society of Mechanical Engineers* **80**, 753.
Kinsloe, H., Ackerman, E. and Reid, J. J. (1954). *Journal of Bacteriology* **68**, 373.
Kirk, J. T. O. (1964). *Journal of Protozoology* **11**, 435.
Kitamura, K., Kaneko, T. and Yamamoto, Y. (1971). *Archives of Biochemistry and Biophysics* **145**, 402.
Kitamura, K., Kaneko, T. and Yamamoto, Y. (1972). *Journal of General and Applied Microbiology* **18**, 57.
Knauf, P. A. and Rothstein, A. (1971). *Journal of General Physiology* **58**, 190.
Konings, W. N. and Freese, E. (1972). *Journal of Biological Chemistry* **247**, 2408.
Krulwich, T. A., Ensign, J. C., Tipper, D. J. and Strominger, J. L. (1967a). *Journal of Bacteriology* **94**, 734.
Krulwich, T. A., Ensign, J. C., Tipper, D. J. and Strominger, J. L. (1967b). *Journal of Bacteriology* **94**, 741.
Kuo, S-C. and Yamamoto, S. (1975). In "Methods in Cell Biology" (D. M. Prescott, ed.), vol. XI, pp. 161–182. Academic Press, New York.
Kusel, J. P. and Storey, B. T. (1972). *Biochemical and Biophysical Research Communications* **46**, 501.
Laufer, J. (1954). Technical Report No. 1174, National Advisory Committee on Aeronautics, United States of America.
Leedale, G. F. (1967). "Euglenoid Flagellates", 242 pp. Prentice-Hall, New Jersey.
Leverett, L. B., Hellum, J. D., Alfrey, C. P. and Lynch, E. P. (1972). *Biophysical Journal* **12**, 757.
Levinthal, C. and Davison, P. F. (1961). *Journal of Molecular Biology* **3**, 674.
Lien, T. and Knutsen, G. (1976). *Archives of Microbiology* **108**, 189.
Lloyd, D. (1965). *Biochimica et Biophysica Acta* **110**, 425.
Lloyd, D. (1974). "The Mitochondria of Microorganisms", 553 pp. Academic Press, London.

Maddy, A. H. (1964). *Biochimica et Biophysica Acta* **88**, 390.
Magnusson, K-E. and Edebo, L. (1976a). *Biotechnology and Bioengineering* **18**, 449.
Magnusson, K-E. and Edebo, L. (1976b). *Biotechnology and Bioengineering* **18**, 865.
Magnusson, K-E. and Edebo, L. (1976c). *Biotechnology and Bioengineering* **18**, 975.
Marchesi, V. T. (1972). *In* "Membrane Research" (C. F. Fox, ed.), pp. 41–52. Academic Press, New York.
Margulis, M. A. (1971). *Soviet Physics Acoustics* **16**, 361.
Marquis, R. E. (1967). *Archives of Biochemistry and Biophysics* **118**, 323.
Marquis, R. E. (1973). *Journal of Bacteriology* **116**, 1273.
Mason, T. L., Boyton, R. O., Wharton, D. C. and Schatz, G. (1973). *Journal of Biological Chemistry* **244**, 1346.
Matin, A. and Konings, W. N. (1973). *European Journal of Biochemistry* **34**, 58.
Milner, H. W., Lawrence, N. S. and French, C. S. (1950). *Science, New York* **111**, 633.
Mirsky, R. and Barlow, V. (1971). *Biochimica et Biophysica Acta* **241**, 835.
Mogren, H., Lindblom, M. and Hedenskog, G. (1974). *Biotechnology and Bioengineering* **16**, 261.
Morell, A. G., Van den Hamer, C. J. A., Scheinberg, I. H. and Ashwell, G. (1966). *Journal of Biological Chemistry* **241**, 3745.
Morgan, N. A., Howells, L., Cartledge, T. G. and Lloyd, D. (1973). *In* "Methodological Developments in Biochemistry" (E. Reid, ed.), vol. 3, pp. 219–232. Longmans, London.
Mowshowitz, D. B. (1976). *Analytical Biochemistry* **70**, 94.
Müller, M. (1975). *Annual Review of Microbiology* **29**, 467.
Munoz, E., Salton, M. R. J., Ng, M. H. and Schor, M. T. (1969). *European Journal of Biochemistry* **7**, 490.
Nečas, O. (1956). *Nature, London* **177**, 898.
Needleman, R. B. and Tzagoloff, A. (1975). *Analytical Biochemistry* **64**, 545.
Nyborg, W. L. (1968). *Journal of the Acoustical Society of America* **44**, 1302.
Nyborg, W. L. and Hughes, D. E. (1967). *Journal of the Acoustical Society of America* **42**, 891.
Ou, L. T. and Marquis, R. E. (1970). *Journal of Bacteriology* **101**, 92.
Ou, L. T. and Marquis, R. E. (1972). *Canadian Journal of Microbiology* **18**, 623.
Ovenall, D. W., Hastings, G. W. and Allen, P. E. M. (1958). *Journal of Polymer Science* **33**, 213.
Oxender, D. L. (1972). *Annual Review of Biochemistry* **41**, 777.
Palek, J., Stewart, G. and Lioetti, F. J. (1974). *Blood* **44**, 584.
Parenti, F., Braurman, G., Preston, J. F. and Eisenstadt, M. (1969). *Biochimica et Biophysica Acta* **195**, 234.
Peacocke, A. R. and Pritchard, N. J. (1968). *In* "Progress in Biophysics and Molecular Biology" (J. A. V. Butler and D. Noble, eds.), vol. 18, pp. 185–208. Pergamon Press, Oxford.
Peberdy, J. F. (1972). *Science Progress, Oxford* **60**, 73.
Pfenniger, W. (1961). *In* "Boundary Layer and Flow Control" (G. V. Lachmann, ed.), vol. 2, pp. 970–980. Pergamon Press, Oxford.
Phillips, D. R. and Morrison, M. (1970). *Biochemical and Biophysical Research Communications* **40**, 284.
Poole, R. K. and Lloyd, D. (1972). *Archives of Microbiology* **88**, 257.
Price, C. A. and Bourke, M. E. (1966). *Journal of Protozoology* **13**, 474.
Reed, P. W. and Lardy, H. A. (1972). *Journal of Biological Chemistry* **247**, 6970.
Richards, O. G. and Boyer, P. D. (1965). *Journal of Molecular Biology* **11**, 327.

Richardson, E. (1974). *Proceedings of the Royal Society, London, Series A* 338, 129.
Richardson, E. (1975). *Biorheology* 12, 27.
Robreau, G. and LeGall, Y. (1974). *Compte Rendus Hebdomadaires des Seances de l'Academie des Sciences, Series D* 279, 1923.
Rodbell, M. and Krishna, G. (1974). *In* "Methods in Enzymology" (S. P. Colowick and N. O. Kaplan, eds.), vol 31, pp. 103–114. Academic Press, New York.
Rodgers, A. and Hughes, D. E. (1960). *Journal of Biochemical and Microbiological Technology and Engineering* 2, 49.
Rogers, H. J. and Perkins, H. R. (1968). "Cell Walls and Membranes", 436 pp. E. & F. N. Spon Ltd., London.
Rooney, J. A. (1972). *Journal of the Acoustical Society of America* 52, 1718.
Rose, J. A. and Warms, J. V. B. (1967). *Journal of Biological Chemistry* 242, 1635.
Rosen, B. P. and McClees, J. S. (1974). *Proceedings of the National Academy of Sciences of the United States of America* 71, 5042.
Schatz, G. and Racker, E. (1966). *Biochemical and Biophysical Research Communications* 22, 579.
Schlösser, U. G., Sachs, H. and Robinson, D. G. (1976). *Protoplasma* 88, 51.
Scully, D. B. and Wimpenny, J. W. T. (1974). *Biotechnology and Bioengineering* 16, 675.
Seeman, P. (1967). *Journal of Cell Biology* 32, 55.
Serrano, R., Gancedo, J. M. and Gancedo, C. (1973). *European Journal of Biochemistry* 34, 479.
Sharpe, J. E. E. (1976). *Laboratory Practice* 25, 28.
Shneyour, A. and Avron, M. (1970). *Federation of European Biochemical Societies Letters* 8, 164.
Sheetz, M. P. and Singer, S. J. (1974). *Proceedings of the National Academy of Sciences of the United States of America* 71, 4457.
Short, S. A. and Kaback, H. R. (1974). *Journal of Biological Chemistry* 249, 4275.
Short, S. A., Kaback, H. R., Kaczorowski, G., Fisher, J., Walsh, C. T. and Silverstein, S. C. (1974). *Proceedings of the National Academy of Sciences of the United States of America* 71, 5032.
Short, S. A., Kaback, H. R. and Kohn, D. (1975). *Journal of Biological Chemistry* 250, 4291.
Singer, S. J. (1974). *Annual Review of Biochemistry* 43, 805.
Singer, S. J. (1975). *In* "Cell Membranes; Biochemistry, Cell Biology and Pathology" (G. Weissman and R. Claiborne, eds.), pp. 35–44. HP Publishing Co. Inc., New York.
Singer, S. J. and Nicholson, G. L. (1972). *Science, New York* 175, 720.
Spoerl, E. (1971). *Journal of Bacteriology* 105, 1168.
Staas, W. H. and Spurlock, L. A. (1975). *Journal of the Chemical Society, Perkin Transactions* 1, Pt II, 1675.
Steck, T. L. (1972). *In* "Membrane Research" (C. F. Fox, ed), pp. 71–93. Academic Press, London and New York.
Steck, T. L. (1974). *In* "Methods in Membrane Biology" (E. D. Korn, ed.), vol. 2, pp. 245–281. Plenum Press, New York.
Steck, T. L. and Kant, J. A. (1974). *In* "Methods in Enzymology" (S. P. Colowick and N. O. Kaplan, eds.), vol. 31, pp. 172–180. Academic Press, New York.
Sutera, S. P. and Mehrjardi, M. H. (1975). *Biophysical Journal* 15, 1.
Svihla, G., Daiko, J. L. and Schlenk, F. (1969). *Journal of Bacteriology* 100, 498.
Taylor, G. I. (1932). *Proceedings of the Royal Society, London, Series A* 138, 41.
Taylor, G. I. (1934). *Proceedings of the Royal Society, London, Series A* 146, 501.
Taylor, G. I. (1935). *Proceedings of the Royal Society, London, Series A* 151, 494.

Taylor, G. I. (1936). *Proceedings of the Royal Society, London, Series A* **157**, 546.
Taylor, C. W., Yoeman, L. C. and Busch, H. (1975). *In* "Methods in Cell Biology" (D. M. Prescott, ed.), vol. IX, pp. 349–376. Academic Press, New York and London.
Thacker, J. (1973). *Biochimica et Biophysica Acta* **304**, 240.
Thienen, V. A. and Postma, A. W. (1973). *Biochimica et Biophysica Acta* **323**, 429.
Toner, J. J. and Weber, M. M. (1967). *Biochemical and Biophysical Research Communications* **28**, 821.
Toner, J. J. and Weber, M. M. (1972). *Biochemical and Biophysical Research Communications* **46**, 642.
Tsuchiya, T. and Rosen, B. P. (1975).*Journal of Biological Chemistry* **250**, 7687.
Tsukagoshi, N. and Fox, C. F. (1971). *Biochemistry, New York* **10**, 3309.
Tzagoloff, A. (1969).*Journal of Biological Chemistry* **244**, 5020.
Villanueva, J. R., Garcia-Acha, I., Gascon, S. and Uruburu, F. (editors) (1973). "Yeast, Mould, and Plant Protoplasts", 381pp. Academic Press, New York and London.
Wallach, D. F. H. (1967). *In* "The Specificity of Cell Surfaces" (B. D. Davies and L. Warren, eds.), pp. 129–163. Prentice-Hall, New Jersey.
Wallach, D. F. H. (1972). "The Plasma Membrane; Dynamic Perspectives, Genetics and Pathology", 186 pp. English Universities Press, London.
Weiner, J. H. (1974).*Journal of Membrane Biology* **15**, 1.
Weiss, L. (1971). *Federation Proceedings* **30**, 1649.
Williams, A. R. (1970). Ph.D. Thesis: University of Wales.
Williams, A. R. (1972).*Journal of the Acoustical Society of America* **52**, 688.
Williams, A. R., Hughes, D. E. and Nyborg, W. L. (1970). *Science, New York* **169**, 871.
Yamamoto, S., Fukuyama, J. and Nagasaki, S. (1974). *Agricultural and Biological Chemistry* **38**, 329.

AUTHOR INDEX

Numbers in italics refer to the pages on which references are listed at the end of each article.

A

Aach, H. G., 333, *336*
Abbo, F. E., 50, *95*
Ackerman, F., 303, *338*
Adams, B. G., 334, 335
Adler, H. I., 120, 121, *150*
Afzelius, B. A., 43, *44*
Aguirre, M. J., 285, *335*
Ahkong, Q. F., 315, *335*
Aiello, E., 10, 20, 30, *44*, *47*
Akai, A., 226, 227, 228, *275*
Alexander, P., 140, *151*
Alfrey, C. P., 293, 294, *338*
Algeri, A., 228, *273*
Allan, D., 313, 314, 322, *335*
Allen. C., 43, *44*
Allen, N. E., 204, *263*
Allen, P. E. M., 309, *339*
Aloni, Y., 167, *273*
Altendorf, K. H., 317, 318, 319, 320, 322, 323, 324, *335*
Anderson, A. W., 108, *150*
Anderson, W. F., 101, *153*
Aplin, R. T., 101, *155*
Arnberg. A. C., 195, 258, *269*
Arwert, F., 130, *152*
Asakura, S., 17, *46*, *47*
Asano, A., 321, 322, *335*
Ashwell, G., 311, *339*
Ashwell, M., 158, *263*
Atkinson, A. W., 333, *335*
Atkinson, M. R., 129, *152*
Attardi, G., 167, *272*, *273*
Aufderheide, K. J., 243, *263*
Augenstein, D. C., 309, 328, *335*
Avers, C. J., 169, 178, 186, 188, 202, *263*, *264*, *266*, *268*
Avner, P. R., 170, 221, 224, 225, 226, 232, 235, 237, 239, 241, 245, *263*, *268*, *275*
Avron, M., 297, 298, 329, *340*

B

Baba, S. A., 71, *44*
Bacilà, M., 179, 187, *269*
Backhaus, B, B., 248, 249, *274*
Bak, A. L., 160, 161, 162, 164, 170, 172, 192, *263*, *264*, *265*
Baker, B., 120, *153*
Baker, R. F., 321, 322, *335*
Balcaddi, G., 167, *263*
Balcavage, W. X., 330, *335*, *337*
Ball, A. J. S., 221, *266*
Banas, E., 333, *336*
Banas, J., 333, *336*
Bandlow, W., 172, 214, 215, *264*
Banks, G. R., 183, 184, *264*
Baranowska, H., 204, *273*
Barnoux, C., 116, *151*
Barolow, V., 313, *339*
Barth, P. T., 53, *97*
Bastus, R. N., 207, 208, 209, 210, 215, *264*, *270*
Bauer, W., 162, *273*
Bech-Hansen, N. T., 205, 209, 220, 221, 231, 233, 257, *264*, *273*
Beck, D., 330, *335*
Beck, J. C., 330, *335*
Bender, W. W., 311, *335*
Bendigkeit, H. E., 50, 70, *96*
Benedict, B., 33, *44*
Bennett, G. M., 145, *155*
Benzon, M. W., 50, 57, *97*
Berends, L. J., 117, *151*
Berg, H. C., 311, *335*
Berliner, M. D., 332, *335*
Bernardi, G., 159, 161, 162, 163, 164, 167, 169, 171, 181, 195, 196, 199, 211, 231, 255, 263, *264*, *266*, *273*
Bernini, F., 30, *45*
Berns, K., 101, *153*
Bhargava, M. M., 329, *335*
Bicknell, J. N., 170, 173, *264*

Biggs, D. R., 214, 217, *269*
Billah. M. M., 313, 315, *335*
Billheimer, F. E., 169, 178, 202, *263*, *264*
Bird, R. E., 51, 76, *95*
Birfelder, E. J., 199, *267*
Birky, C. W., 235, 239, 241, 242, 244, 245, *264*, *272*
Bisschop. A., 321, *335*
Blackshear, P. L., Jr., 294, *335*
Blake, J. R., 18, 19, 20, 21, 22, 23, *44*, *47*
Blamire, J., 159, 160, 183, 189, 202, 249, 259, 263, *264*, *266*, *267*, *271*
Blaurock, A. E., 283, *335*
Bleeg, H. S., 187, 192, *264*
Blum, J. J., 40, *46*
Blumberg, G., 61, 70, 74, *95*, *96*
Bobraw, L., 41, *44*
Boling, M. E., 122, 125, 126, *150*, *154*
Boller, T., 333, *336*
Bollum, F. J., 122, *154*
Bolotin, M., 194, 217, 219, 220, 222, 231, 233, 234, 238, 239, 240, 243, 246, *264*
Bolotin-Fukuhara, M., 194, 200, 202, 204, 219, 222, 235, 237, 238, 245, 246, 248, 249, 252, 253, 254, *264*, *266*, *267*, *272*
Borisy, G. G., 41, 43, *44*, *45*
Borst, P., 158, 160, 161, 162, 165, 166, 167, 169, 171, 172, 179, 181, 187, 195, 196, 198, 199, 205, 207, 215, 216, 219, 220, 222, 252, 253, 254, 258, *264*, *265*, *266*, *267*, *268*, *269*, *271*, *272*, *273*, *274*, *275*
Bostock, C. J., 171, 172, *265*
Bouck, G. B., 9, 30, *44*
Bourke, M. E., 333, *339*
Boyce, R. P., 102, 103, 106, 107, 109, 112, 113, 114, 117, 126, 128, 142, *150*, *152*
Boyer, P. D., 308, *339*
Boyle, J. M., 129, 147, *150*
Boyton, R. O., 330, *339*
Bradfield, J. R. G., 24, *44*
Brandts, J. F., 36, *44*
Braun, A., 112, 125, 127, 128, 131, *150*, *152*
Braun, E., 333, *336*
Braurman, G., 333, *339*
Bremer, H., 57, *95*
Brendel, M., 175, *226*

Brennen, C., 20, 22, *44*, *45*
Bretscher, M. S., 311, 312, *336*
Bridges, B. A., 138, 145, 146, 148, *150*, *151*
Briquet, M., 214, 221, *226*, *267*
Broda, P., 51, *96*
Brodie, A. F., 321, 322, *335*
Brokaw, C. J., 7, 16, 25, 31, 32, 35, 38, 39, 40, 41, *44*, *46*
Bron, S., 130, *152*
Brookman, J. S. G., 297, 299, 300, 305, 306, 329, *336*
Brooks-Low, K., 117, *153*
Brown, R. C., 304, 306, 307, 325, 326, 327, *336*
Brutlag, G., 129, *150*
Buchanan, P. J., 331, *336*
Bucking-Throm, E., 189, *265*
Buetow, D. E., 329, 331, *336*
Bulder, C. J. E. A., 159, 214, *265*
Bunn, C. L., 218, 219, 220, 221, 222, 223, 224, 229, *265*, *270*, *271*, *275*
Bullock, M. L., 118, *151*
Burchiel, K., 168, 169, 221, 225, *271*, *274*
Burczyk, J. H., 333, *336*
Burge, R. E., 7, 17, *45*
Burger, G., 216, 224, *277*
Burgi, E., 309, *336*
Burnasheva, S. A., 8, *44*
Burrell, A. D., 108, 127, 142, *150*
Burnell, J. N., 321, *336*
Burton, A. C., 288, 289, *336*
Busch, H., 331, *336*, *341*
Butow, R. A., 181, 182, 203, *276*
Buttin, G., 115, 147, *150*, *155*
Byers, B., 160, *273*

C

Cabib, E., 334, *336*
Cain, R. F., 108, *150*
Cairns, J., 110, 115, 116, 129, 134, 138, *150*, *151*
Callen, D. F., 220, 221, 225, 231, 233, 239, 240, 242, 243, *265*
Campbell, J. L., 116, *150*
Carnevali, F., 159, 162, 163, 167, 181, 195, 196, *263*, *264*, *265*
Caro, L., 51, 76, *95*
Carrier, W. L., 103, 111, 126, *150*, *154*
Carter, B. L. A., 189, *266*, 334, *336*

AUTHOR INDEX 345

Cartledge, T. C., 329, *339*
Casey, J., 167, 193, 194, 198, 200, 201, 203, 265, 266, *267*, *273*
Casey, J. W., 253, *265*
Castellazi, M., 147, *150*
Cavalieri, L., 54, *97*
Cavalieri, L. F., 308, *336*
Cerutti, P. A., 103, 104, *152*
Chai, N-C., 51, 61, 62, 63, 69, 72, 73, 76, 82, 83, 84, 86, 87, 88, 89, 93, *95*
Chaikoff, I. L., 297, 306, 329, *337*
Chamberlin, M. J., 111, 124, *155*
Champion, J. V., 293, 308, 312, *336*
Chandler, M., 51, 76, *95*
Chanet, R., 175, 184, *265*, *268*
Chase, J., 129, *150*
Cheung, A. T. W., 21, *44*
Chien, J. R., 129, *153*
Choo, K. B., 166, 202, 228, 229, 245, 249, 251, *272*, *275*
Christiansen, C., 160, 161, 162, 164, 170, 172, 187, 192, *263*, *264*, *265*
Christiansen, G., 160, 161, 162, 164, 170, 172, *263*, *265*
Christodoulou, G., 217, 219, 220, *270*
Chung, Y. S., 110, 121, *151*
Chwang, A. T., 7, 16, 17, *44*
Ciferri, O., 185, *265*
Clark, A. J., 110, 113, 114, 133, *150*, *151*, *155*
Clarke, C. H., 145, 151
Clark-Walker, G. D., 162, 169, 171, 172, 204, 217, 255, *265*, *269*, *272*
Coakley, C. J., 36, *44*
Coakley, W. T., 293, 295, 297, 304, 305, 306, 307, 308, 312, 325, 326, 327, 331, *336*, *338*
Cobon, G., 179, *270*
Cobon, G. S., 179, 180, 193, 227, 228, *263*, *271*
Coen, D., 194, 204, 216, 217, 219, 220, 222, 231, 232, 233, 234, 235, 237, 238, 239, 240, 241, 243, 245, 246, 249, 254, 262, *263*, *264*, *265*, *266*, *272*
Cohen, J. A., 119, *154*
Cohen, L. M., 201, *266*
Cohen, N. S., 321, 322, *335*
Cole, R. S., 106, 107, 128, 142, 143, 144, *155*
Collins, J., 53, *97*

Collins, J. F., 61, *95*
Collins, S. B., 293, *336*
Colson, A., 214, *267*
Colson, A.-M., 214, 221, *226*
Commerford, S. L., 297, 298, 306, 329, *338*
Conner, R. L., 334, *336*
Conti, G., 181, *265*
Cooper, P. K., 130, 131, *151*
Cooper, S., 50, 51, 52, 53, 57, 93, *95*, *96*
Copland, H. J. R. 108, 109, 138, 144, 149, *153*
Corneo, 174, 266
Corner, T. R., 288, 319, *336*, *337*
Cosson, J., 225, *275*
Cost, S. O. P., 179, *269*
Costello, D. P., 24, 33, *44*
Cottrell, S., 183, *266*
Cottrell, S. F., 188, *266*
Couder, H., 209, *272*
Court, D., 111, 124, *155*
Cowan, R. M., 298, *337*
Cozzarelli, N. R., 129, *151*
Crandall, M., 214, 215, *266*
Criddle, R. S., 168, 169, 207, 210, 221, 212, 221, 225, 230, 256, 261, 262, 263, *266*, *268*, *271*, *274*, 276
Croft, J. H., 204, *274*
Crowfoot, P. D., 158, 192, 225, *269*
Crum, L. A., 302, *337*
Cryer, D. R., 160, 162, 183, 189, 191, 192, 194, 196, 202, 206, 207, 220, 221, 233, 235, *264*, *266*, *267*, *268*
Culotti, J., 187, *268*
Cumming, D. J., 65, 66, *96*, 331, *336*
Curle, N., 15, *45*
Currie, A., 302, 305, 328, *336*
Cutler, R. G., 56, 69, *95*
Cutting, G. J., 224, 229, 245, 247, 248, 249, 252, *275*

D

Daiko, J. L., 334, *340*
Dallai, R., 30, *45*
Daneo-Moore, L., 95, *95*
Davern, C. I., 76, *97*
Davey, P. J., 222, *266*
Davidson, N., 167, *273*
Davies, D. R., 334, *336*, *338*
Davies, H. J., 15, *45*

Davies, M., 297, 299, 300, 305, *336*
Davis, N. S., 108, *151*
Davison, P. F., 308, 309, *338*
Dawes, I. W., 189, *266*
Dean, C. J., 108, 118, 127, 138, 140, 142, *150*, *151*
DeBoer, E., 315, *336*
Defais, M., 147, *151*
Dellweg, H., 101, *155*
De Lucia, P., 110, 115, 116, 129, 138, *151*
Dennis, P. P., 57, *95*
Dennis, R. E., 146, *150*
Desjardins, R., 331, *336*
Deters, D., 302, 305, 330, *336*
Deuel, T. F., 316, *338*
Deutsch, J., 194, 200, 204, 216, 217, 219, 220, 222, 231, 233, 234, 237, 238, 239, 240, 241, 243, 245, 246, 249, 254, 262, *264*, *265*, *266*, *267*, *272*
De Vries, H., 166, 220, *267*
Di Franco, A., 167, 199, 252, 253, 254 *266*, *274*
Doddema, H., 321, *335*
Donachie, W. D., 53, 57, 94, *95*, *96*
Donch, J., 117, 147, *151*
Donch, J. J., 110, 117, 120, 121, 147, *151*
Dorman, F. D., 294, *335*
Doty, P., 101, *153*, 162, *271*, *274*, 308, *336*
Douglas, G. J., 38, *45*
Douglas, H. C., 170, 173, *264*
Douglass, S., 168, 169, 207, 230, 256, 261, 262, 263, *271*, *274*
Dowben, R. M., 298, *336*
Driedger, A. A., 108, *151*
Duffus, J. H., 329, *336*
Duggan, D., 108, *150*
Dujon, B., 194, 202, 204, 217, 219, 220, 222, 225, 232, 233, 234, 235, 236, 237, 238, 239, 240, 241, 243, 244, 245, 246, 246, 248, 249, 252, 254, *263*, *264*, *266*, *272*, *274*, *275*, *277*
Dunnill, P., 297, 299, 300, 302, 305, 306, 309, 328, 329, *336*, *337*
Duntze, W., 189, *265*
Duran, A., 334, *336*
Durr, M., 333, *336*

E

Eaton, N., 233, 239, 241, *276*
Eaton, N. R., 261, *276*, 329, *337*

Ecker, R. E., 74, *96*
Eckert, R., 40, 41, 42, 43, *45*, *46*
Eddy, A. A., 332, *337*
Edebo, L., 297, 300, 301, 305, *339*
Edelman, A., 119, *154*
Edmiston, S. J., 117, *153*
Edwards, D. G., 299, *337*
Ehrlich, S. D., 163, *266*
Einstein, A., 290, *337*
Eisenberg, A. D., 288, *337*
Eisenstadt, M., 333, *339*
Ekholm, J. E., 282, *337*
Elgar, E. C., 296, *338*
Eller, A. I., 302, *337*
Elliott, J. J., 221, *266*
Ellison, S. A., 112, *151*
El'Piner, I. E., 302, 306, 309, *337*
Emanuel, C. F., 297, 306, 329, *337*
Emmerson, P. T., 114, *151*
English, K. J., 228, 229, 240, 252, *273*
Enns, R., 221, 225, *274*
Ensign, J. C., 284, *338*
Enteric, S., 175, *272*
Ephrussi, B., 159, 204, 213, 256, 257, 258, 259, 260, 262, *266*, *274*
Errera, M., 147, *151*
Errington, F. P., 74, *96*
Evans, J. E., 56, 69, *95*
Eyfjörd, J. E., 148, *151*
Eyring, H., 35, *45*

F

Fangman, W. L., 137, 138, *151*, 160, 184, 189, *272*, *273*
Fath, W. W., 175, *266*
Fauman, M., 198, 199, 200, *266*
Faures, M., 159, 162, 171, 195, 200, 263, *264*, *267*
Fauquet, P., 147, *151*
Faures-Renot, M., 200, *266*
Faye, G., 194, 198, 199, 200, 201, 202, 248, 252, 253, *264*, *266*, *267*, 269, *271*
Feiner, R. R., 112, *151*
Feldberg, R., 125, 131, *152*
Feldman, F., 227, *267*
Feldschreiber, P., 108, 127, 142, *150*
Fennell, D. J., 174, 185, *276*
Ferguson, K. A., 334, *336*
Finean, J. B., 313, 315, 322, *335*

AUTHOR INDEX

Fink, A. M., 183, 277
Finkelstein, D. B., 159, 160, 183, 189, 202, 264, 266, 267
Fisher, D., 315, 335
Fisher, G. W., 42, 45
Fisher, J., 316, 317, 318, 323, 340
Fisher, W. D., 120, 150
Fitz-James, P., 50, 97
Flavell, R. A., 165, 167, 173, 195, 196, 199, 258, 265, 267, 268, 269, 273, 275
Flury, U., 227, 266
Flynn, H., 302, 306, 337
Fogel, S., 189, 274
Follows, M., 297, 299, 300, 305, 306, 309, 328, 329, 337
Fonseca, J. R., 9, 10, 45
Fonty, G., 159, 161, 163, 164, 264, 273
Forrester, I. T., 166, 180, 191, 192, 267, 270
Forro, F., Jr., 65, 71, 96
Forster, J. L., 231, 241, 242, 243, 261, 267
Forstrum, R. J., 294, 335
Foster, J. W., 298, 337
Foury, F., 214, 226, 227, 228, 267, 275, 332, 337
Fowlks, W. L., 106, 154
Fox, C. F., 319, 341
Frank, M. E., 56, 96
Fraser, D., 298, 337
Freedman, M. L., 51, 93, 96
Freese, E., 321, 324, 337, 338
French, C. S., 297, 299, 306, 329, 337, 339
Friedenberg, D. L., 263, 264
Frontali, L., 167, 263
Fukuhara, H., 161, 163, 167, 191, 194, 198, 199, 200, 201, 202, 209, 248, 249, 252, 253, 264, 265, 266, 267, 269, 271, 274
Fukuhara, M., 253, 267
Fukuyama, J., 332, 341
Futai, M., 318, 319, 323, 337

G

Gaertner, F. H., 334, 337
Gaffey, T. A., 298, 336
Gajewska, E., 101, 153
Game, J. C., 184, 267
Gancedo, C., 334, 340

Gancedo, J. M., 334, 340
Gander, J. E., 284, 337
Ganesan, A. K., 135, 137, 138, 151
Ganesan, A. T., 139, 153
Garan, H., 311, 335
Garcia-Acha, I., 332, 341
Garvik, B., 125, 127, 152
Gascon, S., 332, 341
Gates, F. L., 101, 151
Gefter, M. L., 116, 131, 151, 153, 178, 267
Gellert, M., 118, 151
Genin, C., 200, 267
George, D. L., 147, 155
George, J., 147, 150
Getz, G. S., 159, 167, 178, 183, 193, 194, 198, 199, 201, 202, 203, 253, 265, 266, 270, 273, 275
Ghuysen, J. M., 95, 96
Gibbons, B. H., 26, 28, 36, 38, 45
Gibbons, I. R., 4, 5, 25, 26, 28, 30, 33, 36, 38, 39, 44, 45, 47
Gibor, A., 158, 267
Gillham, N. W., 194, 243, 267
Gilula, N. B., 5, 45
Gingold, E. B., 194, 196, 197, 209, 217, 220, 230, 231, 235, 240, 241, 242, 246, 257, 259, 260, 267, 270, 272, 274
Glasstone, S., 35, 45
Glauert, A. M., 283, 337
Glickman, B. W., 129, 138, 151
Goffeau, A., 214, 221, 266, 267, 268, 332, 337
Goldfinger, B., 160, 202, 249, 259, 263, 271
Goldmark, P. J., 115, 131, 151
Goldring, E. S., 162, 170, 180, 182, 190, 192, 194, 196, 202, 206, 207, 267, 268
Goldsmith, H. L., 291, 292, 295, 337
Goldstein, S. F., 28, 32, 37, 45
Goldthwait, D. A., 147, 153
Goldthwaite, C. D., 183, 189, 191, 192, 220, 221, 233, 235, 266, 267
Goodgal, S. H., 121, 151
Gordon, P., 167, 198, 199, 200, 201, 203, 267, 273
Gottlieb, P., 90, 97
Gouhier, M., 205, 206, 221, 267
Gouhier-Monnerot, M., 206, 239, 240, 267

Gould, R. K., 302, 304, 306, 307, 325, 326, 327, 334, *336*, *337*
Grandchamp, C., 169, 194, *268*
Grandchamp, S., 256, 260, 262, *266*
Granick, S., 158, *267*
Gray, J., 7, 15, 16, 31, *45*
Green, D. M., 309, *337*
Green, G. S., 106, *154*
Green, M. H. L., 117, 121, 147, 148, *151*
Greenawalt, J. W., 330, *336*, *337*
Greenberg, J., 110, 111, 117, 120, 121, 147, *151*, *152*
Griffith, D. G., 106, *154*
Griffiths, D. E., 170, 181, 217, 221, 224, 225, 226, *263*, *267*, *269*
Grimes, G. W., 185, 187, 192, *267*
Grimstone, A. V., 5, *45*
Grivell, A. R., 73, *96*
Grivell, L. A., 158, 166, 167, 219, 220, 222, *265*, *267*, *268*, *273*
Groot, G. S. P., 167, 171, 172, 173, *268*, *274*
Groot Obbink, D. J., 221, 223, 224, 226, 227, 228, 236, *265*, *268*
Gross, J. D., 115, 129, 147, *152*, *153*
Gross, M., 115, 129, *152*
Grossman, L., 101, 112, 125, 127, 128, 129, 130, 131, *151*, *152*, *153*
Grossman, L. I., 162, 170, 180, 182, 190, 192, 194, 196, 202, 206, 207, *267*, *268*
Grosveld, F. G., 195, 196, 198, *271*
Grotbeck, R. C., 220, 222, 231, 233, 241, *269*
Grovosky, M. A., 331, *337*
Grunstein, J., 115, *152*
Grzybek, H., 333, *336*
Guarnieri, M., 330, *337*
Gudas, L. I., 70, *96*
Guerineau, M., 169, 170, 209, 226, *268*, *272*
Guidotti, G., 314, *337*
Guisti, F., 30, *45*
Gunge, N., 235, *268*, *276*
Gunning, B. E. S., 333, *335*

H

Halbreich, A., 253, *273*
Hall, D. O., 330, *337*
Hall, R., M., 158, 176, 177, 179, 185, 186, 202, 206, 209, 212, 213, 221, 223, 224, 226, 227, 228, 229, 240, 244, 246, 252, *265*, *268*, *269*, *271*, *272*, *273*
Halvorson, H. O., 166, 171, 185, 187, 189, 190, 192, *271*, *274*, *275*, 329, 334, *335*, *336*
Hamilton, L., 125, 127, *152*
Hamm, L., 118, *154*
Hammond, R. C., 205, 215, *270*, *276*
Hampton, H. L., 324, *337*
Hanahan, D. J., 282, *337*
Hanawalt, P., 130, *154*
Hanawalt, P. C., 65, 66, 73, *96*, 130, 131, 151, *153*
Hancock, G. J., 7, 15, 16, 31, *45*
Handwerker, A., 203, *268*
Hansby, J. E., 181, *269*
Hardigree, A. A., 120, 121, *150*
Hariharan, P. V., 103, 104, *152*
Harm, H., 121, 122, 123, 124, 125, *152*, *154*
Harm, W., 100, 111, 121, 122, 125, *152*
Harold, F. M., 318, 319, 323, 324, *337*
Harrington, R. E., 308, *337*
Harrison, J. S., 170, *273*
Hart, R. W., 125, *152*
Hartman, P. E., 119, *154*
Hartwell, L. H., 170, 182, 183, 184, 187, 189, *263*, *268*
Harvey, E. N., 302, *337*
Harvey, R. S., 61, 74, *96*
Hasegawa, S., 108, *153*
Haslam, J. M., 158, 179, 180, 191, 192, 193, 194, 204, 217, 219, 222, 223, 256, *266*, *270*, *271*
Haslbrunner, E., 159, *274*
Hastings, G. W., 309, *339*
Hatzfeld, J., 175, *268*
Hawthorne, D. G., 170, *271*
Hedenskog, G., 302, 305, 306, 328, *339*
Heijneker, H. L., 129, 130, *151*, *152*
Hellum, J. D., 293, 294, *338*
Helmstetter, C. E., 50, 51, 52, 53, 54, 57, 70, 93, *95*, *96*
Hendler, F., 167, 253, *272*, *273*
Hereford, L. M., 182, 184, *268*
Herriot, R. M., 121, *151*
Hershey, A. D., 309, *336*
Heslot, H., 214, *268*

Hetherington, P. J., 297, 299, 300, 305, 306, 309, 328, 329, *337*
Heude, M., 175, 184, *268*
Heyting, C., 166, 167, 196, 199, 246, 252, 253, 254, 255, *266*, *268*, *274*
Hill, R. F., 111, 112, 116, 145, *152*
Hillebrandt, B., 111, *152*
Hinze, J. O., 290, 291, 293, *337*
Hiramoto, Y., 7, 44
Hirota, Y., 116, *151*
Hirsch, J., 183, 258, *276*, *277*
Hirschberg, R., 183, 189, *266*
Hixon, S. C., 209, 214, *268*
Hoffman, H., 56, 96, 169, 178, 186, *263*, *268*
Hohlfield, R., 51, *96*
Holden, M., 331, *338*
Hollenberg, C. P., 160, 161, 162, 169, 179, 181, 195, 196, 205, 207, 258, *268*, *269*, *275*
Holwill, M. E. J., 7, 8, 9, 10, 15, 16, 17, 22, 24, 30, 32, 34, 36, 37, 38, 40, 41, *44*, *45*, *46*, *47*
Homberger, H., 302, 305, 330, *336*
Hood, J. R., Jr., 50, 53, 55, 59, 74, 77, 79, 80, *97*
Hopkins, J., 43, *47*
Horii, Z., 137, *153*
Houghton, R. L., 221, 225, 226, *267*
Howard-Flanders, P., 102, 103, 106, 107, 109, 112, 113, 114, 117, 119, 120, 125, 126, 128, 133, 134, 135, 136, 137, 142, *150*, *151*, *152*, *154*
Howell, N., 220, 221, 222, 223, 229, 233, 236, 237, 239, 240, 241, 242, 244, 245, 246, *269*, *270*, *271*
Howells, L., 329, *339*
Hsia, D. Y., 298, *336*
Hsu, H., 167, 201, 203, *273*
Hsu, H.-J., 161, 199, 201, 249, 253, 265, *267*, *271*
Huang, M., 217, *269*
Huberman, J. A., 129, *152*
Hughes, D. E., 281, 283, 285, 293, 297, 298, 299, 300, 301, 302, 304, 305, 306, 307, 309, 310, 312, 328, 329, 331, *338*, *339*, *340*, *341*
Hunter, M. J., 297, 298, 306, 329, *338*
Hurwitz, J., 115, *155*
Hyams, J., 334, *338*
Hyams, J. S., 41, *45*

I

Ichikawa, H., 148, *153*
Imamura, A., 204, *275*
Indge, K. J., 333, *338*
Isaac, L., 284, *338*
Ishii, Y., 148, *152*
Iwashima, A., 176, 178, *269*
Iwo, K., 148, *153*
Iyer, V. N., 105, 106, 135, *152*, *155*

J

Jacklet, A. C., 39, *46*
Jacob, F., 147, *153*
Jagger, J., 123, 125, 145, *152*, *153*
Jahn, T. L., 9, 10, 21, *45*
Jakob, H., 195, 260, 262, *266*, *271*
Jakovic, S., 253, *273*
James, A. P., 204, *270*
James, C. J., 295, 304, 305, 306, 307, 325, 326, 327, 331, *336*, *338*
Jayaraman, J., 159, *275*
John, P., 321, *336*
John, P. C. L., 333, *335*
Johnson, B, F., 121, *152*
Joly, M., 309, *338*
Jones, N. C., 94, *96*
Josslin, R., 41, *44*
Juliani, M. H., 179, *269*

K

Kaback, H. R., 316, 317, 318, 321, 322, 323, *338*, *340*
Kaczorowski, G., 316, 317, 318, 323, *340*
Kalf, G. F., 180, *274*
Kamei, S., 249, *276*
Kaneko, H., 41, *46*
Kaneko, T., 332, *338*
Kaneshiro, E. S., 42, *45*, 334, *336*
Kant, J. A., 282, *340*
Kaplan, H. S., 102, 115, 137, 138, 139, 140, 141, *152*, *155*
Kaplan, J. C., 129, 130, *152*
Kapp, D. S., 137, 138, *152*
Kato, T., 115, 145, 148, *152*, *153*
Katz, D., 334, *338*
Katz, D. F., 16, *46*

Kaudewitz, F., 172, 203, 214, 215, 216, 220, 224, 248, 249, 264, 268, 269, 274, 277
Kawakita, M., 331, 338
Kaye, J., 296, 338
Keevert, J. B., 331, 337
Keller, S. R., 20, 45
Kellerman, G. M., 181, 214, 216, 225, 229, 269, 270, 275
Kelley, M. S., 56, 69, 96
Kelly, R. B., 129, 151, 152
Kelner, A., 121, 125, 152, 153
Kendall, D. G., 55, 96
Kiefer, B. I., 4, 45
Kinross, J., 115, 153
Kinsloe, H., 303, 338
Kirby, E. P., 147, 153
Kirk, J. T. O., 333, 338
Kitamura, K., 332, 338
Kjeldgaard, N. O., 75, 96
Kleese, R. A., 220, 222, 231, 233, 236, 241, 242, 243, 261, 267, 269
Kleisen, C. M., 166, 273
Knauf, P. A., 312, 338
Knight-Jones, E. W., 11, 45
Knudsen, J. G., 293, 336
Knutsen, G., 334, 338
Kobatake, M., 108, 153
Koch, A. L., 50, 53, 55, 56, 57, 58, 59, 60, 61, 62, 70, 72, 74, 77, 79, 80, 94, 95, 96, 97
Kohn, D., 317, 340
Kokaisl, G., 74, 96
Kolarov, J., 214, 215, 275
Kondo, S., 115, 145, 146, 148, 152, 153
Konings, W. N., 321, 335, 338, 339
Konrad, E. B., 118, 153
Kopecka, H., 161, 164, 167, 169, 199, 255, 264
Kornberg, A., 129, 150, 151, 52, 153
Kornberg, T., 116, 151
Korylak, Z., 227, 240, 269
Kosel, C., 117, 153
Kovac, L., 181, 205, 214, 215, 269, 275, 276
Krishna, G., 282, 316, 340
Kroon, A. M., 158, 265
Krulwich, T. A., 284, 338
Krupa, M., 225, 275
Krupnick, D., 194, 196, 206, 207, 267
Kruszewska, A., 204, 235, 266

Kubitschek, H. E., 50, 51, 56, 57, 70, 74, 93, 95, 96
Kuenzi, M. T., 190, 192, 269, 271, 275
Kujawa, C., 201, 202, 209, 248, 252, 253, 266, 267
Kuntzel, H., 158, 166, 269, 274
Kuo, S-C., 332, 338
Kuriyama, V., 167, 269
Kung, C., 43, 45, 46
Kusel, J. P., 333, 338
Kushner, S. R., 129, 130, 152
Kutzleb, R., 220, 269
Kuzela, S., 181, 269

L

Laidler, K., 35, 45
Laidler, K. J., 37, 46
Laipis, P. J., 139, 153
Lam, K. B., 183, 189, 266
Lamb, A. J., 181, 200, 217, 219, 220, 269, 270
Lancashire, W. E., 170, 217, 221, 224, 225, 226, 267, 269
Landman, M. D., 9, 10, 45
Landrey, J, T., 334, 336
Landry, Y., 214, 267
Lane, D., 101, 153
Lang, B., 216, 224, 277
Lang, D., 160, 269
Lardy, H. A., 314, 339
Lark, C., 50, 71, 96
Lark, K. G., 51, 61, 62, 63, 69, 71, 72, 73, 76, 82, 83, 84, 86, 87, 88, 89, 93, 95
Laskowski, W., 123, 155, 175, 266
Latarjet, R., 123, 125, 152
Laufer, J., 291, 338
Law, J., 148, 150
Lawrence, C. W., 159, 170, 274
Lawrence, N. S., 297, 299, 329, 339
Lazowska, J., 194, 198, 199, 200, 266, 267, 269, 271
Lederberg, J., 258, 277
Leedale, G. F., 286, 338
Leef, J. L., 334, 337
Le Gall, Y., 334, 340
Lehman, I. R., 118, 129, 153
Lehmann-Brauns, E., 175, 266
Leoni, L., 163, 167, 181, 263, 265
Leverett, L. B., 293, 294, 338
Levinthal, C., 308, 309, 338

AUTHOR INDEX

Lewin, A., 161, 199, 253, *271, 273*
Lewin, R. A., 40, *46*
Lewis, N. F., 108, *153*
Lien, T., 334, *338*
Lighthill, M. S., 15, *46*
Lilly, M. D., 297, 299, 300, 302, 305, 306, 309, 328, 329, *336, 337*
Lindblom, M., 302, 305, 306, 328, *339*
Lindegren, C. C., 185, *272*
Lindegren, G., 185, *272*
Linn, S., 115, 131, *151, 153*
Linnane, A. W., 158, 161, 162, 166, 176, 177, 179, 180, 181, 182, 185, 186, 191, 192, 193, 194, 196, 197, 201, 202, 204, 206, 207, 209, 210, 211, 212, 213, 214, 216, 217, 218, 219, 220, 221, 222, 223, 224, 225, 226, 227, 228, 229, 230, 231, 232, 233, 235, 236, 237, 238, 239, 240, 241, 242, 243, 244, 245, 246, 247, 248, 249, 250, 251, 252, 254, 255, 256, 257, 259, 260, 263, *265*, 266, 267, *268, 269, 270, 271, 272, 273, 274, 275, 276*
Lioetti, F. J., 313, 314, *339*
Lloyd, D., 281, 283, 285, 297, 298, 299, 300, 301, 306, 309, 310, 328, 329, 331, 332, *338, 339*
Locker, J., 167, 194, 198, 199, 201, 202, 203, *266, 70, 273*
Loken, M. R., 50, 70, *96*
Loomis, A. L., 302, *337*
Lorenz, J. R., 106, *155*
Louarn, J., 51, *95*
Louis, C., 214, *268*
Low, B., 114, *155*
Low, M. G., 315, 322, *335*
Lowndes, A. G., 41, *46*
Loyter, A., 315, *336*
Lubliner, J., 40, *46*
Luck, D. J. L., 167, *269*
Lucy, J. A., 315, *335*
Luha, A. A., 171, 205, 215, *270, 276*
Lukins, H. B., 158, 161, 162, 166, 180, 191, 192, 194, 196, 197, 201, 204, 209, 213, 216, 217, 218, 219, 220, 221, 222, 223, 224, 226, 228, 229, 230, 231, 232, 233, 235, 236, 237, 238, 239, 240, 241, 242, 243, 244, 245, 246, 247, 248, 249, 250, 251, 252, 254, 255, 257, 259, 260, *265,* *267, 268, 269, 270, 271, 272, 273, 274, 275*
Lunec, J., 18, *46*
Lusena, C. V., 170, 176, 204, *270, 277*
Lynch, E. P., 293, 294, *338*
Lynch, P. A., 298, *336*
Lynch, P. M., 298, *336*

M

McArthur, C. R., 171, 172, *272*
McClees, J. S., 321, 322, 323, *340*
McFall, E., 50, *96*
McGill, B. B., 308, *336*
McGrath, R. A., 103, 126, 127, 137, *153*
McGregor, J. L., 32, 41, *45*
MacHattie, L. A., 160, *270*
Machemer, H., 10, 11, 12, *46*
Machin, K. E., 32, *46*
Macino, G., 167, *263*
Macintosh, I. J. C., 295, *336*
MacKay, V., 131, *153*
Mackie, G. O., 41, *46*
MacQuillan, A. M., 204, *263*
McVittie, A., 43, *47*
Maag, T. A., 298, *337*
Maaløe, O., 50, 57, 65, 66, 75, *96, 97*
Maddy, A. H., 311, *339*
Magnusson, K-E., 297, 300, 301, 305, *339*
Mahler, H. R., 158, 159, 162, 181, 185, 187, 192, 195, 196, 204, 205, 206, 207, 208, 209, 213, 214, 215, 227, *264, 267, 270, 271, 272, 273, 275*
Mahler, I., 125, 127, 130, 131, *152*
Mallory, F. B., 334, *336*
Manney, T. R., 189, *265*
Marchesi, V. E., 311, *339*
Margerie-Hottingner, H., 256, 257, 258, *266*
Margulies, A. D., 110, 113, *151*
Margulis, M. A., 306, *339*
Markovitz, A., 120, *153, 155*
Marmur, J., 101, *153,* 159, 160, 162, 170, 180, 182, 183, 190, 191, 192, 194, 196, 202, 206, 207, 220, 221, 228, 233, 235, 249, 259, 263, *264, 266, 267, 268, 271, 274, 275*
Maroudas, N. G., 175, 181, 204, *271, 276*

AUTHOR INDEX

Marquis, R. E., 283, 284, 288, 319, *336*, *339*
Marr, A. G., 60, 61, 74, *96*, *97*
Martin, N., 253, *271*
Martin, R., 220, 221, *273*
Martuscelli, J., 51, *95*
Marzuki, S., 177, 179, 180, 193, *268*, *270*, *271*
Masker, W. E., 131, *153*
Mason, D. J., 77, *96*
Mason, S. G., 291, 292, *337*
Mason, T. L., 158, 225, *274*, 330, *339*
Masters, M., 51, *96*
Masurovsky, E. B., 108, *151*
Matin, A., 321, *339*
Matsushita, T., 95, *96*
Mattern, I., 113, *153*
Mattern, I. E., 111, 117, 146, *153*
Matthews, L., 101, *153*
Mattick, J. S., 176, 177, 179, 206, 209, 213, *268*, *272*
Mattingly, A., 149, *153*
Mattoon, J. R., 221, *266*, 330, *335*, *337*
Maxwell, R. J., 227, 228, *265*
Meacock, P. A., 94, *96*
Meadows, P. A., 225, *267*
Mehrjardi, M. H., 293, *340*
Mehrotra, B. D., 162, 181, 195, 196, 204, *271*
Meselson, M., 80, *97*, 174, *271*
Meun, D. H. C., 134, *155*
Mian, F. A., 192, *271*
Michaelis, G., 168, 169, 195, 202, 207, 230, 256, 261, 262, 263, *271*, *276*
Michel, F., 194, 198, 199, 200, 238, 245, *266*, *267*, *269*, *271*
Michell, R. H., 313, 314, 315, 322, *335*
Michels, C. A., 160, 202, 249, 259, 263, *264*, *271*
Miklos, G. L. G., 162, 169, 204, 255, *265*
Miles, C. A., 40, *46*
Miller, R. L., 40, 41, *46*
Milner, H. W., 297, 299, 306, 329, *337*, *339*
Minato, S., 124, *153*
Minckler, S., 185, *272*
Mirsky, R., 313, *339*
Mitchell, C. A., 221, *271*
Mitchell, C. H., 219, 221, 222, 223, 224, *265*, *270*
Mitchison, J. M., 70, 74, *97*

Miura, A., 146, 147, *153*
Mogren, H., 302, 305, 306, 328, *339*
Mohri, H., 43, *46*
Mol, J. N. M., 195, 196, 198, *271*, *273*
Molloy, P. L., 161, 162, 166, 201, 213, 220, 221, 222, 223, 224, 228, 229, 232, 238, 240, 245, 246, 247, 248, 249, 250, 251, 252, 254, 255, *269*, *270*, *271*, *272*, *275*
Molyneux, I. J., 131, *153*
Mondrich, P., 118, *153*
Monk, B. C., 221, 223, 224, 225, 226, 228, 229, *263*, *275*
Monk, M., 115, 147, *153*
Morales, M. F., 37, *46*
Morell, A. G., 311, *339*
Morgan, N, A., 329, *339*
Morimoto, H., 166, *271*
Morimoto, R., 161, 199, 253, *271*, *273*
Morimyo, M., 137, *153*
Morpurgo, G., 181, *265*
Morrison, M., 311, *339*
Mortimer, R. E., 170, *271*
Moseley, B. E. B., 101, 107, 108, 109, 138, 142, 144, 149, *153*, *155*
Mounolou, J., 205, 206, 221, *267*
Mounolou, J. C., 193, 195, *271*
Mount, D. W., 114, 117, *153*, *155*
Moustacchi, E., 169, 175, 184, 188, 191, 206, *265*, *268*, *271*, *272*, *275*, *276*
Mowshowitz, D. B., 334, *339*
Mudd, S., 119, *154*
Muhammed, A., 124, *153*
Müller, M., 329, *339*
Muller, U., 302, 305, 330, *336*
Munoz, E., 313, *339*
Munson, R. J., 145, 146, 148, *150*
Murakami, A., 7, 41, *47*
Muriel, W. J., 148, *151*
Murphy, M., 227, 228, *265*
Nadler, H. C., 298, *336*
Nagai, H., 204, *272*
Nagai, S., 204, *272*
Nagasaki, S., 332, *341*
Nagata, T., 80, *97*

N

Nagley, P., 158, 166, 176, 177, 179, 180, 182, 185, 186, 191, 192, 193, 194, 196, 197, 201, 202, 204, 206, 207,

AUTHOR INDEX

209, 213, 217, 219, 228, 229, 232, 240, 245, 249, 250, 251, 252, 256, 257, 259, 260, 263, 267, 268, 270, 272, 275
Naitoh, Y., 40, 41, 42, 43, 46
Nass, N. M. K., 158, 187, 272
Nathans, D., 161, 272
Nečas, O., 331, 339
Needleman, R. B., 227, 275, 330, 339
Negrotti, T., 203, 273
Nekhorocheff, J., 166, 271
Netter, P., 194, 204, 216, 217, 219, 220, 222, 231, 233, 234, 235, 237, 238, 239, 240, 241, 243, 245, 246, 249, 254, 262, 264, 265, 266, 267, 272
Newlon, C. S., 184, 272
Ng, M. H., 313, 339
Nicolaieff, A., 159, 162, 181, 195, 196, 264
Nicolson, G. L., 310, 340
Nordan, H. C., 108, 150
North, P. F., 293, 308, 312, 336
Nyborg, W. L., 293, 302, 312, 338, 339, 341

O

O'Connor, R. M., 171, 172, 272
O'Connor, R. N., 172, 272
Ogawa, H., 112, 113, 153
Ogawa, K., 43, 46
Oginsky, E. L., 106, 154
Ogur, M., 185, 272
Ojala, D., 167, 272
Okabe, K., 204, 275
Okazaki, R., 133, 135, 154
Okazaki, T., 133, 135, 154
Oosterbaan, R. A., 130, 152
Ormerod, M. G., 140, 151
Orris, S. E., 39, 46
Oschman, J. L., 42, 46
Ostrovskaya, M. V., 8, 44
Ottolenghi, P., 205, 276
Ou, L. T., 283, 339
Oullet, L., 37, 46
Ovenall, D. W., 309, 339
Oxender, D. L., 289, 339

P

Pachler, P. F., 50, 72, 96, 97
Padmanabam, G., 167, 272

Painter, R. P., 51, 60, 61, 74, 96, 97
Palek, J., 313, 314, 339
Palleschi, C., 167, 263
Pannekoek, H., 130, 152
Paoletti, C., 169, 209, 268, 272
Pardee, A. B., 50, 70, 95, 96
Parducz, B., 10, 46
Pareni, F., 333, 339
Parker, J. H., 330, 335
Parks, L. W., 179, 272
Parrish, G., 108, 150
Paterson, M. C., 129, 147, 150
Patterson, M. C., 103, 154
Patzer, J., 167, 272
Paul, D. H., 41, 46
Pauli, R. M., 169, 178, 263
Pauling, C., 118, 138, 151, 154
Payne, C., 330, 337
Payne, J. I., 119, 154
Peacey, M., 147, 153
Peacocke, A. R., 325, 339
Peberdy, J. F., 332, 339
Pedersen, H., 5, 46
Perkins, H. R., 285, 340
Perlman, P. S., 185, 187, 192, 196, 204, 205, 206, 207, 209, 213, 214, 240, 245, 267, 270, 271, 272, 273
Perrodin, G., 193, 204, 271, 274
Person, C., 204, 262, 273
Peters, P. D., 10, 30, 45
Petes, T. D., 160, 170, 189, 273
Petrochilo, E., 194, 195, 204, 216, 217, 219, 220, 222, 230, 231, 233, 234, 237, 238, 239, 240, 241, 243, 245, 246, 249, 254, 262, 264, 265, 266, 271, 272
Pettijohn, D., 130, 154
Pfenniger, W., 290, 339
Phillips, A. W., 119, 154
Phillips, D. M., 43, 46
Phillips, D. R., 311, 339
Pierucci, O., 51, 97
Pinon, R., 190, 273
Piperno, G., 159, 162, 163, 171, 181, 195, 196, 263, 264, 265, 273
Pironneau, O., 16, 46
Pitelka, D. R., 17, 46
Pittman, D., 175, 204, 273, 276
Pivonka, P. R., 28, 45
Plaskitt, A., 334, 336
Plattner, H., 42, 46

Pleger, G. L., 331, *337*
Plummer, D. T., 220, 222, *271*
Poole, R. K., 332, *339*
Postma, A. W., 318, *341*
Pouwels, P. H., 130, *152*
Powell, E. O., 60, 74, *96*, *97*
Powelson, D. M., 77, *96*
Prakash, L., 175, *273*
Prazno, W., 204, *273*
Preston, J. F., 333, *339*
Price, C. A., 333, *339*
Pringle, J. R., 187, *268*
Pritchard, N. J., 325, *339*
Pritchard, R. H., 51, 53, 57, 73, 89, 94, *96*, *97*
Prunell, A., 161, 164, 167, 169, 199, 211, 255, *264*, *273*
Puglisi, P. P., 228, *273*
Putrament, A., 204, *273*

R

Rabinowitz, M., 158, 159, 161, 167, 169, 176, 178, 183, 193, 194, 198, 199, 200, 201, 202, 203, 249, 253, *265*, *266*, *267*, *269*, *270*, *271*, *272*, *273*, *274*, *275*
Racker, E., 331, *340*
Radloff, R., 162, *273*
Radman, M., 147, *151*
Rahn, O., 55, *97*
Randall, Sir John., 43, *47*
Rajanapo. W., 200, *269*
Ranganathan, B., 175, *273*
Rank, G. H., 204, 205, 209, 220, 221, 231, 233, 257, 262, *264*, *273*
Rao, A., 230, 262, *274*
Ray, E., 334, *336*
Reca, M. E., 332, *335*
Reed, P. W., 314, *339*
Reid, B. J., 187, *268*
Reid, J. J., 303, *338*
Reijnders, L., 166, 167, 220, *267*, *273*
Reno, D., 136, 137, *152*
Reno, D. L., 133, 135, *154*
Reynolds, A. J., 19, *46*
Rice, S. A., 308, *336*
Richards, O. G., 308, *339*
Richardson, C. C., 116, 129, *150*
Richardson, E., 293, 295, *340*
Richmond, M. H., 61, *95*

Rikmenspoel. R., 39, *46*
Rintala, D. R., 32, 40, *44*
Robberson, D., 167, *273*
Robinson, D. G., 333, *340*
Robreau, G., 334, *340*
Rodbell, M., 282, 316, *340*
Rodgers, A., 302, *340*
Rodriguez, R. L., 76, *97*
Rogers, H. J., 285, *340*
Roman, H., 256, 257, 258, *266*
Roodyn, D. B., 158, 204, *273*
Rooney, J. A., 302, *340*
Rörsch, A., 111, 113, 114, 117, 119, 121, 129, 138, 146, *151*, *153*, *154*, *155*
Rose, A. H., 170, *273*
Rose, J. A., 313, *340*
Rosen, B. P., 319, 321, 322, 323, *340*, *341*
Rosenberg, B. H., 54, *97*, 308, *336*
Rosenberg, E., 90, 91, 92, *97*
Rosenberger, R. F., 334, *338*
Roth, R., 190, *269*
Rothstein, A., 312, *338*
Rupert, C. S., 121, 122, 123, 124, 125, *151*, *152*, *154*
Rupp, W. D., 128, 133, 134, 135, 136, 137, *152*, *154*
Russel, M., 137, 138, *151*
Ruttenberg, G. J. C. M., 158, *265*
Ryan, R., 167, *272*
Rytka, J., 222, 227, 228, 229, 232, 240, 245, 249, 251, 252, *265*, 270, *272*, *273*

S

Sachs, H., 333, *340*
Sager, R., 158, *273*
Sakabe, K., 133, 135, *154*
Salton, M. R. J., 313, *339*
Salts, Y., 190, *273*
Saltzgaber, J., 181, *274*
Sanders, J. P. M., 161, 162, 166, 167, 171, 172, 173, 196, 199, 246, 252, 253, 254, 255, *266*, *268*, *273*, *274*
Sanghavi, P., 169, *274*
Sarcoe, L. E., 171, 215, *270*
Satir, P., 3, 5, 6, 24, 25, 28, 29, 30, 41, *45*, *46*, *47*
Saunders, G., 217, *276*

AUTHOR INDEX 355

Saunders, G. W., 194, 196, 217, 220, 230, 231, 235, 240, 241, 242, 243, 246, 259, 260, *267*, *270*, *274*
Schaechter, M., 50, 53, 55, 57, 58, 59, 60, 62, 65, 66, 74, 77, 79, 80, 94, *96*, *97*
Schafer, K. P., 166, *274*
Schatz, G., 158, 159, 181, 225, *274*, 330, 331, *339*, *340*
Scheinberg, I. H., 311, *339*
Schildkraut, C. L., 162, *274*
Schlenk, F., 334, *340*
Schlösser, U. G., 333, *340*
Schneller, J. M., 167, *274*
Schooley, C. N., 17, *46*
Schor, M. T., 313, *339*
Schwab, R., 215, *274*
Schweizer, E., 171, 185, 187, *274*, *275*
Schweyen, R. J., 203, 214, 220, 248, 249, *268*, *269*, *274*, *277*
Scragg, A. H., 166, *271*
Scully, D. B., 300, *340*
Sebald, W., 215, *274*
Sebald-Althaus, M., 214, *277*
Sedgwick, S. G., 148, *154*
Seeberg, E., 128, *154*
Seeman, P., 289, *340*
Sena, E. P., 189, *274*
Serianni, R. W., 140, *151*
Serrano, R., 334, *340*
Setlow, J. K., 100, 101, 102, 103, 107, 121, 122, 125, 126, 127, *150*, *154*
Setlow, R. B., 100, 102, 103, 107, 111, 112, 122, 123, 125, 126, 127, 129, 147, *150*, *152*, *154*
Shannon, C., 221, 225, 230, 262, *274*
Sharpe, J. E. E., 297, 298, 328, *340*
Shearman, C. W., 180, *274*
Sheetz, M. P., 315, *340*
Sherman, F., 159, 170, 179, 182, 203, 204, 257, 259, *274*
Shimada, K., 17, *46*, *47*, 112, 113, *153*
Shneyour, A., 297, 298, 329, *340*
Shockman, G. D., 95, *95*
Short. S. A., 316, 317, 318, 321, 323, *340*
Sicurella, N. A., 145, *155*
Siegel, E. C., 113, *154*
Silverman, G. J., 108, *151*
Silverstein, S. C., 316, 317, 318, 323, *340*
Silvester, N. R., 7, 32, 36, 37, *45*, *47*
Simchen, G., 190, *273*
Simpson, L., 214, *274*

Simson, E., 111, 116, 119, 120, *152*
Sinclair, J. H., 169, *274*
Singer, S. J., 289, 310, 313, 314, 315, *340*
Singla, C. M., 41, *46*
Sinsheimer, R. L., 101, *154*
Sinskey, A. J., 309, 328, *335*
Skavronskaya, A. G., 113, *155*
Slater, E. C., 214, *264*
Sleigh, M. A., 2, 7, 9, 10, 13, 16, 20, 22, 23, 25, 30, 41, *44*, *45*, *46*, *47*
Slonimski, P. P., 159, 162, 163, 167, 169, 170, 171, 182, 193, 194, 195, 198, 199, 200, 202, 204, 216, 217, 219, 220, 222, 226, 228, 230, 231, 232, 233, 234, 235, 236, 237, 238, 239, 240, 241, 243, 244, 245, 246, 248, 249, 252, 253, 254, 255, 262, 263, *263*, *264*, *265*, *266*, *267*, *268*, *269*, *271*, *272*, *274*, *277*
Sloof, P., 166, *273*
Smetana, K., 331, *336*
Smigan, P., 181, *269*
Smirnov, G. B., 113, *155*
Smith, D., 171, 187, *275*
Smith, D. G., 243, *275*
Smith, H. O., 161, *272*
Smith, K. C., 101, 102, 115, 133, 134, 135, 137, 138, 139, 140, 141, *151*, *152*, *154*, *155*
Smith, K. E., 187, 192, *264*
Snyder, J. R., 220, 222, 231, 233, 236, 241, *269*
Soll, L., 116, *150*
Somlo, M., 225, *275*
Sora, S., 185, *265*
Spencer, J. H., 195, 196, 198, *271*
Spithill, T. W., 221, 223, 224, 226, 228, *268*
Spoerl, E., 334, *340*
Spurlock, L. A., 307, *340*
Sriprakash, K. S., 166, 202, 222, 224, 228, 229, 232, 240, 245, 247, 248, 249, 250, 251, 252, *270*, *272*, *275*
Srivastava, K. C., 243, *275*
Staas, W. H., 307, *340*
Staehelin, L. A., 317, 318, 319, 320, 322, 323, 324, *335*
Stahl, A., 167, *274*
Stahl, F. W., 174, *271*
Stapleton, G. E., 120, *150*
Starr, P. R., 179, *272*

Steck, T. L., 282, 311, 313, 314, 318, 322, 323, *340*
Stenderup, A., 160, 162, 187, 192, *263*, *264*, *265*
Stent, G. S., 50, *96*
Stephens, R. E., 4, 43, *47*
Stevens, B, J., 169, *274*, *275*
Stewart, G., 313, 314, *339*
Stockenius, W., 283, *335*
Storey, B. T., 333, *338*
Storm, E., 183, 189, *266*
Storm, E. M., 228, *275*
Strauss, F., 161, 164, *264*
Streyrer, U., 249, *274*
Strominger, J. L., 284, *338*
Stuart, K. D., 221, *275*
Subik, J., 214, 215, *275*
Suda, K., 204, 220, 249, *275*
Sugimoto, K., 133, 135, *155*
Sugimura, T., 204, *275*
Sugino, A., 133, 135, *155*
Summers, K. E., 25, *47*
Sutera, S. P., 293, *340*
Sutherland, B. M., 111, 124, *155*
Suzuki, K., 137, *153*
Svihla, G., 334, *340*
Sweet, D. M., 107, 109, 149, *155*
Swift, H., 158, 159, 178, 193, *273*, *275*
Swift, H. H., 200, *266*
Szybalski, W., 105, 106, *152*, *155*

T

Tabak, H. F., 171, 172, *275*
Tait, A., 331, *336*
Takahashi, K., 7, 41, *47*
Tampion, W., 315, *335*
Tanabe, S., 108, *153*
Tate, J. R., 241, 242, 243, *270*
Tauro, P., 171, 187, *275*
Taylor, A. L., 120, *155*
Taylor, C. W., 331, *341*
Taylor, G. I., 19, *47*, 290, 292, 296, *340*, *341*
Teathes, R., 94, *96*
Tecce, G., 159, 162, 181, 195, 196, *264*, *265*
Tewari, K. K., 159, *275*
Thacker, J., 302, 304, *341*
Theriot, L., 112, 113, 119, 120, *152*
Thienen, V. A., 318, *341*

Thiery, J., 163, *266*
Thomas, C. A., 160, *270*
Thomas, D. Y., 219, 220, 221, 230, 233, 241, 243, *275*, *276*
Thompson, N., 74, *96*
Thrasher, K., 309, 328, *335*
Thuring, R. W. J., 160, 161, *268*
Tiboni, O., 185, *265*
Timasheff, S. N., 163, *264*
Tingle, M. A., 190, *275*
Tipper, D. J., 284, *338*
To, K, 123, *154*
Tomizawa, J., 112, 113, 146, 147, *153*
Toner, J. J., 331, *341*
Town, C. D., 102, 115, 137, 138, 139, 140, 141, *155*
Tracy, M. V., 331, *338*
Trembath, M. K., 194, 196, 210, 211, 212, 218, 219, 220, 221, 222, 223, 224, 225, 226, 228, 229, 232, 233, 235, 240, 241, 242, 245, 246, 247, 248, 249, 251, 252, 259, 260, *266*, *268*, *269*, *270*, *272*, *274*, *275*, *276*
Trotter, C. D., 120, *155*
Tsai, M., 168, 169, 207, 256, 261, 263, *271*
Tsuchiya, T., 319, *341*
Tsukagoshi, N., 319, *341*
Tuck, E. O., 19, *47*
Tuppy, H., 159, *274*
Tzagoloff, A., 226, 227, 228, 249, *274*, *275*, 305, 330, *339*, *341*

U

Uchida, A., 204, 220, 249, 259, *275*
Unders, G., 54, *97*
Upholt, W. B., 216, *275*
Uretz, R. B., 120, *155*
Uruburu, F., 332, *341*

V

Van Bruggen, E. F. J., 160, 161, 162, 169, 179, 181, 195, 207, 258, *268*, *269*
Van den Hamer, C. J. A., 311, *339*
Van De Putte, P., 111, 113, 114, 119, 121, *155*
Van Der Kamp, C., 119, *154*
Van Dillewijn, J., 111, *155*
Van Kreijl, C. F., 195, 196, 258, *269*, *275*

Van Ommen, G. J. B., 167, *268*
Van Sluis, C. A., 129, 138, *151*, *155*
Venema, G., 130, *152*
Vidova, M., 205, *276*
Viehhauser, G. L., 181, *276*
Vielmetter, W., 51, *96*
Villa, V., 166, *271*
Villamueva, J. R., 285, 332, *335*, *341*
Vincent, W. S., 70, *97*
Vinograd, J., 162, *273*
Votta, J. J., 21, *45*

W

Wacker, A., 101, *155*
Wakabayashi, K., 236, 237, 239, 243, 249, *276*
Walker, A. C., 117, *153*
Wall, B. J., 42, *46*
Wallach, D. F. H., 298, 316, *341*
Wallis, O. C., 205, *276*
Walsh, C. T., 316, 317, 318, 323, *340*
Walsh, J. M., 263, *264*
Wang, D. I. C., 309, 328, *335*
Ware, G. C., 284, *338*
Waring, M. J., 106, *155*
Warm, J. V. B., 313, *340*
Warner, F. D., 3, 4, 5, 6, 28, 29, 30, *47*
Warr, J. R., 43, *47*
Warr, R., 43, *47*
Waters, R., 175, *276*
Watts, R., 313, 315, *335*
Waxman, M. F., 233, 239, 241, 261, *276*
Webb, R. B., 106, *155*
Weber, M. M., 331, *341*
Wechsler, J. A., 116, *151*
Weigel, P., 221, *266*
Weijers, P-J., 161, 162, 171, 172, 196, 253, *274*, *275*
Weill, L., 204, 235, 236, 237, 238, 241, 243, 244, 245, 246, *266*
Weinblum, D., 101, *155*
Weiner, J. H., 318, 319, 323, *341*
Weinken, A., 333, *336*
Weislogel, P. O., 181, 182, 203, *276*
Weiss, L., 294, *341*
Weiss-Brummer, B., 248, 249, *274*
Welch, J. W., 189, *274*
Wells, J. R. M., 188, *276*
Werbin, H., 124, *153*
Werkheiser, D., 204, *276*

Wertheimer, S. A., 65, 71, *96*
Westenbroek, C., 111, 119, 121, *155*
Weth, G., 202, *276*
Wharton, D. C., 330, *339*
Whatley, F. R., 321, *336*
Wheelis, L., 210, 211, 212, 221, 225, *266*, *268*, *274*, *276*
White, W. E., 209, *268*
Whittaker, P. A., 171, 205, 215, *270*, *276*
Wickner, R. B., 175, *276*
Wilde, C. E., 133, 135, 136, 137, 152, *154*
Wilkie, D., 158, 175, 181, 203, 204, 217, 219, 220, 221, 230, 233, 239, 241, 242, 243, *271*, *272*, *273*, *275*, *276*
Willetts, N., S., 114, *155*
Williams, A. R., 293, 312, 322, *336*, *341*
Williams, D. E., 41, *46*
Williams, R. W., 103, 126, 127, 137, *153*
Williamson, D. H., 170, 174, 175, 181, 182, 184, 185, 188, 191, 204, 263, 265, *272*, *273*, *276*, 332, *337*
Williamson, J P., 50, 53, 55, 59, 74, 77, 79, 80, *97*
Wilson, F., 175, *273*
Wimpenny, J. W. T., 281, 283, 285, 297, 298, 299, 300, 301, 306, 309, 310, 328, *338*, *340*
Winet, H., 21, *44*
Wintersberger, E., 176, 177, 181, *276*
Wintersberger, U., 177, 181, 183, 258, *276*, *277*
Wiseman, A., 299, *337*
Witkin, E. M., 107, 110, 115, 117, 119, 121, 145, 146, 147, 148, *152*, *155*
Woldringh, C. L., 63, 76, 79, *97*
Wolf, K., 202, 203, 214, 216, 224, 232, 236, 237, 239, 241, 245, 248, 252, *264*, *266*, *277*
Wolstenholme, D. R., 158, 159, *275*
Work, T. S., 158, *263*
Wright, L., 7, *44*
Wright, M., 115, *155*, 205, *276*
Wright, R. E., 258, *277*
Wu, T. Y., 7, 16, 17, 20, *44*, *45*
Wyckoff, R. W. G., 101, *155*

Y

Yamamoto, S., 332, *338*, *341*

Yamamoto, Y., 332, *338*
Yanagashima, N., 204, *272*
Yao, M-C., 331, *337*
Yasui, A., 123, *155*
Yielding, K. L., 209, 214, *268*
Yoeman, L. C., 331, *341*
Yoshida, T., 17, *46, 47*
Yotsuyanagi, Y., 178, *277*
Young, E., 50, *97*
Youngs, D. A., 137, 138, *155*
Yurzina, G. A., 8, *44*

Z

Zanders, E. D., 170, 217, 226, *267*
Zaritsky, A., 51, 73, 89, *97*
Zeman, L., 170, 176, *277*
Zimm, B. H., 308, *337*
Zinker, Z., 183, 189, *266*
Zuchowski, C., 51, *97*
Zulch, G., 227, *275*
Zusman, D., 90, 91, 92, *97*
Zwenk, H., 111, 113, 114, 117, 138, 146, *151, 153, 155*

SUBJECT INDEX

A

Abfrontal cilium of *Mytilus* sp., 10
Abrasives, use of to disrupt microbes, 287
Acanthamoeba castellanii, disruption of, 293
Accessibility of enzymes in bacterial membrane vesicles, 318
Accuracy, temporal, of the DNA synthesis cycle in bacteria, 80
Acridine, use of in *petite* induction, 204
Action of ethidium bromide on mtDNA in yeasts, 206
Active bending model of flagellar movement, 33
Activity, enzyme, of microbial extracts, 324
Acylglycerols, presence of in erythrocyte vesicles, 315
Adenine-thymine-rich regions in yeast mtDNA, 164
Adenosine triphosphatase, action in ciliary movement, 35
  activity of bacterial membrane vesicles, 317
  as a source of energy in flagellar action, 37
  ATPase complex in flagellar activity, 37
  concentration, effect on flagellar action, 39
  consumption in ciliary action, 35
  effect of venturicidin on, 225
  hydrolysis of by axonemes in cilia, 39
Adipocytes, disruption of, 316
  preparation of membranes from, 282
Agents, *petite* inducing, types of, 204
Alcohol dehydrogenase, microbial, effect of cavitation on, 307
Algal walls, strength of, 285
Alkaline phosphatase, release of from bacteria by osmotic shock, 289

Allelism test in yeast antibiotic resistance, 218
Allomyces spp., chemotaxis in gametes of, 40
  swimming velocity of gamete of, 23
Alpha factor, yeast, effect of on mtDNA synthesis, 189
Amber nonsense mutations in bacteria, 115
Amino acid, degradation by free radicals, 307
  transport of into bacterial membrane vesicles, 321
Amino-acyl-tRNA genes in *petite* mtDNAs, 201
Amoebae, disruption of, 296
Amplitude as a parameter in flagellar movement, 32
Amplitude of cilia and flagella, 3
Anaerobically irradiated bacteria, DNA repair in, 141
Anaerobiosis, effect of on yeast mtDNA synthesis, 192
Analysis of individual zygotic clones in yeast, 240
Anisotropy in microbial membranes, 312
Anoxic irradiation of bacteria, 140
Antibiotic-resistance genes, in yeasts, location of, 194
Antimycin A, resistance in *Saccharomyces cerevisiae*, 220
  resistance in yeast, 221
Antiplectic metachronal wave patterns in cilia, 12
APV homogenizer, pressure exerted in, 297
Arc-line waveforms in cilia, 7
Arrangement, physical, of sequences in *petite* mtDNA, 202
Arthrobacter crystallopoites, morphology of, 284

Ascites cells, disruption of, 297
Aspergillus oryzae, disruption of, 331
Autoradiography, pulse, of slowly dividing bacterial cells, 63
Axonemal structures in cilia, function of, 24
9 + 2 complex in cilia and flagella, 4
Axoneme, nature of in microbial cilia, 3

B

Bacillus megaterium, disruption of, 284
 stretching of, 284
Bacillus subtilis, membrane vesicles of, 321
 transforming DNA, short patch repair of, 130
Bacteria, engaged in DNA synthesis in a population, distribution of, 83
 halophilic, disruption of, 282
 models for cell division and chromosome replication in, 53
 preparation of vesicles of, 322
 repair of damaged DNA in, 99
Bacterial genome, properties of, 100
Bacterial membrane vesicles, preparation of, 316
Bacterial walls, and disruption, 283
Bactericidal effect of sunlight, 100
Balanced growth of bacteria and the cell cycle, 58
Ball mill, use of in microbial disruption, 328
Ballotini beads, use of in microbial disruption, 330
Basal structures of flagella, need for retention of to obtain activity, 39
Bases, distribution of in yeast mtDNA, 162
 re-insertion of, during repair of bacterial DNA, 130
Batch sonication of micro-organisms, 325
Bead mills, use of to disrupt microbes, 287
Beat frequency, as a parameter in computer simulation of flagellar movement, 32
 dependence of on ATP concentration, 38
 of flagella as a function of viscosity, 4, 39
 of microbial cilia, 3
Beat pattern of gill cilia, of *Elliptio* sp., 30
Bend formation, initiation of in cilia, 27
 mechanism of in cilia, 24
Bends, flagellar, and microbial motility, 2
Berenil, use of in *petite* induction, 204
Bidirectional replication of bacterial chromosomes, 75
Bifunctional alkylaion of bacteria DNA, 105
Bimodal distribution of bacterial division and DNA replication, 85
Binding proteins, release of from bacteria by osmotic shock, 289
Biophysical aspects of ciliary and flagellar motility, 1
Blastocladiella sp., swimming velocity of spore of, 23
Blenders, use of to disrupt microbes, 287
Blood-circulating devices, and erythrocyte breakage, 294
Blue-green algae, disruption of, 287
Boundary layers of microbes, 280
Bracken spermatozoids, ciliary control in, 41
Brain cells, disruption of, 297
Braun homogenizer, use of in microbial disruption, 301
Breakage, controlled, of eukaryotic micro-organisms, 328
 of microbial cells, nature of, 279
 of microbial cells, principles of, 288
Brettanomyces anomalus, respiratory-deficient mutants of, 215
Bromouracil, incorporation of by bacteria, 73
Brucella abortus, disruption of, 299
Bubble motion, and breakage of microbial walls, 303
Bubbles, cavitation, formation of in ultrasonic microbial disruption, 302
Bulgy deformations of membrane, and microbial disruption, 293
Buoyant density of mtDNA from *petite* strains of yeast, 198
 of yeast mtDNA, 159, 163

C

Calcium-affinity sites on cilia, 42
Calcium as an activator of flagellar action, 38
Calcium ions, and flagellar action, 41
  concentration of by bacterial membrane vesicles, 321
Candida parapsilosis, petite-negative strains of, 214
  properties of mtDNA of, 171
C. utilis, disruption of by wet milling, 302
Candida spp., preparation of protoplasts of, 332
Carbomycin resistance in yeast, 218
Carbon source, effect of on induction of *petite* yeasts by ethidium bromide, 205
Caryokinesis in the cell cycle, 50
Catalase, microbial, effect of cavitation on, 307
Cavitation, nitrogen, and microbial disruption, 298
  unloading, and disruption of microbes by ultrasonics, 303
Cell breakage, microbial, techniques for, 279
Cell cycle, bacterial, computer simulation of, 58
  in prokaryotes, 50
  period of occupied by DNA synthesis, 50
  yeast, synthesis of mtDNA during, 187
Cell division and DNA initiation in *Myxococcus xanthus*, 92
  and initiation of chromosome replication in bacteria, 49
  in bacteria, lack of control of DNA initiation over, 86
  in bacteria, models for, 53
  in bacteria, variability between and nuclear division, 76
  relation of to nuclear division in bacteria, 78
  role of DNA initiation in bacterial, 57
Cell extracts, liability of components of in microbial disruption, 305
Cell physiology, effect of on mtDNA synthesis in yeast, 189

Cells, bacterial, engaged in DNA synthesis, distribution of, 83
  microbial, in flow systems, hydrodynamics of, 291
Cell size, bacterial, relation of to rate of protoplasm synthesis, 61
  distributions in a bacterial population, 62
Cellular origin of components involved in yeast mtDNA synthesis, 180
Cellulose in algal walls, 285
Cell walls, microbial, strength of, 281
Central sheath in cilia and flagella, 3
Chaikoff press, use of in microbial disruption, 295
Chemical damage, isolation of bacterial mutants with increased sensitivity to, 111
  to bacterial DNA, 101
Chemical inactivation of microbial cell components, 306
Chemical modification of microbial limiting layers, 333
Chlamydomonas reinhardii, non-motile mutant of, 43
  preparation of protoplasts of, 333
Chlamydomonas sp., ciliary control in, 41
Chemotaxis in gametes, 40
Chloramphenical, effect of on mitochondrial protein synthesis, 217
  effect of on synthesis of mtDNA in yeast, 180
  resistance in yeast, 221
Chlorella pyrenoidosa, disruption of in the French press, 299
  strength of wall of, 285
Chlorella spp., preparation of protoplasts of, 333
Chromophore of reactivating enzyme in bacteria, 123
Chromosomal location of DNA repair genes in *Escherichia coli*, 120
Chromosome regulation in *Myxococcus xanthus*, precision of, 90
Chromosome replication, bacterial, variation in initiation of, 69
  cell division and initiation of in bacteria, 49
  in bacteria, continuity of, 75
  in bacteria, models for, 53

Chromosome replication—*cont.*
  in bacteria, models for relationship of, 82
  in the cell cycle, 51
Cilia, microbial, structure of, 3
Ciliary motility of microbes, biophysical aspects of, 1
Ciliary motion, control of, 40
Ciliary necklace in cilia and flagella, 6
Ciliary reversal in micro-organisms, 41
Cine-photography, use of to study ciliary movement, 7
Circles, omicron, of DNA associated with yeast mitochondria, 169
Circularity of mtDNA in yeasts, 202
  of yeast mtDNA, 161
Clones, zygogitic, analysis of individual in yeast, 240
Coenocytic mycelium of fungi, 285
Colloid mills, use of in microbial disruption, 301
  use of to disrupt micro-organisms, 328
Colonies, bacterial, micromanipulation of cells from, 71
Comb plates, *Pleurobranchia* spp., movement of, 13
Comminution, mechanical, during microbial degradation, 307
Comparative study of the action of ethidium bromide and of other *petite*-inducing agents, 213
Computation of the fraction of cells of a size class engaged in DNA replication, 81
Computer simulation of ciliary movement, 31
  on the bacterial cell cycle, 58
Conformational changes in macromolecules in cilia, 25
Continuous disintegration of microbial suspensions, 304
Controlled breakage of eukaryotic micro-organisms, 328
Control of flagellar motion, 40
Convoluta sp., swimming velocity of, 23
Cooling, importance of in microbial cell disruption, 306
Cooper–Helmstetter model for the cell cycle, 52
Co-ordinate control of chromosome replication and cell division in bacteria, 88

Correlation of events in cell cycles, 54
Correndonucleases, and repair of bacterial DNA, 127
Coupling of chemical and mechanical action in flagella, 40
Courette flow, and microbial disruption, 292
Covalent binding of ethidium bromide to yeast mtDNA, 208
Covalent cross-linking of bacterial DNA, 101
Creeping flow, equations for and ciliary movement, 15
Crithidia fasciculata, disruption of, 297
  removal of ergosterol from, 333
Crithidia oncopelti, beating of isolated flagella of, 32
  effect of calcium ions on flagellar action by, 42
  effect of temperature on ciliary activity of, 36
  reversal of motion of, 40
Crithidia sp., swimming velocity of, 23
Critical size of bacteria for DNA initiation, 81, 85
Cross-bridge cycles in flagellar movement, 33
Cross-link damage, repair of in bacterial DNA, 142
Cross-linking in bacterial wall peptidoglycans, 283
  in cilia, 28
Cross-resistance to antibiotics in yeast, 225
Ctenophores, comb plates of, 15
Curvature, opposite, development of in flagella, 28
Cycle of DNA synthesis in bacteria, temporal accuracy of, 80
Cycloheximide, effect of on synthesis of mtDNA in yeast, 182
  potentiating effect of on induction of *petite* yeasts by ethidium bromide, 205
Cytochrome *b* mutants of *Saccharomyces cerevisiae*, 228
Cytochrome oxidase mutants of *Saccharomyces cerevisiae*, 227
Cytokinesis in the cell cycle, 50
Cytological study of yeast mtDNA, 159
Cytosine-cytosine dimers in bacteria, photoreactivation and, 122

Cytosine-thymine dimers in bacteria, photoreactivation and, 122

D

Damage, ionizing radiation, repair of in bacteria, 137
Damaged DNA, repair of in bacteria, 99
Dark repair of damaged bacterial DNA, 100
 of yeast mtDNA, 174
Decompression, explosive, and microbial disruption, 298
Defective polymerase, DNA, mutants in bacteria, 116
Deformations of droplets, and microbial disruption, 292
Degradation of microbial DNA, 308
 of polymers during microbial disruption, 325
 sonochemical, of microbial enzymes, 307
Degradative enzymes, release of from bacteria by osmotic shock, 289
Dehydrogenase in bacterial membrane vesicles, 318
Deletion mutants, yeast, *petites* as, 193
Denaturable sites in yeast mtDNA, 165
Denaturation maps of yeast mtDNA, 162
 of enzymes at low temperature, possible operation of in flagella, 36
Density of mtDNA in different yeasts, 172
Deoxycholate, use of in microbial disruption, 331
Deoxyribonucleic acid, content of yeasts, 185
 microbial, communition of during disruption, 308
 polymerases in yeast mitochondria, 177
 polymerase I, action of in Type II repair of bacterial DNA, 139
 polymerase II of bacteria, 116
 polymerase III of bacteria, 116
 repair of damaged in bacteria, 99
 synthesis in bacteria, temporal accuracy of cycle of, 80
 synthesis in *Myxococcus xanthus*, 91
 synthesis in the cell cycle, 50
 yeast mitochondrial, structure, synthesis and genetics of, 157

Deoxyribonucleoside triphosphates, synthesis of yeast mtDNA from, 175
Deterministic model for the bacterial cell cycle, 56
Detergent, effect of on spermatozoa flagella, 26
Dexioplectic wave patterns in cilia, 12
Diacylglycerol kinase, presence of in erythrocyte membranes, 315
Diameter of cilia, 16
Diaplectic metachrony in cilia, 11
Didinium sp., swimming velocity of, 23
Digestion of microbial membranes, 311
Dimensions of microbial cilia and flagella, 3
Dimers, DNA, role of in producing cancerous lesions in *Poecelia formosa*, 125
 numbers formed in each bacterial genone by irradiation, 107
Dimethylsulphoxide, use of to disrupt microbial permeability barriers, 334
Discontinuity in the bacterial cell cycle, possible occurrence of, 74
Disintegration rate constant in ultrasonic breakage of microbes, 303
Disruption, cell, mechanical methods for, 295
 of micro-organisms, 279
 of sphaeroplasts, 331
Dissipation of energy in cilia, 13
Distribution of bacteria engaged in DNA synthesis, 83
Divalent cations, need for to obtain flagellar activity, 38
Division, bacterial, variations in cell size at, 63
 cell, and initiation of chromosome replication in bacteria, 49
Donachie model for chromosome replication in bacteria, 53
Dose of radiation required to kill bacteria, 141
Double-stranded breaks, repair of in bacterial DNA, 141
Doublet fibres, movement of in ciliary movement, 25
Doubling time range in bacteria, 57
Dounce homogenizers, use of in microbe disruption, 316
Drag during ciliary movement, 16

Droplet model for disruption of microorganisms, 293
Drug-resistance loci in yeast, 218
Dunaliella spp., disruption of, 329
Dynamic behaviour of a sliding filament system of ciliary movement, 31
Dynein arms, in ciliar axonemes, 4
  role of in ciliary movement, 26
Dyno mill, use of to disrupt microbes, 328
  use of to disrupt *Saccharomyces cerevisiae*, 305
Dyskinetoplastic strains of trypanosomes, 214

E

Eaton press, use of to disrupt microbes, 329
Effective stroke of cilia, 10
Electron microscopy, denaturation mapping of mtDNA by, 198
Elliptio spp., ciliary structure in, 5
  movement of cilia of, 25
Endocytosis, and erythrocyte vesicles, 313
Endonuclease, action of yeast mtDNA, 208
  effect of on damaged bacterial DNA, 126
  mitochondrial, stimulation of by ethidium bromide in yeast, 212
  effect of on damaged bacterial DNA, 127
Energy dissipation in flagella, 17
Energy requirements for microbial motility, 2
Enthalpy changes in flagellar action, 38
Entropy changes in flagellar action, 38
Envelope technique for study of ciliary movement, 18
Enzyme activity of microbial extracts, 324
Enzymes, accessibility of in bacterial membrane vesicles, 318
  microbial, inactivation by ultrasonics, 326
Ergosterol, effect of on mtDNA synthesis by yeast, 193
  removal of from *Crithidia fasciculata*, 333
  role of in yeast mitochondria, 179

Erythrocytes, critical flow stress needed to disrupt, 293
  membrane vesicles from, 312
  preparation of membranes from, 282
  preparation of vesicles of, 322
Erythromycin, effect of on mitochondrial protein synthesis in yeast, 217
  effect of on mtDNA synthesis in yeast, 180
  resistance in *Saccharomyces cerevisiae*, 219
Escherichia coli, chromosomal location of DNA repair genes in, 120
  contribution of different repair mechanisms to recovery of irradiated, 140
  disruption of, 284, 298
  DNA synthesis in, 63
  doubling time range in, 57
  energy required for pyrimidine dimer formation in, 134
  formation of photoreactivating enzyme-substrate complex in, 125
  gamma irradiation survival curve of, 107
  inducible DNA repair in, 156
  inducible nature of ultraviolet radiation-induced mutagenesis in, 147
  mapping of *phr* gene in, 124
  nuclear bodies in division of, 76
  photoreactivation repair of DNA in, 121
  preparation of membrane vesicles of, 316
  preparation of membrane vesicles of, 323
  radiation damage in, 100
  radiation-resistant mutants of, 110
  rate of repair of ionizing radiation damage in, 137
  removal of cross links from damaged DNA of, 143
  repair-proficient strains of, 103
  uvr mutants of, 112
Ethidium bromide, action of on mtDNA in yeasts, 206
  activated endonuclease in yeast, 209
  effect of high concentrations of on yeast, 210
  permeability of yeast cells to, 212
  use of in *petite*-induction, 204

SUBJECT INDEX

Ethylene diaminetetraacetic acid, effect of on strand mending of bacterial DNA, 140
Ethyl methane sulphonate, differential effects of on bacteria, 109
Eucaryotes, cell cycle in, 50
Euflavine, induction of *petite* yeasts by, 209
Euglena spp., disruption of, 329
  swimming velocity of, 23
Euglena viridis, hairs on flagellum of, 17
Euglenoid flagellates, strength of walls of, 286
Eukaryotic micro-organisms, controlled breakage of, 328
Evagination of erythrocytes, role of spectrin in, 314
Excision-defective mutants of bacteria, 127
Excision, in repair of damaged bacterial DNA, 128
  mechanisms in *uvr* mutants of bacteria, 112
  of cross-links from damaged bacterial DNA, 142
  of pyrimidine dimers in bacteria, role of polymerases in, 116
  repair, and formation of bacterial mutants, 145
  repair of damaged DNA in bacteria, 125
  repair of damaged DNA, study of in bacteria, 126
Exonuclease, ATP-dependent, coding for in *rec* mutants, 115
  role of in repair of damaged bacterial DNA, 128
Exonuclease V, ATP-dependent, coded for by *rec* genes in bacteria, 137
Explosive decompression, and microbial disruption, 298
exr genes in bacteria, 116
Extent of repair of DNA damage in bacteria, 106
Extranuclear genetic systems in yeast, 159

F

Fatty acid desaturase mutants of *Saccharomyces cerevisiae*, 179
Fibrils, central, stiffening of in cilia, 34

Fidelity of repair to bacterial DNA, 144
Filamentous fungi, disruption of, 287
  preparation of protoplasts of, 332
Filamentous mutants of bacteria, 119
fil mutants of bacteria, 119
Fish gonad cells, disruption of, 297
Flagella, coupling of chemical and mechanical activity in, 40
  microbial, structure of, 3
Flagellar, motility of microbes, biophysical aspects of, 1
  motion, control of, 40
Flagellates, disruption of, 329
Flipping of proteins in microbial membranes, 319
Flow systems of microbial disruption, 326
Fluctuation in DNA synthesis in a bacterial population, 84
Fluid-flow, patterns in *Ochromonas* flagella, 10
  round cilia, 20
Fluidity of plasma membranes, 315
Forces required to disrupt micro-organisms, magnitude of, 280
Fraction of bacteria of a size class engaged in DNA replication, computation of, 81
Fragmentation of mtDNA, by ethidium bromide, 206
Fragments, small, of bacterial DNA, 75
Free radicals, inactivation of microbial components by, 306
Freeze presses, use of to disrupt microbes, 287
Freeze thawing, of membrane vesicles, 324
  use of to disrupt microbes, 287
French press, basis for action of, 299
  pressure exerted in, 297
Frequency as a parameter in flagellar movement, 32
Freshwater mussel, movement of cilia of, 25
Functions of axonemal structures in cilia, 24
Fungal lytic enzymes, use of to prepare protoplasts, 332
Fungal walls, and strength of, 285
Fusogenic action of diacylglycerols in membranes, 315

G

Galactose residues in microbial membranes, labelling of, 311
Gamma-irradiation survival curves of bacteria, 107
Gamma radiation, differential effects of on bacteria, 109
Gap formation opposite pyrimidine dimers in repair of bacterial DNA, 133
Gap repair of bacterial DNA, 135
Gelatin, effect of on microbial disruption, 301
Gene purification, mitochondrial, in *petite* yeasts, 196
Genetic control, of motility, 43
  of post-replication repair in bacteria, 135
Genetic crosses of yeast, transmission and recombination of mitochondrial genes during, 230
Genetic determinants on yeast mtDNA, 217
Genetic factors affecting transmission and recombination of mitochondrial genes in yeast, 239
Genetic factors influencing transmission and recombination in yeast, 238
Genetic map, of *Saccharomyces cerevisiae* mitochondrial genome, 229
  of yeast mtDNA, 245
Genetics, mitochondrial, in yeast, 216
Genetics of yeast mitochondrial DNA, 157
Genome, bacterial, properties of, 100
  mitochondrial, of *Saccharomyces cerevisiae*, genetic and physical map of, 229
  size of *grande* mtDNA of yeast, 160
Geometry of outlet, effect of on microbial disruption, 301
Gill cilia of *Elliptio* sp., mechanism of movement of, 30
Glass beads, use of in microbial disruption, 302
Glucans in fungal walls, 285
Glucosamine resistance in *Saccharomyces cerevisiae*, 220
Glucose repression, effect of on mtDNA synthesis in yeast, 191
Glusulase, use of to prepare yeast protoplasts, 332
Glyceraldehyde 3-phosphate dehydrogenase activity of erythrocytes, 323
Gradient centrifugation, use of to separate size classes of bacteria, 70
Grains, autoradiographic, in study of DNA synthesis in bacteria, 64
Gram-positive, wall structure of, 283
Grande strains of yeast, nature of, 159
Grinders, use of to disrupt microbes, 287
Grinding, use of in microbial disruption, 331
Growth rate, and interval between nuclear division in bacteria, 58
  effect of on nuclear division in the cell cycle, 52

H

Haemolysis of erythrocytes, 289
Hairs on flagella, 9
Halobacterium halobium, disruption of, 282
Halophilic bacteria, disruption of, 282
Hansenula wingei, properties of mtDNA of, 171
Haploid *Saccharomyces cerevisiae*, DNA content of, 185
Helix pomatia juice, use of to prepare yeast protoplasts, 332
Heterosexual crosses in yeast, 236
Highly polar mitochondrial genetic crosses in yeast, 234
Homogenizers, use of to disrupt microbes, 287
Hughes pressing of *Pseudomonas aeruginosa*, 324
Hughes press, use of to disrupt microorganisms, 300
Hybridization, of yeast mtDNA, 168
  use of to study mtDNA from *petite* yeast strains, 198
Hydrodynamic considerations of ciliary movement, 13
Hydrodynamic efficiency of ciliary movement, 24
Hydrodynamics, basic, of microbial disruption, 289
Hydrodynamic shearing forces, and microbial disruption, 280

Hydroperoxyl radicals, inactivation of microbial components by, 306
Hydroxyapatite columns, use of to resolve yeast mtDNA, 163
Hydroxylamine, differential effects of on bacteria, 109

I

Inactivation, chemical, of microbial cell, components, 306
Inactivation of microbial enzymes by ultrasonics, 326
Incision of damaged bacterial DNA, 127
Increased sensitivity of bacteria to radiation damage, isolation of mutants with, 111
Inducible nature of ultraviolet radiation-induced mutagenesis in bacteria, 147
Induction, of *petite* yeasts, 203
  of *petite* yeasts by ethidium bromide, mechanism of, 205
Inertial forces on a moving cylinder, and ciliary movement, 14
Informational content of mtDNA from *petite* yeast strains, 200
Informational sequences in yeast mtDNA, 165
Inheritance, mitochondrial, criteria for in *Saccharomyces cerevisiae*, 216
Inhibition of mtDNA synthesis by euflavine, 213
Initiation, of bacterial chromosome replication, variation in, 69
  of chromosome regulation in *Myxococcus xanthus*, precision of, 90
  of chromosome replication and cell division in bacteria, 49
  of DNA synthesis in bacteria, lack of control of cell division, 86
  of DNA synthesis in the cell cycle, 52
  of mtDNA synthesis in yeast, 186
Inner arm in cilia and flagella, 3
Intact organelles, isolation from disrupted micro-organisms, 281
Interciliary co-ordination, 11
Interdoublet links, function of in ciliary movement, 29
Internal resistances to motion in cilia, 31

Interstrand cross-links, removal of from damaged bacterial DNA, 143
Intracellular transport of RNA in yeast, 158
Iodination of microbial membrane proteins, 311
Ionizing radiation, damage in bacteria, repair of, 137
  damage to bacterial DNA, 103
Ionophore A 23187, effect of on ion permeability of erythrocyte vesicles, 314
  use of to study flagellar action, 41
Ions, effect of on erythrocyte vesicles, 313
Irradiated bacterial cell, contributions of different repair mechanisms in recovery of, 140
Irregular spacing of chromosome initiation in bacteria, 71
Isolated yeast mitochondria, DNA synthesis in, 175
Isolation of mutants defective in DNA repair in bacteria, 109

K

Kidney cells, disruption of, 297
Kinetics of synthesis of yeast mtDNA, 176
Kluyveromyces fragilis, effect of ethidium bromide on, 215
  homology of DNA of, 170
K. lactis, homology of DNA of, 170
  mtDNA synthesis during cell cycle of, 187
Koch and Schaechter model for chromosome replication in bacteria, 53
Kontes mini-bomb, pressure exerted in, 297
Kornberg polymerase, DNA, in bacteria, 115

L

Labelling of microbial membrane proteins, 311
Lability of cell extract components in microbial disruption, 305
Lactoperoxidase-catalysed labelling of microbial membrane proteins, 311

368 SUBJECT INDEX

Laminar flow velocity, and disruption of micro-organisms, 289
Large bacterial cells, DNA synthesis in, 68
Laser microbeam, use of to isolate flagella, 32
Length of cilia and flagella, 3
Leptomonas sp., swimming velocity of, 23
Leptospira pomona, disruption of, 299
Leucyl-tRNA, retention of in some *petite* mtDNAs, 201
lex genes in bacteria, 116
Life lengths of mother cells of bacteria, 55
Ligase-deficient mutants of bacteria, 118
Ligase, polynucleotide, mutants of bacteria, 118
lig genes in bacteria, 118
Limited defects in respiratory function, loci conferring in yeast, 227
Linkage maps for *Escherichia coli*, recalibration of, 156
Linomycin, effect of on mitochondrial protein synthesis in yeast, 217
Lipid composition of yeast mitochondria, 179
Lipids, intracellular transport of in yeast, 158
Lipid supplements, effect of on yeast mtDNA synthesis, 193
Liquid extrusion presses, use of to disrupt microbes, 287
Liver cells, disruption of, 297
Loci, conferring limited defects in respiratory function in yeast, 227
  genetic, in yeast conferring antibiotic resistance, 217
Locus for X-ray sensitivity mutants of bacteria, 117
Long-form mutants of bacteria, 119
Long-patch repair of bacterial DNA, 129
lon mutants of bacteria, 119
Low-stress devices for microbial disruption, 295
Low-temperature, denaturation, possible operation of in flagella, 36
  instability of *rho* factor in yeast, 203
Lysis, hypotonic, of bacterial sphaeroplasts, 322
  of microbial cells, 282
Lysosomes, isolation of, 329

Lysozyme, action of on bacteria, 284
  effect of cavitation on, 307

M

Macromolecule synthesis and regulation of the bacterial cell cycle, 55
Magnesium ions, effect of on polarity of erythrocyte vesicles, 314
Magnesium, need for to obtain flagellar activity, 38
Mammalian cells, mtDNA synthesis during cell cycle of, 187
Manganese as an activator of flagellar action, 38
Manganese ions, use of in *petite* induction, 204
Mannans in fungal walls, 285
Manton–Gaulin homogenizer, use of in microbial disruption, 305, 330
Mapping of *uvr* genes in bacteria, 112
Map position, of *uvr* mutants in *Escherichia coli*, 113
  or *rec* genes in bacteria, 114
Marker enzymes and bacterial membrane vesicles, 317
Mass, bacterial, and cell division, 59
Mastigoneme action in cilia, 9
Mastigonemes in cilia and flagella, 5
Mating, yeast, mtDNA synthesis during, 189
Meander forms in cilia, 7
Mechanical breakage of eukaryotic micro-organisms, 328
Mechanical comminution, during microbial degradation, 307
Mechanical disruption of microbial DNA, 308
Mechanical methods for cell disruption, 295
Mechanical strength of microbial walls, 281
Mechanism of *petite* induction by ethidium bromide, 205
Mechanochemical aspects of ciliary movement, 34
Megahertz ultrasonic waves, use of in microbial disruption, 302

SUBJECT INDEX

Meiosis, yeast, mtDNA synthesis during, 190
Membrane asymmetry in micro-organisms, 310
Membrane potential and ciliary action, 43
Membrane surrounding cilia and flagella, 5
Membrane synthesis and regulation of the bacterial cell cycle, 54
Membrane vesicles, microbial, 310
Mending of strands in repair of damaged bacterial DNA, 139
Menoidium sp., swimming velocity of, 23
Mercapto compounds, use of to prepare protoplasts, 332
Metabolism and ciliary movement, 35
Metabolism and regulation of the bacterial cell cycle, 54
Metachronal patterns in cilia, 11
Metachrony in cilia, 11
Methods for preparing membrane vesicles, 322
N-Methyl-N-nitro-N-nitrosoguanidine, differential effects of on bacteria, 109
Microbial cells, principles of breakage of, 288
Micrococcal nuclease, action of on yeast mtDNA, 164
Micrococci, resistance of to radiation damage, 108
Micrococcus denitrificans, membrane vesicles of, 321
M. luteus, endonuclease from, 130
M. radiodurans, amount of DNA in each cell of, 108
  chromatographic separation of dimers from DNA of, 126
  cross-link formation in DNA of, 144
  gamma irradiation survival curves of, 107
  repair of double-stranded breaks in DNA of, 142
  repair to ionizing radiation damage in, 138
  thymine dimer formation in DNA of, 104
Microfibrillar arrays in fungal walls, 285
Micromanipulation of bacterial cells from colonies, 71

Micro-organisms, disruption of, 279
Microtubule bending in the active bending model of ciliary movement, 34
Microtubules, central, role of in ciliary movement, 30
  in cilia and flagella, 3
Mikamycin resistance, in *Saccharomyces cerevisiae*, 220
  in yeast, 221
  in yeast, link with antimycin A resistance, 224
Mills, colloid, use of in microbial disruption, 301
Mini-mill, use of in microbial disruption, 330
Minimization of damage during disruption of micro-organisms, 281
Mit^- mutants of yeast, 227
Mitochondrial DNA, in *petite* yeasts, properties of, 195
  number of molecules of in mitochondria, 184
  polymerases in yeast, 177
  physical arrangement of sequences in, 202
  synthesis in yeast, cellular origin of components involved in, 180
  synthesis of during the yeast cell cycle, 187
  yeast, properties of, 170
  yeast, repair in, 174
  yeast, structure, synthesis and genetics of, 157
  yeast, synthesis of, 173
Mitochondrial gene, purification mutants in *petite* yeasts, 196
  influence of on *petite* induction in yeasts, 203
  of yeast, transmission and recombination of during genetic crosses, 230
Mitochondrial genetics in yeast, 216
Mitochondrial genome of *Saccharomyces cerevisiae*, genetic and physical map of, 229
Mitochondrial inheritance in *Saccharomyces cerevisiae*, criteria for, 216
Mitochondrial membranes and mtDNA replication, 178
Mitochondrial ribosomes, yeast, DNA for, 165

SUBJECT INDEX

Mitochondria, isolated yeast, electron microscope study of, 178
number of mtDNA molecules in yeast, 184
osmotically shocked, of *Saccharomyces cerevisiae*, 160
preparation of, 330
Mitomycin C, differential effects of on bacteria, 109
radiation damage in bacteria, 105
Models for chromosome replication in bacteria, 53
Models for ciliary movement, 14
Models for regulation of cell division in bacteria, 53
Models for the relationship of chromosome replication in bacteria, 82
Molecular models of recombination in yeast, 243
Mono-adduct formation on bacterial DNA, 144
Motility, microbial, biophysical aspects of, 1
Mouse ascites carcinoma cells, disruption of, 293
Movement, patterns of in cilia and flagella, 7
Mucoid-colony mutants of bacteria, 119
Multiple flash photography, use of to study ciliary movement, 7
Multiple forks in bacterial chromosomes, 71
Muscular activity, activation parameters associated with, 37
Mutagenesis of yeast mtDNA, 175
Mutagenic consequences of repair to bacterial DNA, 144
Mutagenic effect of ethidium bromide on yeast mtDNA, 209
Mutagens, bacterial, role of repair in action of, 148
comparative effects of on bacteria, 109
Mutants, non-motile, of microbes, 43
Mutants of bacteria defective in DNA repair, isolation of, 109
phr Mutants of bacteria, isolation of, 111
Mutants of yeast defective in respiratory function, 227
Mutations leading to increased sensitivity of bacteria to radiation damage, 111

Mycobacterium phlei, membrane vesicles of, 321
M. tuberculosis, disruption of in the French press, 229
Mycoplasma gallinarum, disruption of, 299
Mycoplasmas, disruption of, 283
Mytilus sp., abfrontal cilium of, 10
Myxococcus xanthus, precision of initiation of chromosome regulation in, 90

N

Nalidixic acid, potentiating effect of on induction of *petite* yeasts by ethidium bromide, 205
Near-field effects and ciliary movement, 20
Netzch–Moleux mill, use of to disrupt microbes, 328
use of in microbial disruption, 302
Neurospora crassa, wall-less variants of, 334
Nexin linkages in flagella, effect of trypsin on, 26
Nitrogen cavitation, and microbial disruption, 298
Non-motile mutants of microbes, 43
Non-polar mitochondrial genetic crosses in yeast, 234
Nuclear division, in bacteria, variability of time between and cell division in bacteria, 76
in the cell cycle, 50, 52
relation of to cell division in bacteria, 78
Nuclear DNA, yeast, role of in synthesis of mtDNA, 174
Nuclear location of genes involved in yeast mtDNA synthesis, 183
Nuclear yeast genes, influence of on *petite* induction, 203
Nuclease, presence of in yeast mitochondria, 176
Nuclei, isolation of, 329
Nucleotidase, release of from bacteria by osmotic shock, 289
Number of mtDNA molecules in each yeast cell, 204

O

Ochromonad flagellates, movement of, 9
Ochromonas flagella, fluid-flow patterns in, 10
Ochromonas sp., role of central microtubules in ciliary movement of, 30
swimming velocity of, 23
Okazaki fragments in bacterial DNA synthesis, 133
Oleandomycin, effect of on mitochondrial protein synthesis in yeast, 217
Oligomycin, resistance in Saccharomyces cerevisiae, 220
resistance in yeast, 224
Ome^+ allele in yeast, 238
Omega locus in yeast, nature of, 231
Omicron DNA circles associated with yeast mitochondria, 169
Opalina sp., changes in beat frequency of, 40
ciliary movement in, 11
metachronal waves on cilia of, 19
speed of movement of, 20
Organelles, intact, isolation from disrupted micro-organisms, 281
motility, of microbes, 1
Organization of mtDNA in petite yeasts, 193
Osmotic lysis of micro-organisms, 282
Osmotic shock of Gram-negative bacteria, 289
Osmotic stress, effect of on microbes, 288
Oxidases, inactivation of microbial by chemicals, 307
Oxygen, effect of on irradiation of bacteria, 140
effect of on mtDNA synthesis by yeast, 192

P

Pancreatic deoxyribonuclease, effect of on bacterial DNA, 129
Pantoyl lactone, effect of on growth of filamentous mutants of bacteria, 121
Paramecium aurelia, disruption of, 331
P. cilia, effect of calcium ions on, 42
Paramecium sp., changes in beat patterns of cilia on, 40
ciliary movement in, 11
ciliary reversal in, 41
direction of movement of, 19
metachronal wave patterns in cilia of, 12
swimming speeds of, 22
Parameters of the size distribution for balanced bacterial growth, 62
Paromomycin resistance in Saccharomyces cerevisiae, 220
Patterns of ciliary and flagellar movement, 7
Peptidoglycan, bacterial wall, and mechanical strength, 283
Peranema sp., different motion forms of, 41
swimming velocity of, 23
Periplasmic binding proteins, release of from Gram-negative bacteria, 289
Permeability, barriers, microbial, destruction of, 44, 334
of microbial plasma membranes to solutes, 288
of yeast cells to ethidium bromide, 212
Petite deletion analysis, use of to map yeast mtDNA, 162
Petite-inducing agents, types of, 204
Petite mutants of Saccharomyces spp., nature of, 158
Petite-negative yeasts, 170
Petite negativity in yeasts, 214
Petite yeasts, induction of, 203
Pet series of yeast mitochondrial mutants, 182
Phasing of bacterial DNA synthesis with growth, 63
Phosphatidate phosphohydrolase activity of erythrocyte membranes, 315
Phosphatidylcholine, distribution of in membranes, 312
Phosphodiester strand, breakage of in bacterial DNA by ultraviolet radiation, 102
Phospholipases, effect of on erythrocytes, 315
effect of on microbial membranes, 312

Photons, and repair of damaged DNA in bacteria, 123
Photoreactivating enzyme, in bacteria, 122
  in bacteria, substrate for, 122
  substrate complex in bacteria, 124
Photo-reactivating mutants of bacteria, isolation of, 111
Photoreactivation, and excision of thymine-thymine dimers in bacteria, 122
  as an analytical tool in bacteria, 125
  in formation of bacterial mutants, 145
  of damaged yeast mtDNA, 175
  repair of DNA in bacteria, 121
Photosensitive azide derivative of ethidium bromide, use of in *petite* induction, 209
Physarum spp., mtDNA synthesis during cell cycle of, 187
Physical arrangement of sequences in *petite* mtDNA, 202
Physical map of *Saccharomyces cerevisiae* mitochondrial genome, 229
Physical size of yeast mtDNA, 160
Physiological changes, effect of on yeast mtDNA synthesis, 189
Pichia spp., preparation of protoplasts of, 332
Piston-pressure extrusion, and microbial disruption, 299
Plasmid replication in bacteria, 72
Pleurobranchia comb plates, movement of, 13
Pleurobrachia sp., fluid flow around bomb plates of, 20
Pneumatic disrupter, use of to break micro-organisms, 298
Poecelia formosa, role of DNA dimers in producing cancerous lesions in, 125
Poiseuille flow, and microbial disruption, 291
polA mutants of *Escherichia coli*, DNA repair in, 138
Polarity of recombination in homosexual crosses in yeast, 235
pol genes in bacteria, 115
Polidy, effect of on content of mtDNA in yeast, 185

Polyamines, protective action of in microbial disruption, 331
Polymerase-defective mutants, and repair of DNA, 129
Polymerase-deficient mutants of bacteria, 115
Polymerases, DNA, role of in repair of damaged bacterial DNA, 129
  DNA, in yeast mitochondria, 177
Polymer degradation during microbial disruption, 325
Polynucleotide ligase mutants of bacteria, 118
Polysaccharides in fungal walls, 285
Polytomella spp., disruption of, 329
Population of bacteria, distribution of cells engaged in DNA synthesis in, 83
Pores, formation of in erythrocyte membranes by imposition of shear stress, 294
  in bacterial plasma membranes, 319
Post-radiation incubation, effect of on damaged bacterial DNA, 104
Post-replication, recombination repair of bacterial DNA, 131
  repair in bacteria, genetic control of, 135
Potter–Elvehjem homogenizers, and microbial disruption, 296
Power dissipation in cilia, 18
Precision of initiation of chromosome regulation in *Myxococcus xanthus*, 90
Pre-excision mutants of bacteria, 128
Pressure, atmospheric, effect of on ciliary activity, 36
  effect of on bacterial membranes, 319
  effect of on cilia of *Crithidia oncopelti*, 36
  exerted in Chaikoff press, 297
Pritchard model for chromosome replication in bacteria, 53
Probes for DNA repair in bacteria, 125
Procaryotes, cell cycle in, 50
Projections, arrangements of in central microtubules in cilia and flagella, 6
Prophage induction and mutagenesis in bacteria, 147
B-Propiolactone, differential effects of on bacteria, 109
Propulsive speed of colia and flagella, 3

SUBJECT INDEX 373

Protease digestion of microbial membranes, 311
Protein-DNA cross-links, formation of in bacterial DNA, 102
Proteins, effect of mechanical forces on, 309
  surrounding euglenoid flagellates, 286
  synthesis by bacteria, rate of, 74
  intracellular transport of in yeast, 158
Proteus vulgaris, nuclear to cell division interval in, 79
Protofilaments in cilia and flagella, 3
Protoplasm synthesis in bacteria, rate of in relation to cell size, 61
Prototheca zopfi, disruption of, 331
Protozoa, contractile vacuole of, 282
  diameter of cilia of, 16
  disruption of, 282, 329
  motility of, 1
Pseudomonas aeruginosa, Hughes pressing of, 324
  membrane vesicles of, 321
Psoralen-plus-near ultraviolet radiation damage in bacteria, 105
Pulse autoradiography of slowly dividing bacterial cells, 63
Pyrimidine dimers, energy required for formation in DNA of *Escherichia coli*, 134
  formation of in bacterial DNA, 102
  separation of from damaged DNA, 126

Q

Quartz wind effect in microbial disruption by ultrasonics, 304

R

Radial migration of spheres, and microbial disruption, 292
Radial spokes, in cilia and flagella, 3
  role of in initiation of bend formation in cilia, 27
Radiation damage to bacterial DNA, 101
Radiation-resistant, mutants of bacteria, 110
  strains of *Escherichia coli*, 107

Radiodecomposition of label in bacteria, 72
Radiolysis of thymine residues in bacterial DNA, 103
Random diploid analysis of yeasts, 231
Random initiation of DNA replication in bacteria, 89
Rarefaction wave, use of to disrupt micro-organisms, 299
Ratio of bases in yeast mtDNA, 163
Reactivation by host cells of irradiated bacteriophage, 117
rec genes in bacteria, 113
rec mutants of bacteria, 136
Reckless degradation of DNA in bacteria, mutants, 114
Recombination, and formation of bacterial mutants, 146
Recombination, and transmission, of mitochondrial genes in genetic crosses of yeast, 230
Recombination-deficient genes in bacteria, 113
Recombination-deficient mutants of bacteria, 135
Recombination-deficient mutants of *Escherichia coli*, isolation of, 110
Recombination repair, post replication, of bacterial DNA, 131
Recovery stroke of cilia, 10
Regulation of bacterial cell division, role of DNA initiation in, 57
Regulation of synthesis of mtDNA in yeast, 186
Re-insertion of bases during repair of bacterial DNA, 130
Rejoining action of Type I DNA repair in bacteria, 139
Relationship between microbial walls strength and disruption method, 286
Removal of bases from bacterial DNA, 101
Renaturation of mtDNA from yeast, 161
Renaturation of mtDNA of *Saccharomyces carlsbergensis*, kinetics of, 161
Repair-dependent mutagens and bacteria, 148
Repair, excision, of damaged DNA in bacteria, 125

Repair-independent mutagens and bacteria, 148
Repair-mediated induced mutation in bacteria, 145
Repair of bacterial DNA by polymerases, 115
Repair of bacterial DNA, fidelity of, 145
Repair of cross-link damage in bacterial DNA, 142
Repair of damaged DNA in bacteria, 73, 99
Repair of DNA, and resistance to ethidium bromide in yeasts, 206
damage, extent of in bacteria, 106
isolation of mutants of bacteria defective in, 109
Repair of ionizing radiation damage in bacteria, 137
Repair of yeast mtDNA, 174
Repair photoreactivation, of DNA in bacteria, 121
Repair-proficient strains of *Escherichia coli*, 103
Repair to bacterial DNA, mutagenic consequences of, 144
Repeating patterns in sequences in yeast mtDNA, 203
Replication of chromosomes in *Escherichia coli*, 63
Replication of DNA, computation of fraction of bacteria of a size class engaged in, 81
Replication of yeast mtDNA, 174
Replication of yeast mtDNA and the mitochondrial membrane, 178
Repression, glucose, effect of on yeast mtDNA synthesis, 191
Resistance to ethidium bromide in yeasts, 205
res mutants of bacteria, 115
Respiratory-competent yeast, structure of mtDNA in, 158
Restriction endonucleases, effect of on yeast mtDNA, 161
Reversal of direction of wave propagation in cilia, 8
Reynolds numbers, and microbial disruption, 290
Reynolds numbers, and movement of cilia, 14
Rhabdomonas sp., swimming velocity of, 23

Rhodamine G resistance in yeast, 226
Ribonuclease, release of from bacteria by osmotic shock, 289
Ribonucleic acid, intracellular transport of in yeast, 158
Ribosomal RNA genes, in yeast, location of, 166
retention of in some mtDNAs, 201
Ribosome population and regulation of the bacterial cell cycle, 54
Right side-out nature of bacterial membrane vesicles, 316
Rods, solid, movement of as a model in microbial disruption, 291
Role of DNA initiation in regulating bacterial cell division, 57
Rolling circle replication of mtDNA in yeasts, possibility for, 203
Rotation, rate of in cell bodies, 17

S

Saccharomyces cerevisiae, action of ethidium bromide on mtDNA of, 206
content of DNA in haploid, 185
disruption of, 304, 305
S. carlsbergensis, disruption of by wet milling, 302
S. cerevisiae, disruption of in the Braun homogenizer, 302
disruption of in the French press, 299
effect of glucose on mtDNA synthesis by, 191
effect of mechanical forces on enzymes of, 309
fatty acid desaturase mutants of, 179
genetic and physical map of mitochondrial genome of, 229
mtDNA synthesis during cell cycle of, 187
petite mutants of, 159
photoreactivating enzyme in, 123
strength of wall of, 285
Saccharomyces spp., preparation of protoplasts of, 332
Salmonella typhimurium, doubling time range in, 57, 77
Salt, effect of on microbial disruption, 301
Sarcina lutea, disruption of, 284

SUBJECT INDEX 375

Sarcinalutea, disruption of in the French press, 299
Scale up of mechanical methods for microbial disruption, 304
Scenedesmus obliquus, disruption of by wet milling, 302
Scenedesmus spp., preparation of protoplasts of, 333
Schizosaccharomyces pombe, effect of ethidium bromide on, 215
  preparation of protoplasts of, 332
  properties of mtDNA of, 171
Sciara sperm, mechanism of movement of, 30
Scission of mtDNA by ethidium bromide action, 207
Screening of *Saccharomyces cerevisiae* for mutants defective in respiratory function, 228
Sea-urchin, spermatozoa, movement of, 26
  sperm flagella, ATP and movement of, 35
Secondary structure of yeast mtDNA, 162
Sedimentation velocity of bacterial DNA, effect of ionization radiation on, 103
Semiconservative replication of bacterial DNA, 116
Sequence relationships of different mtDNAs with that from *Saccharomyces cerevisiae*, 172
Septation of cells of *Myxococcus xanthus*, 90
Sequence differences between mtDNAs of different *grande* yeasts, 167
Sequence organization of yeast mtDNA, 162
Sequences in *petite* mtDNA, *physical* arrangement of, 202
Serratia marcescens, disruption of, 299
Shearing, random, of bacterial DNA by X-rays, 142
Shear stress, development of and microbial disruption, 289
  effect of on microbial DNA, 308
Shock-tube techniques, and microbial disruption, 298
Short patch repair of bacterial DNA, 130
Sialidase activity of erythrocytes, 323
Sidedness of bacterial membrane vesicles, 317

Sidedness of microbial membranes, 310
Single-strand breaks in bacterial DNA caused by ionizing radiation, 103
Single-strand gaps, formation of in repair of bacterial DNA, 134
Sinusoidal waves in cilia, 16
Sister cells, bacteria, lifelengths of, 55
Size, class, computation of fraction of bacteria of a engaged in DNA replication, 81
  classes of bacteria, separation from an asynchronous culture, 70
  critical, in bacteria for DNA initiation, 85
  distributions in a bacterial population, 62
  effect of on microbial movement, 22
  of bacteria at cell division, 59
  of cells and initiation of DNA replication, 52
  physical, of yeast mtDNA, 160
Sliding filaments, in cilia and possible action of calcium ions, 42
  model of ciliary movement, 25
Small bacterial cells, growth rate of, 59
Snail juice, use of to prepare yeast protoplasts, 332
  use of to remove fungal walls, 285
Snake-form mutants of bacteria, 119
Solid-shear force, use of to disrupt bacteria, 284
Solid spheres, movement of as a model for microbial disruption, 291
Solvated electrons, inactivation of microbial components by, 306
Sonication of membrane vesicles, 324
Sonochemical modification of microbial components, 306
Sorvall-Ribifractionator, use of in microbial disruption, 306
SOS repair of bacterial DNA, 156
Spacers in yeast mtDNA, 164
Specificity of bacterial endonucleases, 127
Spectrin, effect of on erythrocyte morphology, 313
Spectrin, nature of in erythrocyte membranes, 313
Spermatozoa, marine invertebrate, effect of temperature on cilia of, 36
Spermatozoa, sea-urchin, movement of, 26

Sphaeroplasts, preparation of, 331
  release of mitochondria from, 202
  yeast, use of to isolate mitochondria, 160
Spiramycin, effect of on mitochondrial protein synthesis in yeast, 217
  resistance in *Saccharomyces cerevisiae*, 210
Spirillum spp., stretching of, 284
Spirostomum sp., swimming velocity of, 23
  velocity profile of fluid surrounding, 21
Spleen cells, disruption of, 297
Spoke heads in cilia and flagella, 3
Spontaneous formation of *petite* yeast strains, 203
Sporulation, yeast, mtDNA synthesis during, 190
Staphylococcus aureus, disruption of, 284, 299
  membrane vesicles of, 321
Steady-state distribution of cell sizes in a bacterial population, 61
Stentor, sp., swimming velocity of, 23
Stiffness, control of in flagella, 28
Strand breakage in bacterial DNA, 101
Strand exchange in removal of cross-links from damaged bacterial DNA, 143
Streptococcus faecalis, disruption of, 300
S. griseus conidia, photoreactivation repair of DNA in, 121
Stress-time regimes for haemolysis of erythrocytes, 294
Stretching of cross-links in flagellar movement, 33
Stroboscopic methods to study ciliary movement, 7
Structure and organization of mtDNA in *petite* yeasts, 193
Structure of microbial walls, 281
Structure of yeast mitochondrial DNA, 157
Subcellular fractionating of microbial organelles, 281
Substrate for photoreactivating enzyme in bacteria, 122
Supercoiled DNA from yeast mitochondria, 161
Survival value of DNA repair mechanisms in bacteria, 100
Symplectic metachrony in cilia, 11

Synchronized cells and the bacterial cell cycle, 56
Synchronous cultures of bacteria, chromosome replication in, 69
Synthesis of DNA in isolated yeast mitochondria, 175
Synthesis of mtDNA during the yeast cell cycle, 187
Synthesis of yeast mitochondrial DNA, 157, 173

T

Target analysis in *petite* induction, 204
Taylor number, relation to Reynolds number in microbial disruption, 296
  and low-stress microbial disruption, 296
Temperature, effect of on ciliary activity, 36
  effect of on irradiation of bacteria, 106
  effect of on lability of microbial cell components, 305
Temperature-sensitive, mutants of *Escherichia coli*, and mutagenesis, 147
  petite mutants of yeast, 203
  yeast mutants for DNA synthesis, 183
Temporal accuracy of the DNA synthesis cycle in bacteria, 80
Tensile strength of boundary layers of microbes, 280
Termination of DNA synthesis in the cell cycle, 52
Tetracycline, effect of on mitochondrial protein synthesis in yeast, 217
Tetrahymena pyriformis, disruption of, 297
Tetrahymena sp., effect of ethidium bromide on, 216
  mtDNA synthesis during cell cycle of, 187
  swimming velocity of, 23
Thermal lability of microbial cell components, 305
Threshold mass and initiation of chromosome replication in bacteria, 53
Thymidine diphosphorhamnoside, labelling of in bacteria, 72

Thymidine incorporation into bacteria, 69
Thymidine labelling of *Myxococcus xanthus*, 91
Thymine auxotrophs, use of to study DNA synthesis by bacteria, 73
Thymine dimers, formation of in bacterial DNA, 103
 structural formulae of, 104
 structure of in bacterial DNA, 103
Thymine, incorporation into bacteria, 69
 requirement by bacteria, variations in, 73
Timing of mtDNA synthesis in the yeast cell cycle, 188
Toluene, effect of on bacterial membrane vesicles, 318
Tonsate movement of cilia, 7
Topological studies on microbial membranes, 311
Tracer experiments in study of DNA synthesis in bacteria, objections to, 72
Transforming DNA, photoreactivation repair of damaged, 121
Translational velocity of microbes, 22
Translocation of membrane proteins, 319
Transmission and recombination, in yeast, genetic factors influencing, 238
 of mitochondrial genes in genetic crosses of yeast, 230
Tricarboxylic acid cycle enzymes in *Escherichia coli*, effect of mechanical force on, 309
Triethyl tin resistance in yeast, 224
Trigger in the bacterial cell cycle, cell size as a, 56
Trimethylpsoralen, effect of on bacteria, 105
Tritium labelling of bacterial cells, 72
Triton X100, effect of on bacterial membrane vesicles, 318
 effect of on flagella of spermatozoa, 26
Trypanosoma ranarum, swimming velocity of, 23
Trypanosomes, disruption of, 329
 initiation of bend formation in cilia of, 27
 movement of, 8
 reversal of movement in, 41

Trypsin, effect of on spermatozoa flagella, 26
Tubularia sperms, chemotaxis in, 40
Tubulin, role of in ciliary action, 43
Turbulence, and disruption of microbial suspensions, 290
Turbulent flow, and microbial disruption, 290
Type I repair of damaged bacterial DNA, 139
Type II repair to damaged bacterial DNA, 138
Type III repair of damaged bacterial DNA, 137
Types of *petite*-inducing agents, 204

U

Ultrasonic inactivation of microbial enzymes, 326
Ultrasonics, use of in microbial disruption, 302
Ultraviolet-induced mutagenesis of yeast mtDNA, 175
Ultraviolet irradiation, and photoreactivation in bacterial DNA, 122
Ultraviolet radiation, damage to bacteria, 102
 effect of on bacterial DNA, 106
 induced mutagenesis in *Escherichia coli*, inducible nature of, 147
 isolation of mutants of bacteria resistant to, 110
 use of to induce *petite* yeasts, 213
Ultraviolet-resistant genes in bacteria, 111
Uncouplers, inhibiting effect of on ethidium bromide binding to yeast mtDNA, 208
Undulatory movement of cilia, 7
Unidirectional movement of microbes, 43
Unidirectional replication of bacterial DNA, 75
Unsaturated fatty acids, effect of on mtDNA synthesis by yeast, 193
Unsaturated fatty-acyl residues, role of in yeast mitochondria, 179
Unwinding of bacterial DNA, 75

Unwinding protein, action of in repair of bacterial DNA, 131
uvr genes, in bacteria, 111, 127
  role of in removal of cross-links from damaged bacterial DNA, 143
Uronema sp., movement of, 22

V

Vacuole, contractile, of protozoa, 282
  microbial, preparation of, 333
Variability of the time between nuclear and cell divisions in bacteria, 76
Variation-free protoplasm growth in bacteria, 55
Variation in initiation of bacterial chromosome replication, 69
Venturicidin resistance, in *Saccharomyces cerevisiae*, 220
  in yeast, 224
Vernier arrangement of spokes in cilia, 29
Vesicles, bacterial membrane, heterogeneous nature of, 319
  membrane, methods for preparing, 322
  microbial, nature of, 310
Vesicularization of microbial membranes, 310
Visco-elasticity of cell membranes, and microbial disruption, 293
Viscosity, effect of on flagellar action, 38
  of microbial suspensions, 290
Viscous forces on a moving cylinder, and ciliary movement, 14

W

Wall-defective micro-organisms, disruption of, 334
Wall-less forms of micro-organisms, 282
Walls, microbial, strength of, 281
Wall turbulence, and disruption of microbial suspensions, 290
Waring blender, use of in microbial disruption, 330
Wavelength, as a parameter in flagellar movement, 32
  of cilia and flagella, 3
  of flagella action as a function of viscosity, 39
Wave patterns, metachronal, in cilia, 12
Wet milling techniques in microbial disruption, 301

X

X press, continuous modification of, 305
  use of to disrupt micro-organisms, 300
X-Ray-induced damage to bacterial DNA, repair of, 137
X-Ray-sensitive mutants of bacteria, 113, 117
X-Rays, effect of on double-stranded breaks in bacteria, DNA, 108
Xylans in algal walls, 285

Y

Yeast, chromophore of reactivating enzyme in, 124
  enzymes, effect of mechanical forces on, 309
  mitochondrial genetics, 216
  mitochondrial DNA, structure, synthesis and genetics of, 157
  mtDNA, genetic determinants on, 217
  mtDNA, synthesis of, 173
  photoreactivation repair of DNA in, 121
  protoplasts, preparation of, 331
Yeasts, disruption of, 299, 330
  induction of petite, 203
Yeda press, pressure exerted in, 297
  use of to disrupt eukaryotic micro-organisms, 329

Z

Zygote formation, fate of mitochondrial genomes during, 231
Zygotic cell lineages in yeast, 242
Zygotic clones, analysis of individual in yeast, 240
Zymolase, use of to prepare yeast protoplasts, 332

68329